MW01001377

Introduction to Laser Technology

IEEE Press
445 Hoes Lane
Piscataway, NJ 08855

Technical Reviewer
Joe Falk, *Professor, Dept. of Electrical and Computer Engineering*
University of Pittsburgh

Introduction to Laser Technology

Fourth Edition

C. Breck Hitz
J. Ewing
Jeff Hecht

IEEE PRESS

A JOHN WILEY & SONS, INC., PUBLICATION

For general information on our other products and services please contact our Customer Care Department within the United States at (800) 762-2974, outside the United States at (317) 572-3993 or fax (317) 572-4002.

Wiley also publishes its books in a variety of electronic formats. Some content that appears in print, however, may not be available in electronic formats. For more information about Wiley products, visit our web site at www.wiley.com.

Library of Congress Cataloging-in-Publication Data:

Hitz, C. Breck.
Introduction to laser technology / C. Breck Hitz, J. J. Ewing, Jeff Hecht.—4th ed.
p. cm.
ISBN 978-0-470-91620-9 (hardback)
1. Lasers. I. Ewing, J. J. (James J.), 1942– II. Hecht, Jeff. III. Title.
TA1675.H58 2012
621.36'6—dc23 2011037608

Printed in the United States of America.

10 9 8 7 6 5 4 3 2 1

Contents

Preface

HOW DOES A LASER WORK AND WHAT IS IT GOOD FOR?

Answering this question is the goal of this textbook. Without delving into the mathematical details of quantum electronics, we examine how lasers work as well as how they can be modified for particular applications.

THE BOOK'S APPROACH

You should have some feeling for the overall organization of this textbook before you begin reading its chapters. The book begins with an introductory chapter that explains in unsophisticated terms what a laser is and describes the important applications of lasers worldwide.

Lasers produce light, and it is essential to understand how light works before you try to understand what a laser is. Chapters 2 through 5 are dedicated to light and optics, with lasers rarely mentioned. The subjects discussed in these chapters lead naturally to the laser principles in the following chapters, and the laser chapters themselves will not make much sense without the optics concepts presented in Chapters 2 through 5.

The heart of this text is contained in Chapters 6 through 9 because these are the chapters that explicitly answer the question, How does a laser work? As you read these chapters, you will find that two fundamental elements must be present in any laser: some form of optical gain to produce the light, and some form of feedback to control and amplify the light.

Having covered the fundamentals, the book turns to more sophisticated topics in Chapters 10 through 20. Chapters 10 to 13 describe how a laser can be modified for particular applications. Lasers can be pulsed to produce enormously powerful outputs, or their beams can be limited to a very narrow portion of the optical spectrum. And the color of the light produced by a laser can be altered through nonlinear optics.

Finally, the last seven chapters of the book apply the principles developed in the first 13 chapters to explain the operation and engineering of today's commercial lasers. All important lasers—gas lasers, optically pumped solid-state lasers, fiber lasers, and semiconductor lasers—are explicitly covered in these chapters.

Breck Hitz
Photonetics Associates

J. J. Ewing
Ewing Technology Associates, Inc.

Jeff Hecht
Laser Focus World

Acknowledgments

Earlier editions of this book have been used by one of us (BH) in teaching this material to numerous scientists and technicians over the years. Many of those individuals have made suggestions that have been incorporated into the current version of the text, and we wish to acknowledge the assistance of Professor Joel Falk of the University of Pittsburgh with the original manuscript, and of the late Professor Anthony Siegman of Stanford University for helpful suggestions about explaining the subtleties of quantum mechanics on an intuitive level.

Chapter 1

An Overview of Laser Technology

The word laser is an acronym that stands for "light amplification by stimulated emission of radiation." In a fairly unsophisticated sense, a laser is nothing more than a special flashlight. Energy goes in, usually in the form of electricity, and light comes out. But the light emitted from a laser differs from that from a flashlight, and the differences are worth discussing.

You might think that the biggest difference is that lasers are more powerful than flashlights, but this concept is more often wrong than right. True, some lasers are enormously powerful, but many are much weaker than even the smallest flashlight. So power alone is not a distinguishing characteristic of laser light.

Chapter 5 discusses the uniqueness of laser light in detail. But for now it is enough to say that there are three differences between light from a laser and light from a flashlight. First, the laser beam is much narrower than a flashlight beam. Second, the white light of a flashlight beam contains many different colors of light, whereas the beam from a laser contains only one, pure color. Third, all the light waves in a laser beam are aligned with each other, whereas the light waves from a flashlight are arranged randomly. The significance of this difference will become apparent as you read through the next several chapters about the nature of light.

Lasers come in all sizes, from tiny diode lasers small enough to fit in the eye of a needle to huge military and research lasers that fill multistory buildings. And different lasers can produce many different colors of light. As we will explain in Chapter 2, the color of light depends on the length of its waves. Listed in Table 1.1 are some of the important commercial lasers. In addition to these fixed-wavelength lasers, several important tunable lasers are discussed in Chapter 20.

The "light" produced by carbon dioxide lasers and neodymium lasers cannot be seen by the human eye because it is in the infrared portion of the spectrum. Red light from a ruby or helium–neon laser, and green and blue light from an argon laser, can be seen by the human eye. But the krypton-fluoride laser's output at 248 nm is in the ultraviolet range and cannot be directly detected visually.

Table 1.1 is by no means a complete list of the types of lasers available today; indeed, a complete list would have dozens, if not hundreds, of entries. It is also incomplete in the sense that many lasers can produce more than a single, pure color. Nd:YAG lasers, for example, are best known for their strong line at 1.06 μm, but

Introduction to Laser Technology, Fourth edition. C. Breck Hitz, J. Ewing, and J. Hecht

Table 1.1. Some important commercial lasers

Laser	Wavelength	Average power range
Carbon dioxide	10.6 μm	Milliwatts to tens of kilowatts
Nd:YAG	1.06 μm	Milliwatts to hundreds of watts
	532 nm	Milliwatts to watts
Nd:glass	1.05 μm	Watts[1]
Diodes	Visible and IR	Milliwatts to kilowatts
Argon-ion	514.5 nm	Milliwatts to tens of watts
	488.0 nm	Milliwatts to watts
Fiber	IR	Watts to kilowatts
Excimer	Ultraviolet	Watts to hundreds of watts[2]

[1]Although glass lasers produce relatively low average powers, they almost always run in pulsed mode, where their peak powers can reach the gigawatt levels. Peak powers are explained at the beginning of Chapter 11.

[2]Excimers, like the glass lasers discussed in the note above, are pulsed lasers, capable of peak powers in the tens of megawatts.

these lasers can also lase at perhaps a dozen other wavelengths. Or, with the aid of nonlinear optics, Nd:YAG lasers can produce wavelengths in the visible, such as the green line of laser pointers, and even in the ultraviolet. Diode lasers produce beams throughout the infrared spectrum and in the short- and long-wavelength regions of the visible spectrum.

The yttrium–aluminum–garnet (YAG) and glass lasers listed are solid-state lasers. The light is generated in a solid, crystalline rod that looks much like a cocktail swizzlestick. The ytterbium-doped fiber laser is also a solid-state laser, but the solid is a thin glass fiber. Diode lasers are also solid-state devices, but through the fickleness of human terminology, the term "solid-state laser" is usually understood to include lasers such as Nd:YAG and glass, but not diode lasers. Diode lasers are based on semiconductors, and in many ways resemble high-powered light-emitting diodes.

All the other lasers listed are gas lasers that generate light in a gaseous medium, in some ways like a neon sign. If there are solid-state lasers and gaseous lasers, it is logical to ask if there is such a thing as a liquid laser. The answer is yes. The most common example is the organic dye laser, in which dye dissolved in a liquid produces the laser light.

1.1 WHAT ARE LASERS USED FOR?

We have seen that lasers usually do not produce a lot of power. By comparison, an ordinary 1200 W electric hair dryer is more powerful than 99% of the lasers in the world today. And we have seen that some types of lasers do not even produce power very efficiently, often wasting at least 99% of the electricity they consume.* So

*An important exception is the diode laser, whose efficiency can sometimes be 70%.

what is all the excitement about? What makes lasers so special, and what are they really used for?

The unique characteristics of laser light are what make lasers so special. The capability to produce a narrow beam does not sound very exciting, but it is the critical factor in most laser applications. Because a laser beam is so narrow, it can read the minute, encoded information on a CD or DVD, or on the bar-code patterns in a grocery store. Because a laser beam is so narrow, the comparatively modest power of a 200 W carbon-dioxide laser can be focused to an intensity that can cut or weld metal. Because a laser beam is so narrow, it can create tiny and wonderfully precise patterns in a laser printer.

The other characteristics of laser light—its spectral purity and the way its waves are aligned—are also important for some applications. And, strictly speaking, the narrow beam could not exist if the light did not also have the other two characteristics. But from a simple-minded, applications-oriented viewpoint, a laser can often be thought of as nothing more than a flashlight that produces a very narrow beam of light.

One of the leading laser applications is materials processing, in which lasers cut, drill, weld, heat-treat, and otherwise alter both metals and nonmetals. Lasers can drill tiny holes in turbine blades more quickly and less expensively than mechanical drills. Lasers have several advantages over conventional techniques of cutting materials. For one thing, unlike saw blades or knife blades, lasers never get dull. For another, lasers make cuts with better edge quality than most mechanical cutters. The edges of metal parts cut by lasers rarely need be filed or polished because the laser makes such a clean cut.

Laser welding can often be more precise and less expensive than conventional welding. Moreover, laser welding is more compatible with robotics, and several large machine-tool builders offer fully automated laser-welding systems to manufacturers.

Laser heat-treating involves heating a metal part with laser light, increasing its temperature to the point where its crystal structure changes. It is often possible to harden the surface in this manner, making it more resistant to wear. Heat-treating requires some of the most powerful industrial lasers, and it is one application in which the raw power of the laser is probably more important than the narrow beam.

Have you purchased a quart of milk or anything else with a "Use by" date on it recently? Odds are, that date was put there with a laser. Laser marking is the largest market for materials-processing lasers, in terms of the number of lasers sold.

1.2 LASERS IN TELECOMMUNICATIONS

One of the more exciting applications of lasers is in the field of telecommunications, in which tiny diode lasers generate the optical signal transmitted through optical fibers. Because the bandwidth of these fiber-optic systems is so much greater than that of conventional copper wires, fiber optics is playing a major role in enabling the fast-growing Internet.

Many modern fiber-optic telecommunication systems transmit multiple wavelengths through a single fiber, a technique called wavelength-division multiplexing.

The evolution of this technology, together with erbium-doped fiber amplifiers to boost the signal at strategic points along the transmission line, is a major driving force in today's optoelectronics market.

Any time you pick up a telephone or connect to the Internet, it is likely that, somewhere along the line, you are transmitting information across a fiber-optic link. But today, optical technology is starting to transmit data over much smaller distances, even from one point to another inside your computer. As electronic devices—computers, phones, and readers—get smaller and smaller, eventually they run into a roadblock. Because electrons push each other around (repel each other, actually), there is a limit on how close together they can be. One beam of light, however, exerts no force on another beam. Hence, transmitting information with light instead of electrons avoids the roadblock to further miniaturization.

1.3 LASERS IN RESEARCH AND MEDICINE

Lasers started out in research laboratories, and many of the most sophisticated ones are still being used there. Chemists, biologists, spectroscopists, and other scientists count lasers among the most powerful investigational tools of modern science. Again, the laser's narrow beam is valuable, but in the laboratory the other characteristics of laser light are often important too. Because a laser's beam contains light of such pure color, it can probe the dynamics of a chemical reaction while it happens or it can even stimulate a reaction to happen.

In medicine, the laser's narrow beam has proven a powerful tool for therapy. In particular, the carbon dioxide laser has been widely adopted by surgeons as a bloodless scalpel because the beam cauterizes an incision even as it is made. Indeed, some surgeries that cause profuse bleeding had been impossible to perform before the advent of the laser. The laser is especially useful in ophthalmic surgery because the beam can pass through the pupil of the eye and weld, cut, or cauterize tissue inside the eye. Before lasers, any procedure inside the eye necessitated cutting open the eyeball. The LASIK procedure, described in Chapter 18, can correct vision so people no longer need glasses or contract lenses. Lasers also are used for gum surgery, to selectively destroy cancer cells, and even to treat toenail fungus.

1.4 LASERS IN GRAPHICS AND GROCERY STORES

Laser printers are capable of producing high-quality output at very high speeds. Twenty-five years ago, they were also very expensive, but good, PC-compatible laser printers can now be obtained for less than a hundred dollars. In a laser printer, the laser "writes" on an electrostatic surface, which, in turn, transfers toner (ink) to the paper.

Lasers have other applications in graphics as well. Laser typesetters write directly on light-sensitive paper, producing camera-ready copy for the publishing industry. Laser color separators analyze a color photograph and create the information a printer needs to print the photograph with four colors of ink. Laser platemakers produce the printing plates, or negatives in some cases, so that national newspapers such as the *Wall Street Journal* and *USA Today* can be printed in locations far from their editorial offices.

And everyone has seen the laser bar-code scanners at the checkout stand of the local grocery store. The narrow beam of the laser in these machines scans the bar-code pattern, automatically reading it into the store's computer.

1.5 LASERS IN THE MILITARY

So far, lasers have been found to make poor weapons, and many scientists believe that engineering complexities and the laws of physics may prevent them from ever being particularly useful for this purpose. Nonetheless, many thousands of lasers have found military applications, not in weapons, but in range finders and target designators.

A laser range finder measures the time a pulse of light takes to travel from the range finder to the target and back. An on-board computer divides this number into the speed of light to find the range to the target. A target designator illuminates the target with infrared laser light, and then a piece of "smart" ordnance, a rocket or bomb, equipped with an infrared sensor and some steering mechanism, homes in on the target and destroys it.

Diode lasers are sometimes used to assist in aiming small arms. The laser beam is prealigned along the trajectory of the bullet, and a policeman or soldier can see where the bullet will hit before firing.

Diode lasers are used as military training devices in a scheme that has been mimicked by civilian toy manufacturers. Trainees use rifles that fire bursts of diode-laser light (rather than bullets) and wear an array of optical detectors that score a hit when an opponent fires at them.

1.6 OTHER LASER APPLICATIONS

There seems to be no end to the ingenious ways a narrow beam of light can be put to use. In sawmills, lasers are used to align logs relative to the saw. The laser projects a visible stripe on the log to show where the saw will cut it as the sawman moves the log into the correct position. On construction projects, the narrow beam from a laser guides heavy earth-moving equipment. Laser light-shows herald the introduction of new automobile models and rock concerts. And laser gyroscopes guide many commercial aircraft (an application that depends more on a laser's spectral purity than on its narrow beam).

Chapter 2

The Nature of Light

What is light? How does it get from one place to another? These are the questions that are addressed in this chapter. But the answers are not all that easy. The nature of light is a difficult concept to grasp because light does not always act the same way. Sometimes it behaves as if it were composed of tiny waves, and other times it behaves as if it were composed of tiny particles. Let us take a look at how light waves act and at how light particles (photons) act, and then we will discuss this so-called "duality" of light.

2.1 ELECTROMAGNETIC WAVES

"Light is a transverse electromagnetic wave" is a simple sentence. But what does that mean? Let us take the phrase "transverse electromagnetic wave" apart and examine it one word at a time.

Figure 2.1 is a schematic of a wave. It is a periodic undulation of something—maybe the surface of a pond, if it is a water wave—that moves with characteristic velocity, v. The wavelength, λ, is the length of one period, as shown in Figure 2.1. The frequency of the wave is equal to the number of wavelengths that move past an observer in one second. It follows that the faster the wave moves—or the shorter its wavelength—the higher its frequency will be. Mathematically, the expression

$$f = v/\lambda$$

relates the velocity of any wave to its frequency f and wavelength λ.

The amplitude of the wave in Figure 2.1 is its height, the distance from the center line to the peak of the wave. The phase of the wave refers to the particular part of the wave passing the observer. As shown in Figure 2.1, the wave's phase is 90° when it is at its peak, 270° at the bottom of a valley, and so on.

So much for waves. What does transverse mean? There are two kinds of waves: transverse and longitudinal. In a transverse wave, whatever is waving is doing so in a direction transverse (perpendicular) to the direction in which the wave is moving. A water wave is an example of a transverse wave because the thing that is waving (the surface of the water) is moving up and down, while the wave itself is moving horizontally across the surface. Ordinary sound, on the other hand, is an example of a longitudinal wave. When a sound wave propagates through air, the compressions

Introduction to Laser Technology, Fourth edition. C. Breck Hitz, J. Ewing, and J. Hecht

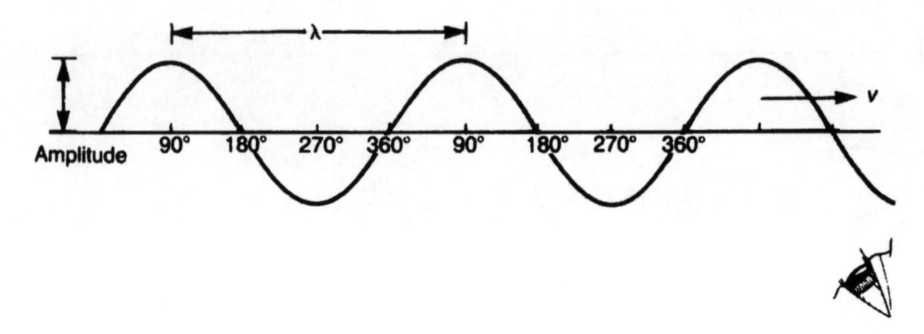

Figure 2.1 A wave and an observer.

and rarefactions are caused by gas molecules moving back and forth in the same direction that the wave is moving. Light is a transverse wave because the things that are waving—electric and magnetic fields—are doing so in a direction transverse to the direction of wave propagation.

Light is an electromagnetic wave because the things that are waving are electric and magnetic fields. Figure 2.2 is a diagram of the fields of a light wave. It has an electric field (E) undulating in the vertical direction and a magnetic field (B) undulating in the horizontal direction.* The wave can propagate through a vacuum because, unlike sound waves or water waves, it does not need a medium to support it. If the light wave is propagating in a vacuum, it moves at a velocity $c = 3.0 \times 10^8$ m/s, the speed of light.†

Visible light is only a small portion of the electromagnetic spectrum diagrammed in Figure 2.3. Radio waves, light waves, and gamma rays are all transverse electromagnetic waves, differing only in their wavelength. But what a difference that is! Electromagnetic waves range from radio waves hundreds or thousands of meters long down to gamma rays, whose tiny wavelengths are on the order of 10^{-12} m. And the behavior of the waves in different portions of the electromagnetic spectrum varies radically, too.

But we are going to confine our attention to the "optical" portion of the spectrum, which usually means part of the infrared, the visible portion, and part of the ultraviolet. Specifically, laser technology is usually concerned with wavelengths between about 10 μm (10^{-5} m) and 100 nm (10^{-7} m). The visible portion of the spectrum, roughly between 400 and 700 nm, is shown across the bottom of Figure 2.3.

The classical (i.e., nonquantum) behavior of light—and all other electromagnetic radiation—is completely described by an elegant set of four equations called Maxwell's equations, named after the nineteenth century Scottish physicist James Clerk Maxwell. Maxwell collected the conclusions of several other physicists and then modified and combined them to produce a unified theory of electromagnetic

*Although it would make sense to use M to designate a magnetic field, that is not the letter physicists have chosen. B is the accepted letter to designate a magnetic field.

†It is convenient to remember that the speed of light is about 1 ft/ns. Thus, when a laser produces a 3 ns pulse, the pulse is about 3 ft long.

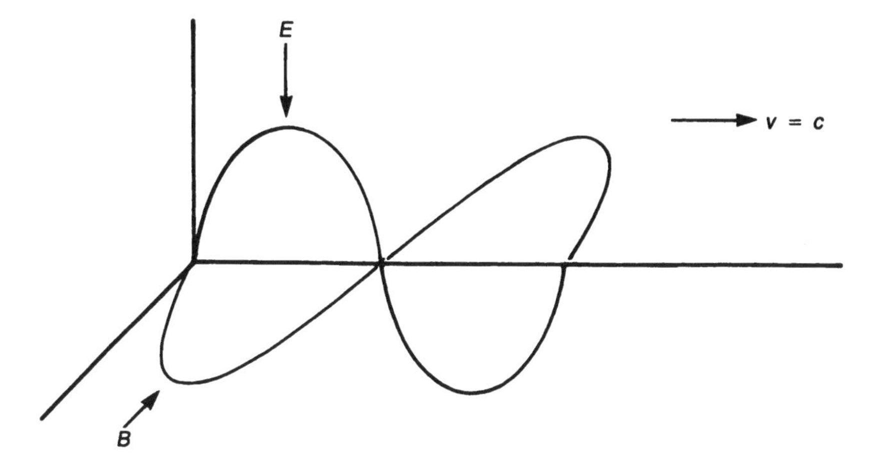

Figure 2.2 The electric (E) and magnetic (B) fields of a light wave.

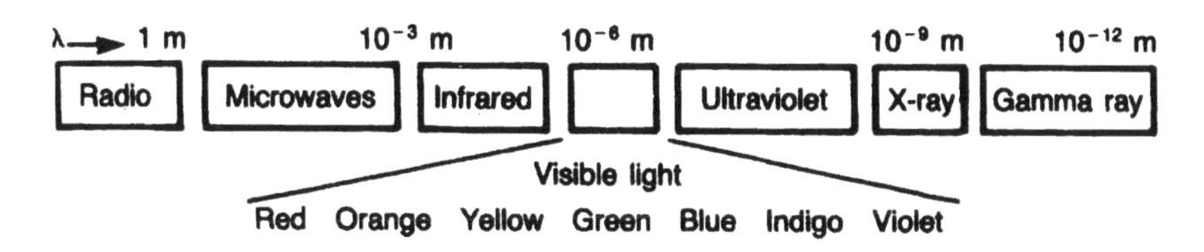

Figure 2.3 The electromagnetic spectrum.

phenomena. His equations are among the most important in physics. Here is what they look like in a vacuum:

$$\nabla \times E = 0$$

$$\nabla \times B = 0$$

$$\nabla \times E + \frac{\partial B}{\partial t} = 0$$

$$\nabla \times B = c^{-2} \frac{\partial E}{\partial t}$$

where c is the speed of light.

Now, these are differential equations, but you do not have to understand differential calculus to appreciate their simplicity and beauty.* The first one—Gauss's law

*$\nabla \times E$ is read "divergence of E"; $\nabla \times E$ is read "curl of E"; and $\partial E/\partial t$ is read "partial time derivative of E."

for electricity—describes the shape of an electric field (E) created by electric charge (ρ). The second equation—Gauss's law for magnetism—describes the shape of a magnetic field (B) created by a magnet. The fact that the right side of this equation is zero means that it is impossible to have a magnetic monopole (e.g., a north pole without a south pole).

An electric field is created by electric charge, as described by Gauss's law, but an electric field is also created by a time-varying magnetic field, as described by Faraday's law (the third equation). Likewise, a magnetic field can be created by a time-varying electric field and also by an electric current, J.* The shape of this magnetic field is described by Ampere's law, the fourth equation.

The fame of these four little equations is well justified, for they govern all classical electrodynamics and their validity even extends into the realm of quantum and relativistic phenomena. We will not be dealing directly with Maxwell's equations any more in this book, but they have been included in our discussion to give you a glimpse at the elegance and simplicity of the basic laws that govern all classical electromagnetic phenomena.

There are two special shapes of light waves that merit description here. Both of these waves have distinctive wavefronts. A *wavefront* is a surface of constant phase. An example is the plane wave in Figure 2.4. The surface sketched passes through the wave at its maximum. Because this surface that cuts through the wave at constant phase is a plane, the wave is a *plane wave.*

The second special shape is a spherical wave, and, as you might guess, it is a wave whose wavefronts are spheres. A cross-sectional slice through a spherical wave in Figure 2.5 shows several wavefronts. A spherical wavefront is the three-dimensional analogy of the two-dimensional "ripple" wavefront produced when you drop a pebble into a pond. A spherical wave is similarly produced by a point source, but it spreads in all three dimensions.

2.2 WAVE–PARTICLE DUALITY

Let us do a thought experiment with water waves. Imagine a shallow pan of water 3 ft wide and 7 ft long. Figure 2.6 shows the waves that spread out in the pan if you strike the surface of the water rapidly at point A. Now look at what happens at points X and Y. A wave crest will arrive at Y first because Y is closer to the source (Point A) than X is. In fact, if you pick the size of the pan correctly, you can arrange for a crest to reach X just as a trough arrives at Y, and vice versa.

On the other hand, if you strike the water at point B, the wave crest will arrive at X first. But (assuming you are still using the correct-size pan) there will still always be a crest arriving at X just as a trough arrives at Y, and vice versa.

*Because there are not enough letters in the English (and Greek) alphabets to go around, some letters must serve double duty. For example, in Maxwell's equations E represents the electric-field vector, but elsewhere in this book it stands for energy. In Maxwell's equations, J represents an electric-current density and B represents the magnetic-field vector, but elsewhere J is used as an abbreviation for joules and B for brightness. The letter f is used to mean frequency and to designate the focal length of a lens. The letters and abbreviations used in this book are consistent with most current technical literature.

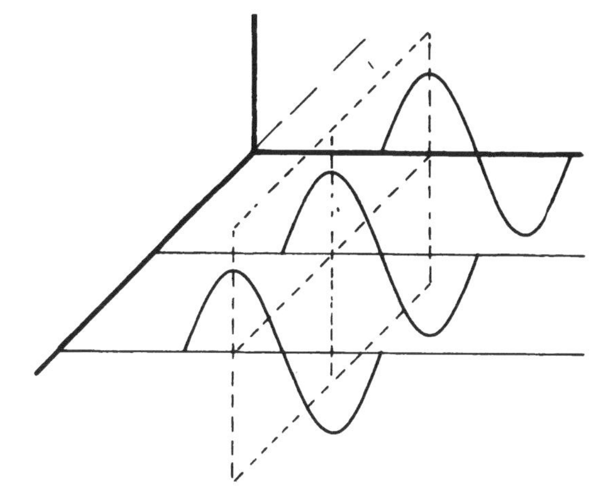

Figure 2.4 A plane wave.

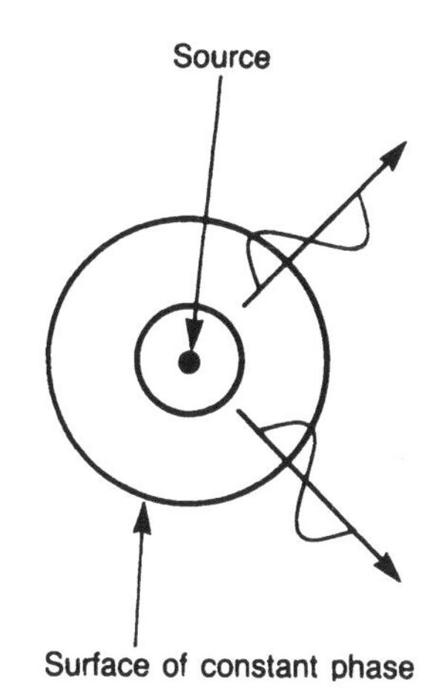

Figure 2.5 A spherical wave.

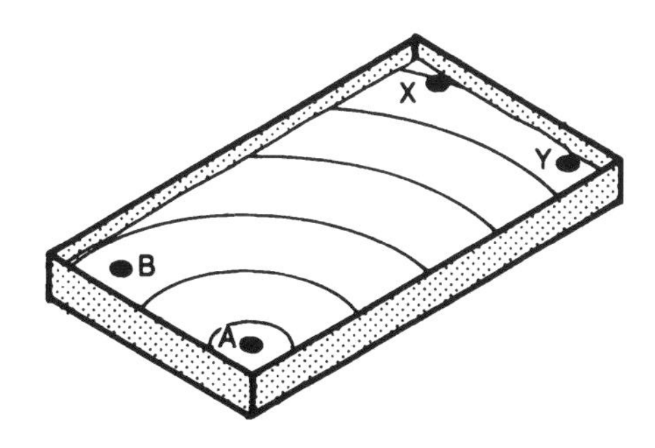

Figure 2.6 Wave experiment in a shallow pan of water.

What happens if you strike the water at A and B simultaneously? At point X, a crest from A will arrive at exactly the same time as a trough arrives from B. Likewise, a crest from B will be canceled out by a trough from A. At point X, the surface of the water will be motionless. The same argument holds for point Y. But at a point halfway between X and Y, where crests from A and B arrive simultaneously, there will be twice as much motion as there was before.

This is the phenomenon of interference: waves interacting with each other to cancel or enhance each other.

A similar situation can be observed with light, as diagrammed in Figure 2.7. Here, two slits correspond to the sources, and dark stripes on a viewing screen correspond to motionless water at points X and Y. This arrangement, called Young's double-slit experiment, is analyzed in detail in Chapter 5. But here is the point for now: the only way to explain the observed results is to postulate that light is behaving as a wave. There is no possible way to explain the bright spot at the center of the screen if you assume that the light is made up of particles. However, it is easily explained if you assume light is a wave.

During most of the nineteenth century, physicists devised experiments like this one and explained their results quite successfully from the assumption that light is a wave. But near the turn of the century, a problem developed in explaining the photoelectric effect.

A photoelectric cell, shown schematically in Figure 2.8, consists of two electrodes in an evacuated tube. When light strikes the cathode, the energy in the light can liberate electrons from the cathode, and these electrons can be collected at the anode. The resulting current is measured with an ammeter (A). It is a simple experiment to measure the current collected as a function of the voltage applied to the electrodes, and the data look like the plot in Figure 2.9.

There is a lot of information in Figure 2.9. The fact that current does not change with positive voltage (voltage that accelerates the electrons toward the anode) im-

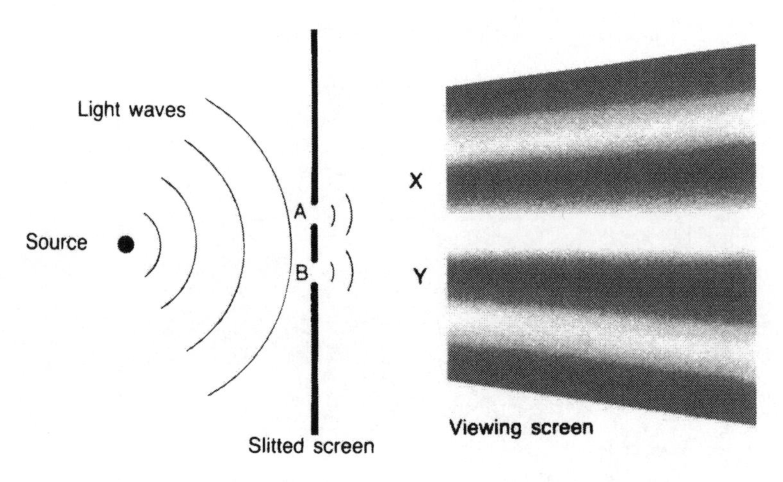

Figure 2.7 Optical analogy to wave experiment in Figure 2.6.

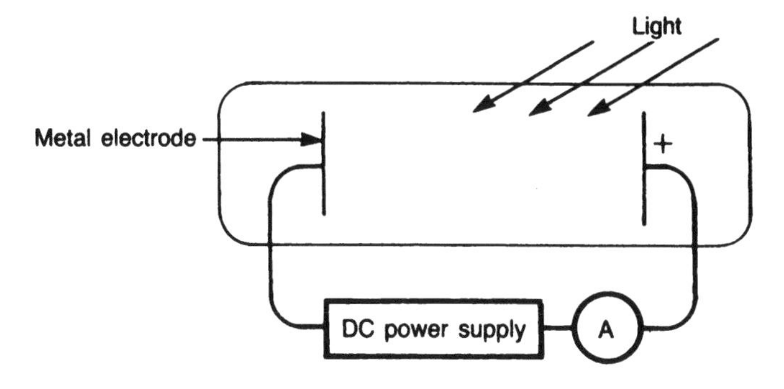

Figure 2.8 A photoelectric cell.

plies that every electron emitted from the cathode has at least some kinetic energy. In other words, an electron emitted from the cathode does not need any help to get to the anode. But as soon as the voltage starts to go negative, the current decreases. This implies that some of the electrons are emitted with very little energy; if they have to climb even a small voltage hill, they do not make it to the anode. The sharp cutoff of current implies that there is a definite maximum energy with which electrons are emitted from the cathode.

Thus, some electrons are emitted with high energy, and some barely get out of the cathode. This makes sense if you assume that the high-energy electrons came from near the surface of the cathode whereas the low-energy ones had to work their way out from farther inside the cathode. What is hard to understand is why the maximum energy for emitted electrons does not depend on the intensity of the light illuminating the cathode.

Think about it for a moment. The electric field in the light wave is supposed to be exerting a force on the electrons in the cathode. The field vibrates the electrons,

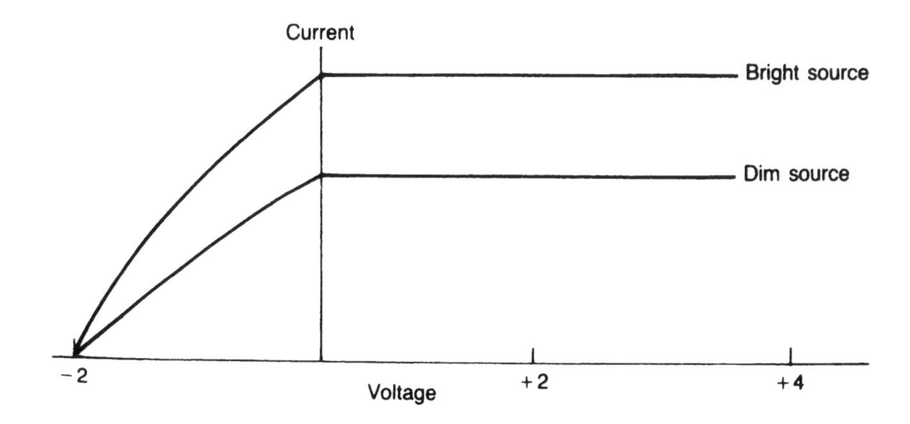

Figure 2.9 Current versus voltage for a photoelectric cell.

shaking them so they can break free from the cathode. As the illumination intensifies—that is, the electric-field strength increases—the energy of vibration should increase. An electron right on the surface of the cathode should break free with more energy than it did before the illumination intensity was increased. That is, the current from the bright source in Figure 2.9 should go to zero at a greater negative voltage than it does from the dim source. But that is not what happens.

There are other problems. For example, you can easily figure out how much energy the most energetic electrons have when they leave the cathode. (For the experiment whose data appear in Figure 2.9, those electrons would have 2 eV of energy.) If all the energy falling on an atom can somehow be absorbed by one electron, how long does it take that electron to accumulate 2 eV of energy?

The rate at which energy hits the whole surface is known from the illumination intensity. To calculate the rate at which energy hits a single atom, you have to know how big the atom is. Both now and back at the turn of the nineteenth century, when all this confusion was taking place, scientists knew that an atomic diameter was on the order of 10^{-8} cm. For typical illumination intensities of a fraction of a microwatt per square centimeter, it takes a minute or two for an atom to absorb 2 eV. But in the laboratory, the electrons appear immediately after the light is turned on with a delay of much less than a microsecond. How can they absorb energy that quickly?

In 1905, Albert Einstein proposed a solution to the dilemma. He suggested that light is composed of tiny particles called photons, each photon having energy

$$E = hf$$

in which f is the frequency of the light, and h is Planck's constant ($h = 6.63 \times 10^{-34}$J-s). This takes care of the problem of instantaneous electrons. If light hits the cathode in discrete particles, one atom can absorb one photon while several million of its neighbors absorb no energy. Thus, the electron from the atom that was hit can be liberated immediately.

Einstein's theory also explains why the maximum energy of electrons emitted from the cathode does not depend on illumination intensity. If each liberated electron has absorbed the energy of one photon, then the most energetic electrons (those that came right from on the surface of the cathode) will have energy almost equal to the photon energy. But increasing the illumination intensity means more photons, not more energy per photon. So a brighter source will result in more electrons but not more energy per electron. That is exactly the result shown in Figure 2.9.

On the other hand, changing the color of light—that is, changing its wavelength and, therefore, its frequency—will change the energy per photon. In subsequent experiments, other physicists changed the color of light hitting the cathode of a photocell and observed data like those shown in Figure 2.10. As the energy of the incident photons increases, so does the maximum energy of the photoelectrons.

Thus, Einstein's photons explained not only the photoelectric effect but also other experiments that were conducted later that defied explanation from the wave theory. But what about experiments like Young's double-slit experiment, which absolutely cannot be explained unless light behaves as a wave? How was it possible to resolve the seemingly hopeless contradiction?

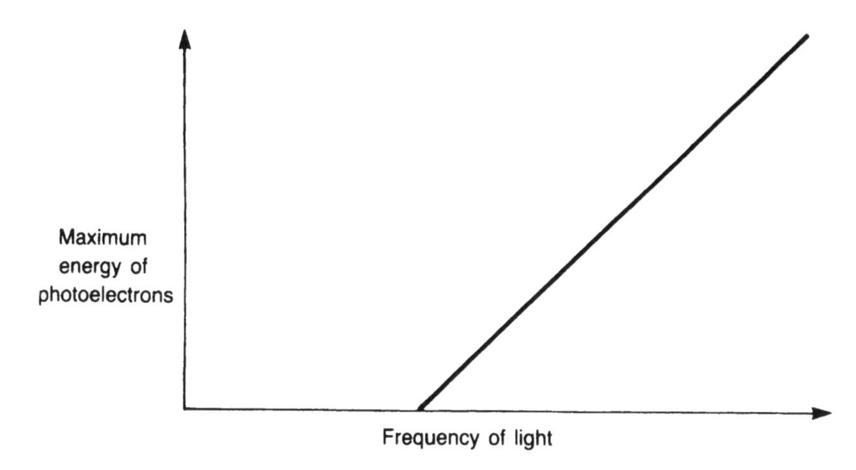

Figure 2.10 Energy of liberated electrons increases with photon energy.

The science of quantum mechanics developed during the early years of the twentieth century to explain this and other contradictions in classical physics. Quantum mechanics predicts that when nature is operating on a very tiny scale (an atomic scale or smaller), it behaves much differently than it does on a normal, "people-sized" scale, so intuition has to be reeducated to be reliable on an atomic scale.

As a result of quantum mechanics, physicists now believe that the dual nature of light is not a contradiction. In fact, quantum mechanics predicts that particles also have a wavelike property, and experiments have proven that this property exists. By reeducating their intuitions to deal reliably with events on an atomic scale, physicists have found that the duality of light is not a contradiction of nature but a manifestation of nature's extraordinary complexity.

If a laser produces a 1 ns, 1 J pulse of light whose wavelength is 1.06 μm, there are two ways you can think of that light. As shown in Figure 2.11, you can think of that pulse as a foot-long undulating electric and magnetic field. The period of the undulation is 1.06 μm, and the wave moves to the right at the speed of light. On the other hand, you could think of the laser pulse as a collection of photons, as shown in Figure 2.12. All the photons are moving to the right at the speed of light, and each photon has energy $E = hf = hc/\lambda$.

Either way of thinking of the pulse is correct, provided that you realize neither way tells you exactly what the pulse is. Light is neither a wave nor a particle, but it

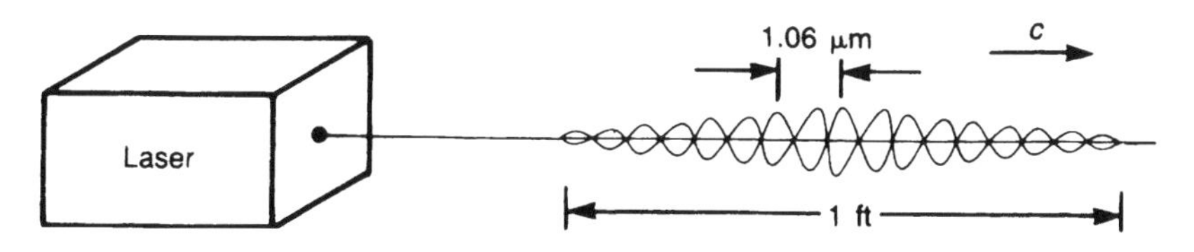

Figure 2.11 A 1 J, 1 ns pulse of 1.06 μm laser light pictured as a wave.

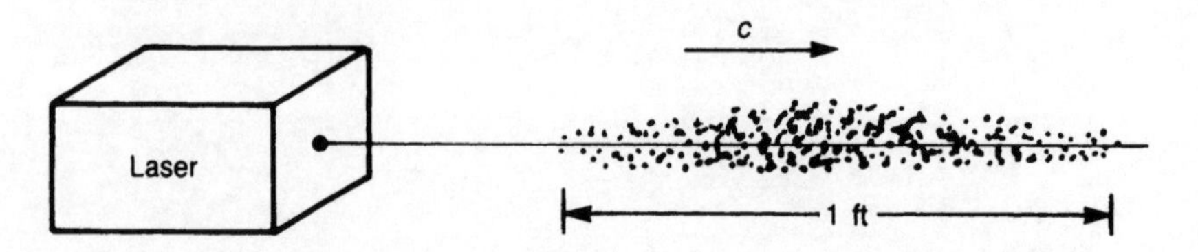

Figure 2.12 A 1 J, 1 ns pulse of 1.06 μm laser light pictured as photons.

is often convenient to think of light as one or the other in a particular situation. Sometimes, light can act as both a wave and a particle simultaneously. For example, you could envision illuminating the cathode of a photocell with stripes of light from Young's experiment. Electrons would still be liberated instantaneously in the photocell, proving the particle-like nature of light despite the stripes, which prove light's wavelike nature.

QUESTIONS

1. What is the frequency of green light whose wavelength is $\lambda = 530$ nm? Roughly how many nanoseconds does it take this light to travel from one end of a 100-yd-long football field to the other?
2. Sketch Figure 2.3 on a piece of paper. Beneath the figure, add the frequencies of the electromagnetic radiation that correspond to the wavelengths given above the figure.
3. A compass will not work properly underneath a high-voltage power line. Which of Maxwell's equations accounts for this? Which of Maxwell's equations describes the earth's magnetic field?
4. Calculate the frequency of the light wave emerging from the laser in Figure 2.11. Calculate the number of photons emerging from the laser in Figure 2.12.

Chapter 3

Refractive Index, Polarization, and Brightness

In Chapter 2, we talked about what light is. In the next several chapters we talk about some of the properties of light. In this chapter, we begin with a discussion of how light propagates in a transparent medium such as glass or water. Next, we talk about the polarization of light. It is an important characteristic that deals with the orientation of the electric and magnetic fields that make up the light wave. We conclude with a discussion of what is meant by the brightness of an optical source.

3.1 LIGHT PROPAGATION—REFRACTIVE INDEX

The speed of light in a vacuum is 3×10^8 m/s, but it moves less rapidly in a transparent medium like glass or water. The electrons in the medium interact with the electric field in the light wave and slow it down. This reduction in velocity has many important consequences in the propagation of light. The refractive index of a material is determined by how much light slows down in propagating through it. The index is defined as the ratio of light's velocity in a vacuum to its velocity in the medium. Table 3.1 gives the refractive indices for some common transparent materials. The values listed are approximate at best because the index of refraction of a material depends slightly on the wavelength of light passing through it. That is, red light and blue light travel at exactly the same velocity in a vacuum, but red light will travel a little faster in glass. This effect is called dispersion.

The change in velocity that light experiences in moving from one medium to another accounts for the bending, or refraction, of light at the interface. Figure 3.1 shows wavefronts passing through the interface. Consider wavefront AB. In Figure 3.1a, the light at both A and B is moving at $c = 3 \times 10^8$ m/s. In Figure 3.1b, the light at B has entered the medium and has slowed down, whereas the light at A has not yet slowed. The wavefront is distorted as shown. In Figure 3.1c, the light at A has also entered the medium and the planar wavefront has been restored, propagating in a different direction than it had been outside the medium. If you think about the way a bulldozer turns, by slowing one tread relative to the other, you will have a pretty good analogy.

We can understand several important phenomena from this model of light bending when it moves from one medium to another. Note in Figure 3.1 that light is always bent toward the normal in the higher-index material. (The normal is an imagi-

Table 3.1 Refractive indices for common transparent materials

Material	n
Dry air	1.0003
Crown glass	1.517
Diamond	2.419
YAG	1.825
Ice (–8°C)	1.31
Water (20°C)	1.33

nary line perpendicular to the surface.) Likewise, light is bent away from the normal when it passes into a lower-index material.

Figure 3.2 illustrates the basic focusing capability of a lens. Rays emerging from the rectangular block of glass on the left side are traveling in the same direction as they were before they passed through the glass. But rays passing through the lens on the right side have been deviated because the shape of the surface is different for the

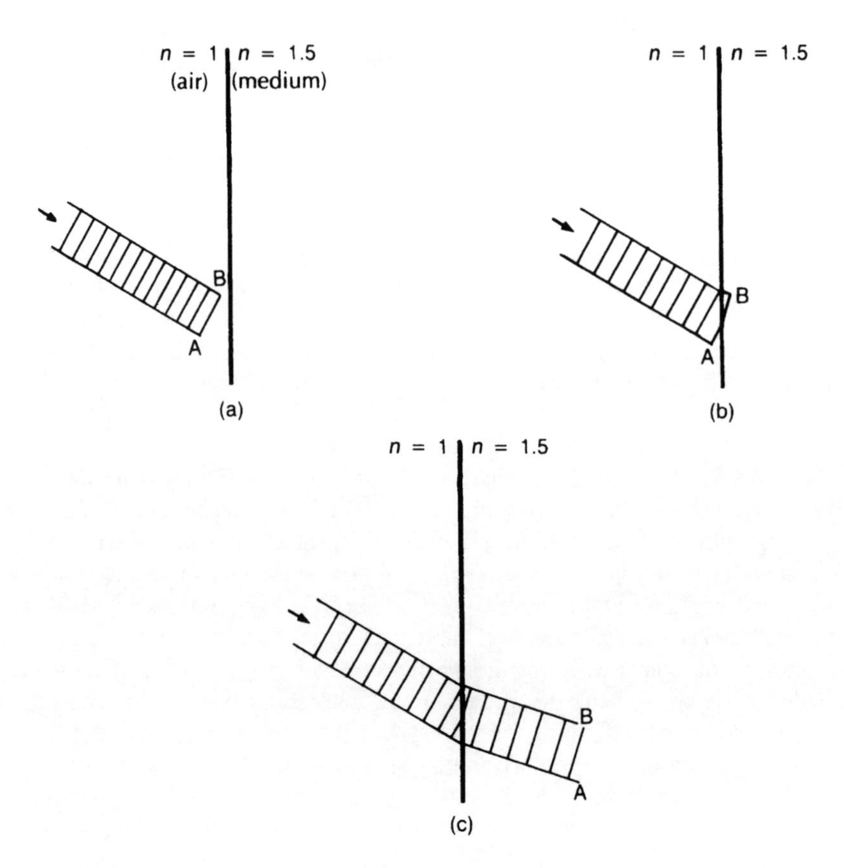

Figure 3.1 Refraction of a wavefront at an interface between optical media.

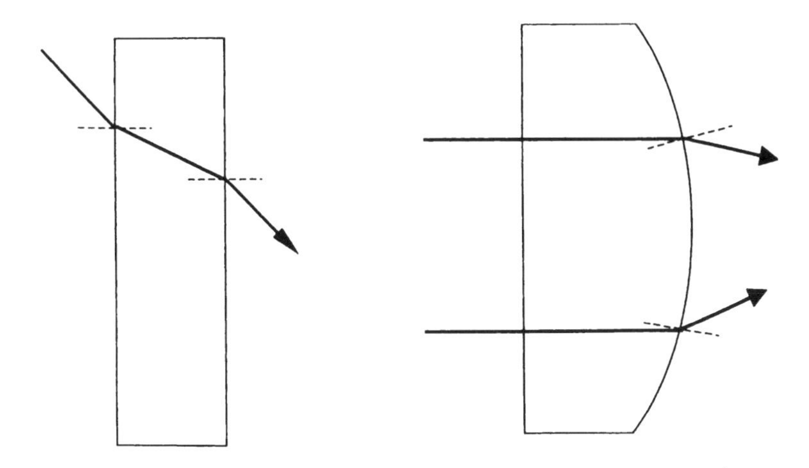

Figure 3.2 Refraction of light accounts for the focusing capability of a lens.

two rays. Notice that both rays, when they enter the low-index material (air), are bent away from the normal.

A ray that just grazes the surface will be bent toward the normal when it enters the high-index material, as illustrated in Figure 3.3. Inside the glass, the ray travels at an angle ϕ to the normal. An interesting question is, What happens to light that hits the inside surface of the glass at an angle greater than ϕ? The answer is, This light is totally reflected from the surface—none of the light emerges. This phenomenon is known as total internal reflection.

Part of the light that hits any interface between materials of different refractive index is reflected. This reflected light is not shown in Figure 3.2 nor on the left side

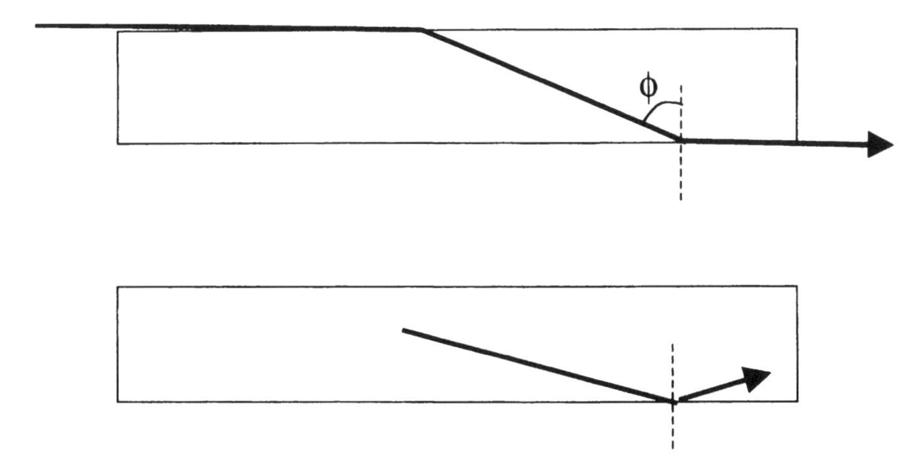

Figure 3.3 A grazing ray of light (top) defines the *critical angle,* ϕ, of a refractive material. An internal ray incident on the surface at an angle greater than the critical angle (bottom) is totally internally reflected.

of Figure 3.3. For more information about how much light is reflected, see the discussion of Brewster's angle later in this chapter.

Now you can understand how a prism separates white light into its component colors. In Figure 3.4, red and blue wavefronts approach the prism together. But because the blue light slows down a little more when it enters the glass, it is bent at a slightly greater angle than the red light, Thus, the two colors emerge from the prism at slightly different angles and will separate from each other as they travel away from the prism.

The frequency of light is an absolute measure of the energy of the light. Because energy is conserved, the frequency of light cannot change as the light moves from one medium to another. But the wavelength depends on the velocity, according to the equation introduced in the beginning of Chapter 2:

$$\lambda = v/f$$

Hence, when light moves from one refractive medium to another, its wavelength changes by an amount proportional to the ratio of refractive indices of the two media. This is analogous to what happens when a small child is bouncing up and down in the backseat of a car. Figure 3.5a shows the path his nose will follow as the car moves along at 50 mph. On the other hand, if the child does not gain or lose any energy (i.e., if he keeps bouncing at the same frequency), the path his nose follows will have half its former wavelength when the car slows to 25 mph, as shown in Figure 3.5b. Likewise, when the speed of light decreases as it moves from one medium to another, the wavelength decreases proportionally.

This wavelength-changing phenomenon gets more interesting when you consider dispersion. For example, take two light waves, one of which has twice the wavelength of the other in a vacuum. As shown in Fig. 3.6, when these two waves enter an optical medium, their wavelengths will change. But because the refractive indices for the two waves are different, the changes in fractional wavelength will not be the same, and one wavelength will no longer be twice that of the other. Does this

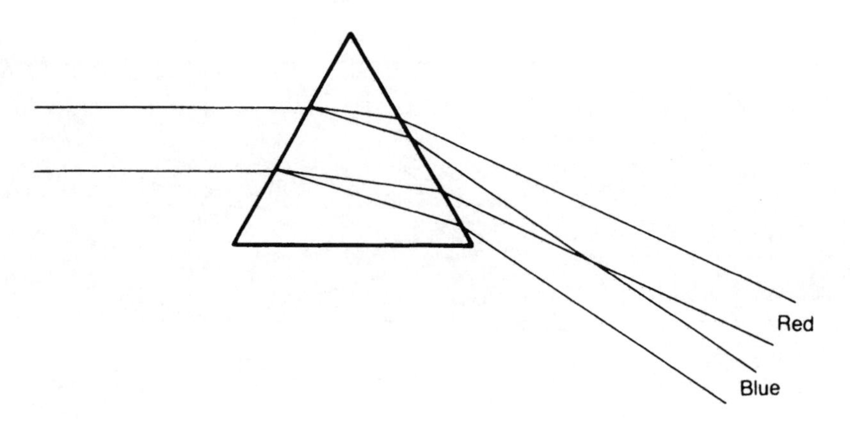

Figure 3.4 A prism refracts different colors at different angles because it is dispersive.

Figure 3.5 The path followed by the child's nose at (a) 50 mph and (b) 25 mph. The wavelength is shorter as the car slows.

strike you as a rather esoteric and perhaps meaningless observation? As you will discover in Chapter 13, this effect becomes very important in the world of nonlinear optics.

3.2 HUYGENS' PRINCIPLE

In 1678, Dutch physicist Christian Huygens formulated a way of visualizing wave propagation that has become known as Huygens' principle. This concept is still useful today for gaining an intuitive "feel" for how light waves behave.

Huygens' principle lets you predict where a given wavefront will be later if you know where it is now. This can be useful because it lets you understand how a light wave diverges. Simply stated, Huygens' principle holds that all points on the given wavefront can be considered sources that generate spherical secondary wavelets. The new position of the original wavefront is described by the surface of tangency to these wavelets.

To see how this works, let us look at the trivial case of a plane wave. Figure 3.7 shows a plane wavefront and some of the points along that wavefront that can be considered sources of the secondary wavelets. These spherical secondary wavelets

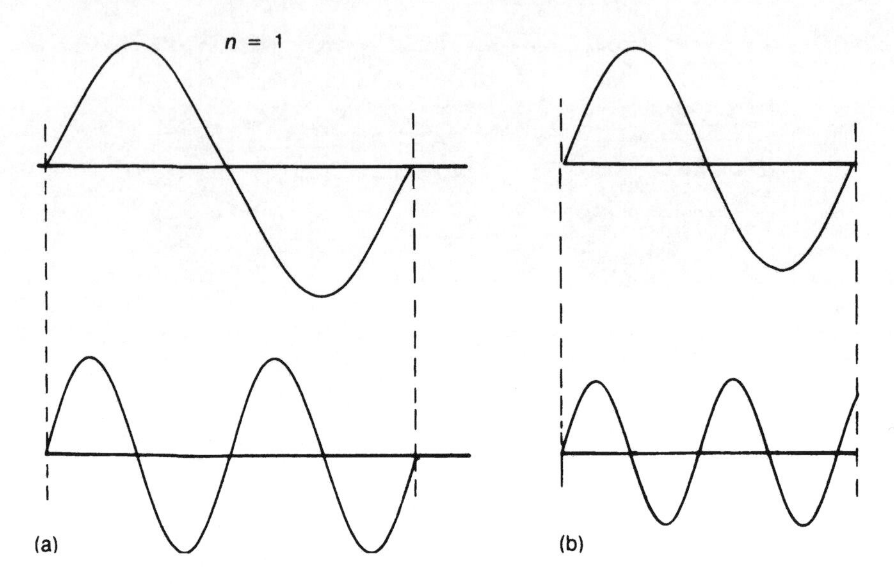

Figure 3.6 Although one wavelength is twice that of the other in a vacuum (a), dispersion in a transparent medium destroys that relationship (b).

spread out as shown, and a short time later the new position of the wavefront can be deduced by constructing the dotted surface tangent to each wavelet.

You may be asking yourself, "What happens to the wavefront that would be tangent to the back surfaces of the wavelets? Is Huygens trying to tell me there is another wavefront going backward?" Of course there isn't, but Huygens did not have a very good response to that question.

Usually, it is assumed that the intensity of the secondary wavelets diminishes to zero in the backward direction, thus getting rid of the backward-moving wavefront. Huygens' principle is not a rigorous law of physics (after all, when Huygens formulated it in the seventeenth century, he had no idea what light really was), but it is of-

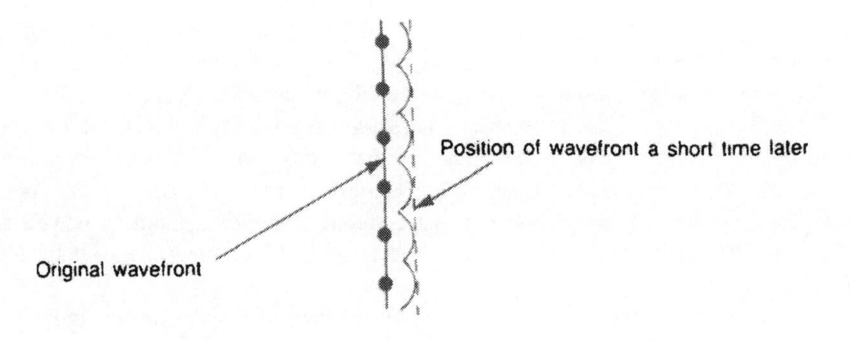

Figure 3.7 Huygens' principle applied to a plane wave.

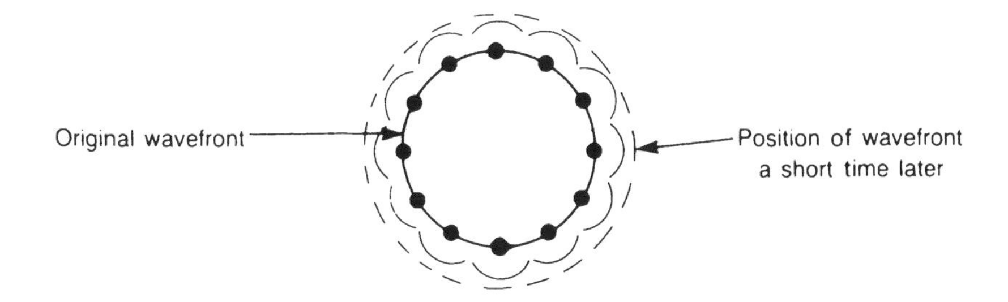

Figure 3.8 Huygens' principle applied to a spherical wave.

ten a useful aid to the imagination. The only truly rigorous explanation of how light behaves depends on solving Maxwell's equations, but often we can gain some intuitive insight on a less-formal level.

As another example of Huygens' principle, Figure 3.8 shows a spherical wave and some of the points on that wavefront that can be considered sources of secondary wavelets. These wavelets move away from their sources, and a surface of tangency to them is the new spherical wavefront. That is, Huygens' principle predicts that the solid wavefront in Figure 3.8 will develop into the wavefront represented by the broken line. As in the case with the plane wave, you must assume that the secondary wavelets diminish to zero in the backward direction.

You can even use Huygens' principle to understand bending of light, the same idea as explained in Figure 3.1. Instead of showing many point sources in Figure 3.9, only the two at the edge of the original wavefront are shown. The wavefront expanding in air is bigger than the one in glass, because the velocity of light is reduced inside the glass. Thus, the wavefront inside the glass, shown as a tangent to the Huygens wavelets, is propagating at a different angle than the wavefront in air; the beam has been bent by refraction at the air–glass interface.

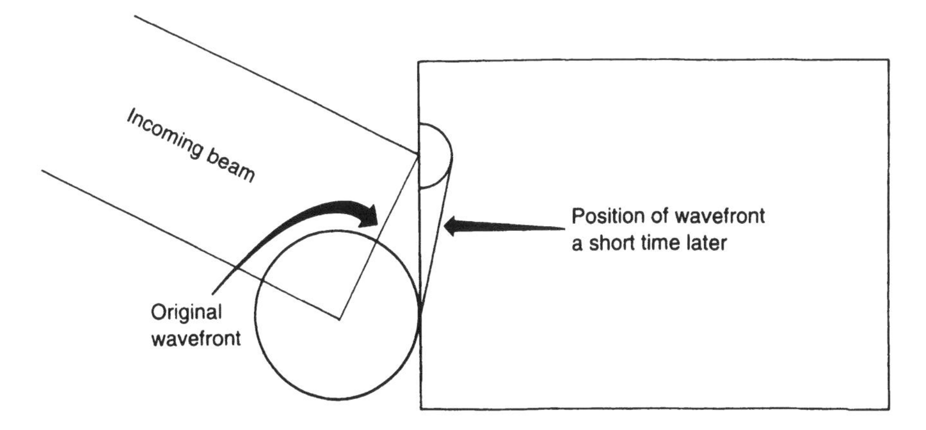

Figure 3.9 Huygens' principle can be used to explain bending of light by refraction.

3.3 POLARIZATION

Remember that light is composed of orthogonal electric and magnetic waves, as shown in Figure 3.10. The polarization of light is the direction of oscillation of the electric field. For example, the light in Figure 3.10 is plane polarized because the electric field oscillates only in one plane (the *y–z* plane). And since this plane is vertical, the light is vertically polarized. Horizontally polarized light is shown in Figure 3.11.

Figure 3.11 is the last representation of the magnetic field in a light wave you will see in this book. It is the electric field that determines the polarization of the light, and that is the only field with which we are concerned. Not that the magnetic field is not important. Indeed, it is, for light could not exist without the magnetic field. But as a matter of convenience, we will only show the electric field in future diagrams.

The light by which you're reading this book is a collection of many waves: some polarized vertically, some horizontally, and some in between. The result is unpolarized light, light in which the electric field oscillates in all random directions.

Suppose the light wave in Figure 3.11 were coming directly at you. What would you see as the wave hit your retina? At the beginning, you would see no electric field at all. Then you'd see it growing and diminishing to the left and then to the right, as shown in Figure 3.12. Over several cycles, you would see the electric field behaving as represented in Figure 3.14b.

One way to understand Figure 3.12 is to recall the "paper-pad movies" children make. A moving picture is made by drawing a slightly different picture on each page of a tablet and then quickly flipping the pages. By putting each drawing in Figure 3.12 on a different page of a tablet, you could create a moving picture of an ar-

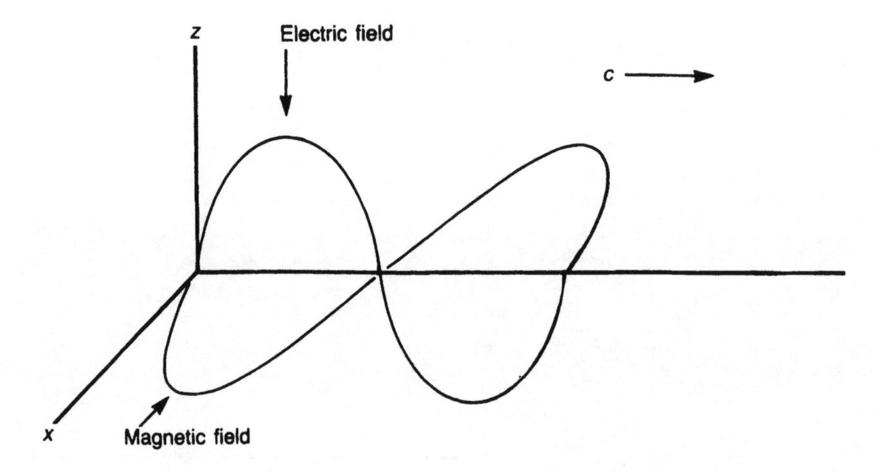

Figure 3.10 Light is composed of orthogonal electric and magnetic waves. This light is vertically polarized because the electric field oscillates in a vertical plane.

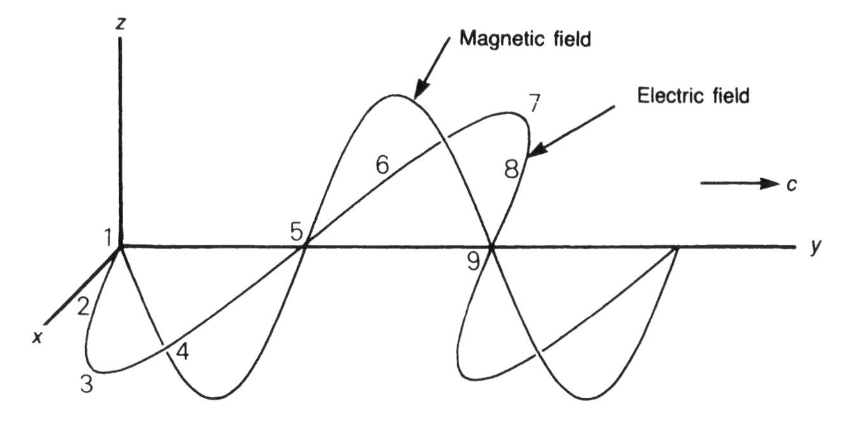

Figure 3.11 A horizontally polarized light wave.

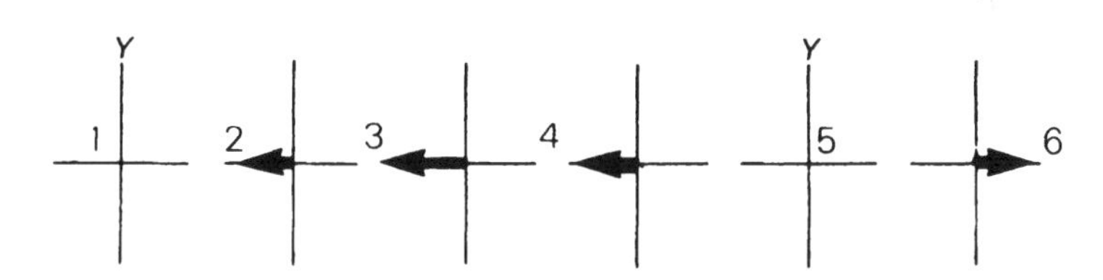

Figure 3.12 What the electric field of Fig. 3.11 looks like as it moves past you; each picture is an instant in time later.

row that grows and diminishes to the left, then to the right. Figure 3.14b is a "time exposure" of this movie over several cycles.

On the other hand, unpolarized light coming directly at you would look like Figure 3.13. The direction and amplitude of the electric field at any instant would be completely random. Over several cycles, you would observe the electric field behaving as represented in Figure 3.14c: a collection of random field vectors going off in random directions.

There is another type of polarization that is important in laser optics. The polarization vector (which is the same as the electric-field vector) describes a circle as circularly polarized light moves toward you, as shown in Figure 3.15. Over several cycles, you would see the electric field behaving as represented in Figures 3.16 and

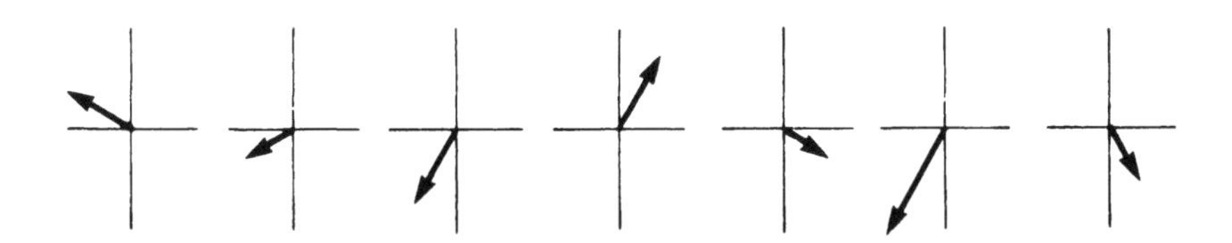

Figure 3.13 What the electric field of unpolarized light looks like as it moves past you.

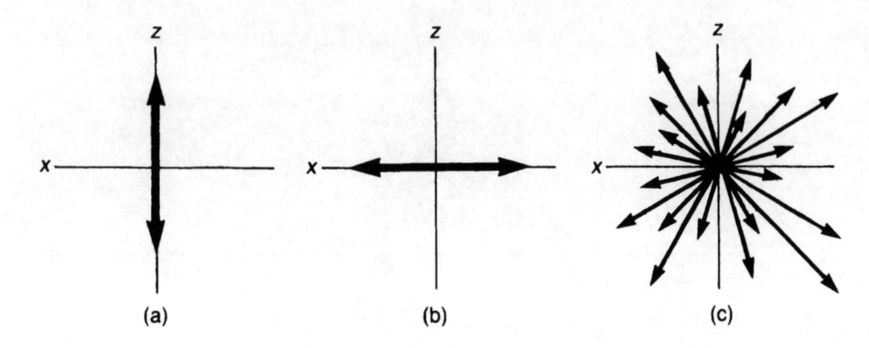

Figure 3.14 Orientation of the electric field for (a) vertically polarized light coming directly at you, (b) horizontally polarized light, and (c) unpolarized light.

3.17. Just as plane-polarized light can be vertically or horizontally polarized, circularly polarized light can be clockwise or counterclockwise polarized.

3.4 POLARIZATION COMPONENTS

To understand how the polarization of light can be manipulated by devices like waveplates, Pockels cells, and birefringent filters, it is necessary to understand how light can be composed of two orthogonally polarized components.

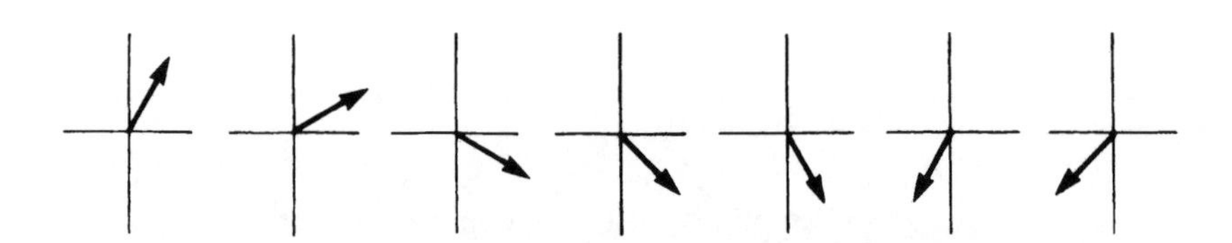

Figure 3.15 What the electric field of (clockwise) circularly polarized light looks like as it moves past you.

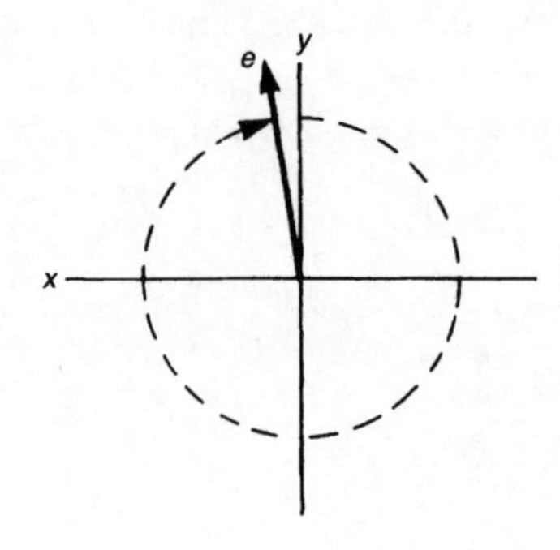

Figure 3.16 Circularly polarized light coming directly at you.

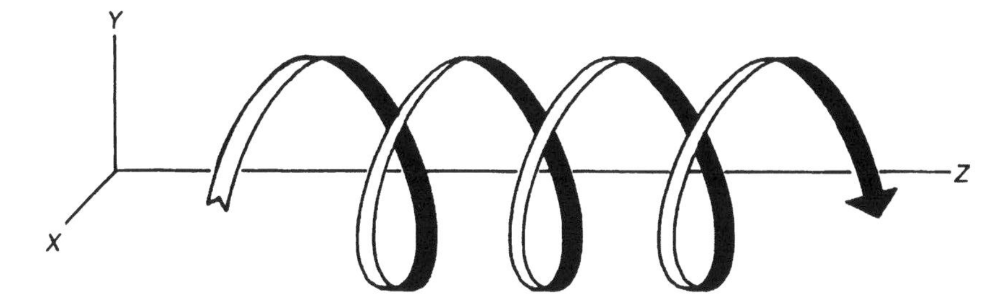

Figure 3.17 The path traced by the tip of the electrical field vector in circulatory polarized light.

Suppose at some point in space there are two electric fields, as shown in Figure 3.18. These two fields are equivalent to a single electric field, which is called the vector sum of the two original fields. You can construct the vector sum by lining up the individual vectors, tip to tail, without changing the direction in which they point (Figure 3.19). It is meaningless to try to say whether it is the two original fields or their vector sum that "really" exist at that point in space; the two pictures are exactly equivalent. Moreover, two different fields could describe the situation equally

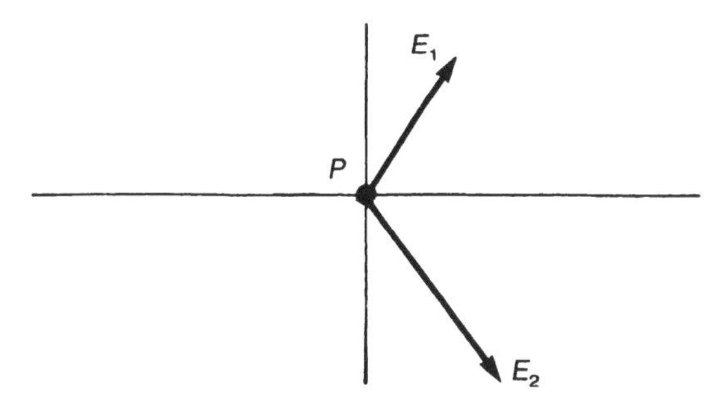

Figure 3.18 Two electric fields at a point in space.

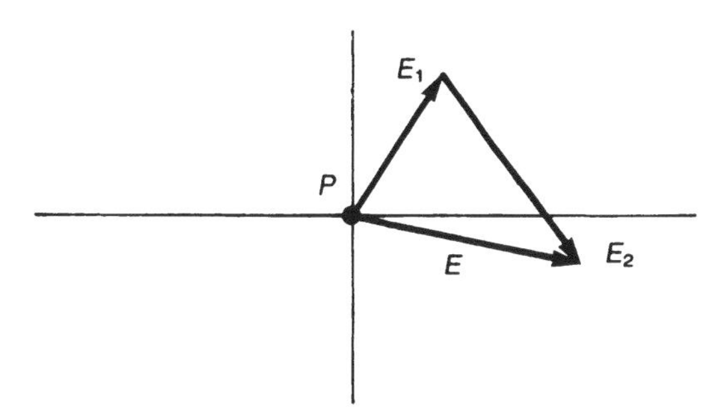

Figure 3.19 To find E, the vector sum of E_1 and E_2 in Figure 3.18, line up E_1 and E_2 tip to tail without changing their direction.

well if their vector sum is the same as that of the first two. Figure 3.20 shows two such fields that might exist at the point. In other words, Figures 3.18, 3.19, and 3.20 are three different ways of describing the same physical situation.

Now take a look at Figure 3.21, which shows two electric fields. (Do not confuse this drawing with Figure 3.11, which shows the electric and magnetic fields of a light wave. Figure 3.21 shows the electric fields of two light waves.) At any point along the *y*-axis, the two fields are equivalent to their vector sum. And what does the vector sum look like as the waves of Figure 3.21 move past you? That is shown in Figure 3.22; the two waves in Figure 3.21 are exactly equivalent to the single, plane-polarized light wave shown in Figure 3.23. Figures 3.21–3.23 are different ways of describing exactly the same physical situation.

If you go backward through this explanation, you will see that any plane-polarized light wave can be thought of as the vector sum of two orthogonal components that are in phase with each other.

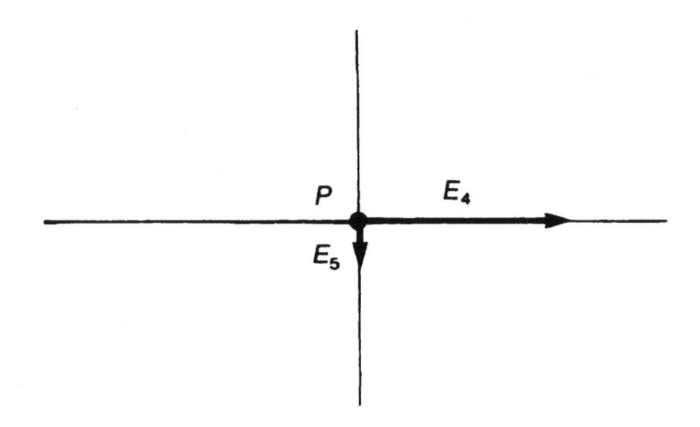

Figure 3.20 These two electric fields produce the same vector sum as E_1 and E_2 in Figure 3.18.

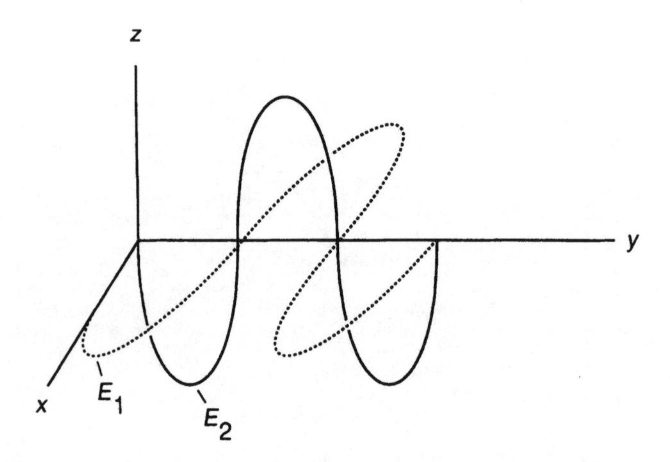

Figure 3.21 Two electric waves whose vector sum is a plane-polarized wave.

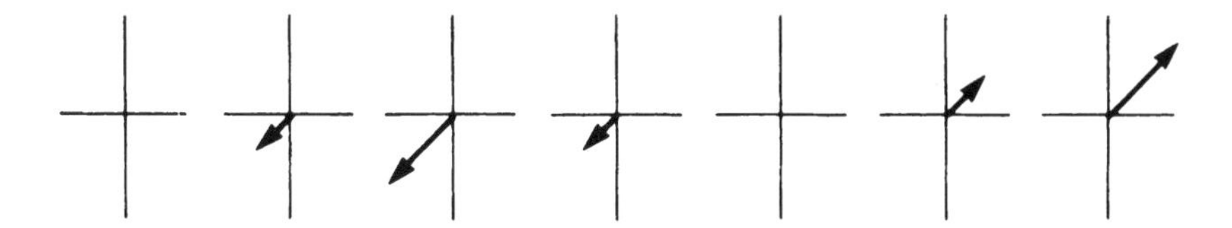

Figure 3.22 What the vector sum of the waves in Figure 3.21 looks like as it moves past you.

The phrase "in phase with each other" is crucial. What happens if the two components are out of phase with each other? At this point, the three-dimensional visualization gets a little tricky. It is not conceptually difficult, but you have to think carefully. Figure 3.24 shows two orthogonal waves that are out of phase with each other. Compare this drawing with Figure 3.21 and be sure you understand the difference.

What is the polarization of the light in Figure 3.24? To figure this out, look at the total electric field at several places along the wave, as in Figure 3.25. At point a in Figure 3.24 there is zero z-component of field, and the only field is in the negative x-direction. This is shown in Figure 3.25a. (In Figure 3.25 the y-axis is coming straight out of the book.) At point b in Figure 3.24, there is a field in both the negative x-direction and the negative z-direction. The sum of these two fields is shown in Figure 3.25b. And Figure 3.25c shows the electric field at point c in Figure 3.24. Suppose you are standing at the right-hand edge of Figure 3.24 and the wave is coming directly at you. First, the electric fields at point a will enter your eye, then point b, then point c. That is, Figure 3.25 is exactly what your eye sees at three different instants. And what is the electric field doing in Figure 3.25? It is going around in a circle. Figure 3.24 is a representation of the components of circularly polarized light.

Now compare Figure 3.21 with Figure 3.24 again. Figure 3.21 is plane-polarized light; Figure 3.24 is circularly polarized. Do you understand why the light in Figure

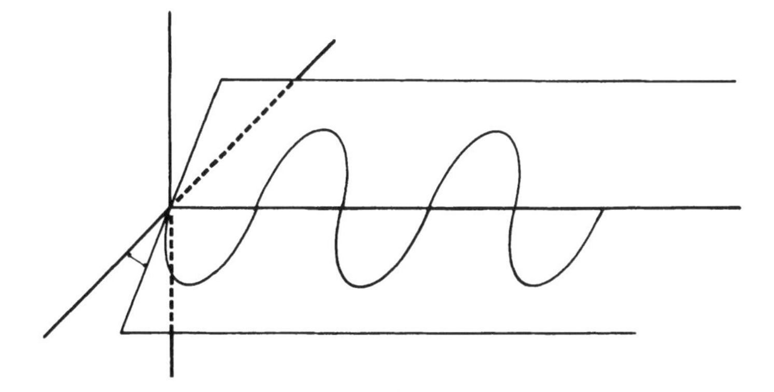

Figure 3.23 The two waves in Fig. 3.21 add at every point along the axis to produce a single field that looks like this.

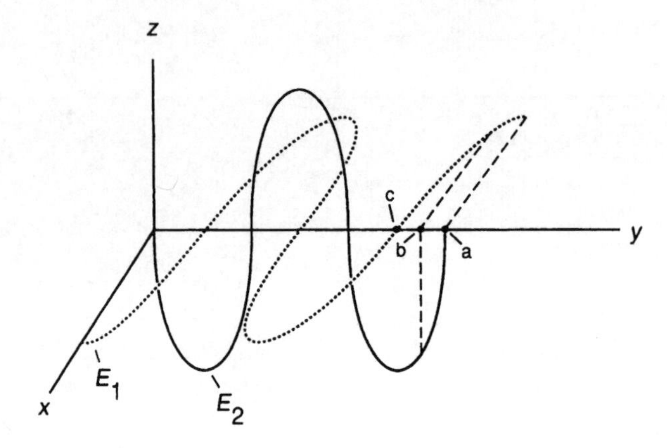

Figure 3.24 Two electric waves whose vector sum is a circularly polarized wave.

3.25 is called "clockwise" circular polarization? Ask yourself how you would change Figure 3.24 to represent counterclockwise circular polarization.

3.5 BIREFRINGENCE

Remember from the earlier discussion that the refractive index of a material determines how much light slows down when it travels through the material. In a fundamental sense, light moves more slowly in a material than in a vacuum because light interacts with the electrons in the material. Most materials have only one refractive index (for a particular wavelength), but a birefringent material has two.

These two indices are experienced by light in two orthogonal polarizations. In Figure 3.26, horizontally polarized light will behave as if the crystal's refractive index is 1.7, and vertically polarized light will behave as if the crystal's refractive index is 1.4. How can a crystal have two refractive indices at the same time? To understand that, think about the internal structure of a birefringent crystal, shown in Figure 3.27. The forces that hold electrons at their equilibrium positions in the crystal are represented here by springs. In an ordinary crystal, one that's not birefringent, all the springs would be the same; the asymmetry shown in Figure 3.27 is what causes birefringence.

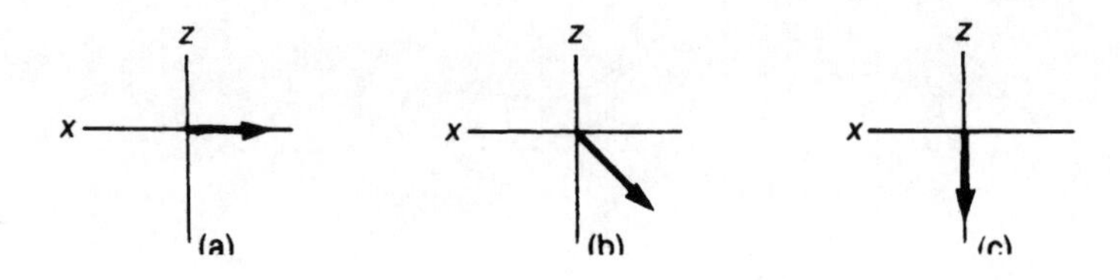

Figure 3.25 What the vector sum of the waves in Figure 3.24 looks like as it moves past you.

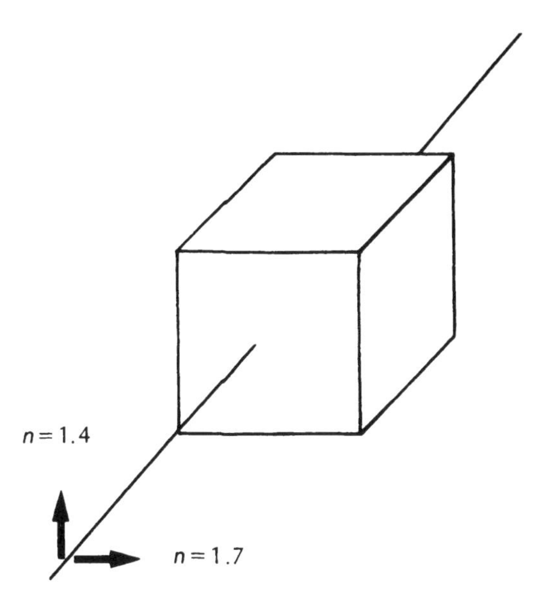

Figure 3.26 The refractive index of a birefringent material depends on polarization.

Remember that an electric field exerts a force on a charged particle, so the light wave passing through this crystal will vibrate the electrons. But here is the important thing to understand: a horizontally polarized wave will vibrate the electrons horizontally and a vertically polarized wave will vibrate them vertically. (Why? Because the force on the electrons is in the same direction as the light wave's electric field.) But because electrons respond differently to horizontal and vertical forces—

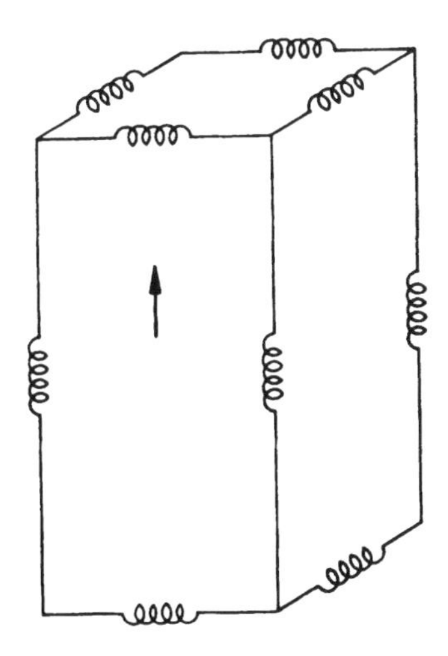

Figure 3.27 The internal structure of a birefringent material.

horizontal forces excite the short springs in Figure 3.27, and vertical forces excite the long ones—horizontally polarized light will interact differently with the crystal than vertically polarized light: that is, the horizontally polarized light will experience a different refractive index.

In the previous section, you read about the two orthogonal cases of plane polarization and the two orthogonal cases of circular polarization (clockwise and counterclockwise). Next, we explain how you can use birefringent crystals to convert one of these to another. You can, for example, convert plane polarization to circular or horizontal polarization to vertical.

Let us start by converting plane-polarized light to circularly polarized light. Recall that plane-polarized light is shown in Figure 3.21, and circularly polarized light is shown in Figure 3.24. Thus, the problem becomes: What do you do to the light in Figure 3.21 to make it look like the light in Figure 3.24?

Look carefully at the two figures. In both cases the vertical field (E_2) is the same. But in Figure 3.24 the horizontal field (E_1) is shifted to the left by one-quarter of its length. Or, since the wave is moving left to right, the horizontal component has been retarded by a quarter wave.

How can you retard one component of polarization with respect to the other? The same way you retard one racehorse with respect to another: you make it move slowly. How do you do that? If you send the light wave pictured in Figure 3.21 through the birefringent crystal in Figure 3.26, what happens? The horizontal component (E_1 in Figure 3.21) moves more slowly than the vertical component. And what emerges looks like the light in Figure 3.24. Plane-polarized light has been converted to circularly polarized light! The entire process is depicted in Figure 3.28.

The crystal in Figure 3.28 has to be exactly the right length or the retardation will be something other than exactly a quarter wave. But if the crystal is long enough to produce a quarter wave of retardation, then the device is called a quarter-wave plate, and it converts plane-polarized light to circularly polarized light.

What happens if you put another quarter-wave plate to the right of Figure 3.28? How do you describe the polarization of the light that emerges from it? The horizontal component has been retarded by another quarter-wave, so now the light looks

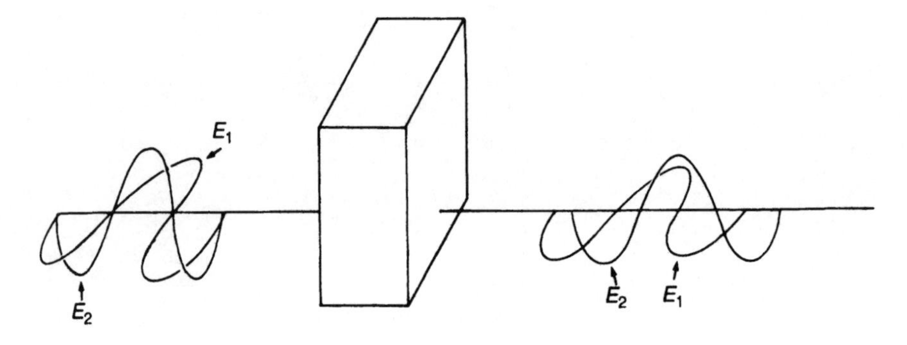

Figure 3.28 A birefringent crystal converts a plane-polarized light to circular.

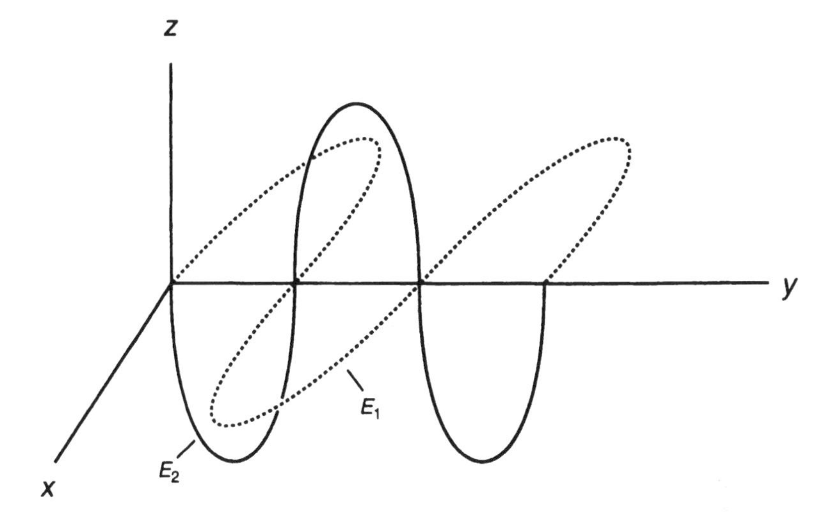

Figure 3.29 An exercise for the student.

like that in Figure 3.29. It is left as an exercise for the student to show that this is plane-polarized light and that it is orthogonal to the plane-polarized light we started with in Figure 3.21. So two quarter-wave plates, or a single half-wave plate, convert plane-polarized light to the orthogonal plane polarization.

What would come out of a third quarter-wave plate? Out of a fourth? It is left as another exercise for the student to show that Figure 3.30 correctly summarizes the discussion of wave plates.

Note that in relation to the previous figures, the world has been rotated by 45 degrees in Figure 3.30. That is, in Figure 3.28 light was polarized diagonally and the crystal axes were horizontal and vertical. In Figure 3.30, light is polarized horizontally or vertically, and the crystal axes are diagonal. It all works out the same since the only thing that matters is the relative orientation of the polarization components to the crystal axes.

So far, we have been looking at only a special case of light propagation in a birefringent material: the case in which light goes straight into the material (i.e., light traveling parallel to the short springs in Figure 3.27). That is all you need understand if you want to explain wave plates and birefringent filters (see Chapter 10) and electrooptic modulators (see Chapters 11 and 12). But when we talk about nonlinear optics

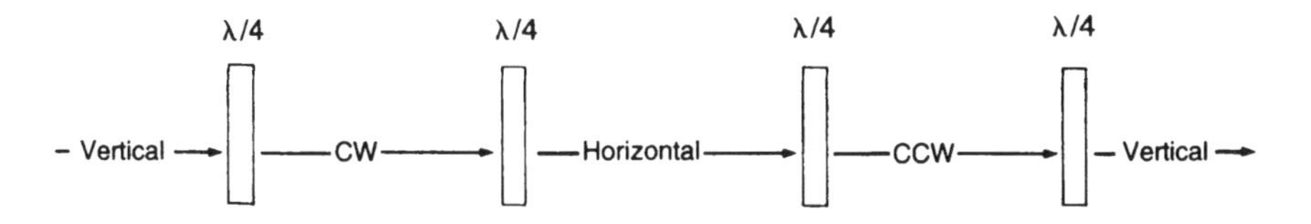

Figure 3.30 A summary of the effect of wave plates on polarization.

in Chapter 13, we will have to consider cases in which light propagates through a birefringent crystal at an arbitrary angle. The remainder of this section deals with that problem.

There is an arrow in Figure 3.27 that has gone unmentioned thus far. It indicates the direction of the optic axis in the birefringent material. The optic axis is always perpendicular to a square of springs, as shown in Figure 3.27.

We need a couple of definitions now. The "ordinary" light is polarized perpendicular to the optic axis. The "extraordinary" light is polarized perpendicular to the ordinary. No matter in what direction the light is going, those are the only two possibilities. Any light in the crystal can be broken down to its ordinary and extraordinary components. In the sequence of Figures 3.26 through 3.29, the horizontally polarized light is ordinary, and the vertically polarized light is extraordinary.

The special case that we have already considered is when light propagates perpendicular to the optic axis. We have seen that the light divides itself into ordinary and extraordinary components that propagate at different velocities. The question to ask now is: What happens when the light is not propagating perpendicular to the optic axis?

First consider another special case: light propagating along the optic axis. In Figure 3.27, light going straight up (or straight down) will vibrate the short springs no matter how it is polarized because its electric field will always be in the same direction as the short springs. Hence, light traveling parallel to the optic axis will experience only one refractive index, independent of polarization. There is no birefringence for light traveling along the optic axis.

In the general case, light propagates at an oblique angle to the optic axis. The ordinary light, polarized perpendicular to the optic axis, still vibrates only the short springs in Figure 3.27, so its refractive index does not depend on the propagation angle. But the extraordinary light now vibrates both kinds of springs. Which springs it vibrates more—and, hence, the extraordinary refractive index—does depend on the propagation angle. Our problem is to understand how the extraordinary light behaves as a function of propagation angle.

A good way to do this is to visualize the Huygens wavelets expanding in the birefringent medium. Next, we develop *The Baseball in the Gouda Cheese Model** of birefringent propagation.

First visualize a point source in an ordinary, nonbirefringent medium, and ask yourself what the Huygens wavelet produced looks like. It is a sphere. It is like the circular wavelet you produce when you drop a pebble into a pond, except that a point source in three-dimensional space produces a three-dimensional sphere.

Next visualize the extraordinary Huygens wavelet produced by a point source in a birefringent medium. This is not easy to do the first time you try. You know that the part of this wavelet moving along the optic axis experiences the same refractive index as ordinary light. And you know that the part of this wavelet moving perpen-

*Okay, the authors made that name up. If you Google "Baseball and Gouda Cheese," you will find nothing. But it is a descriptive name for the model.

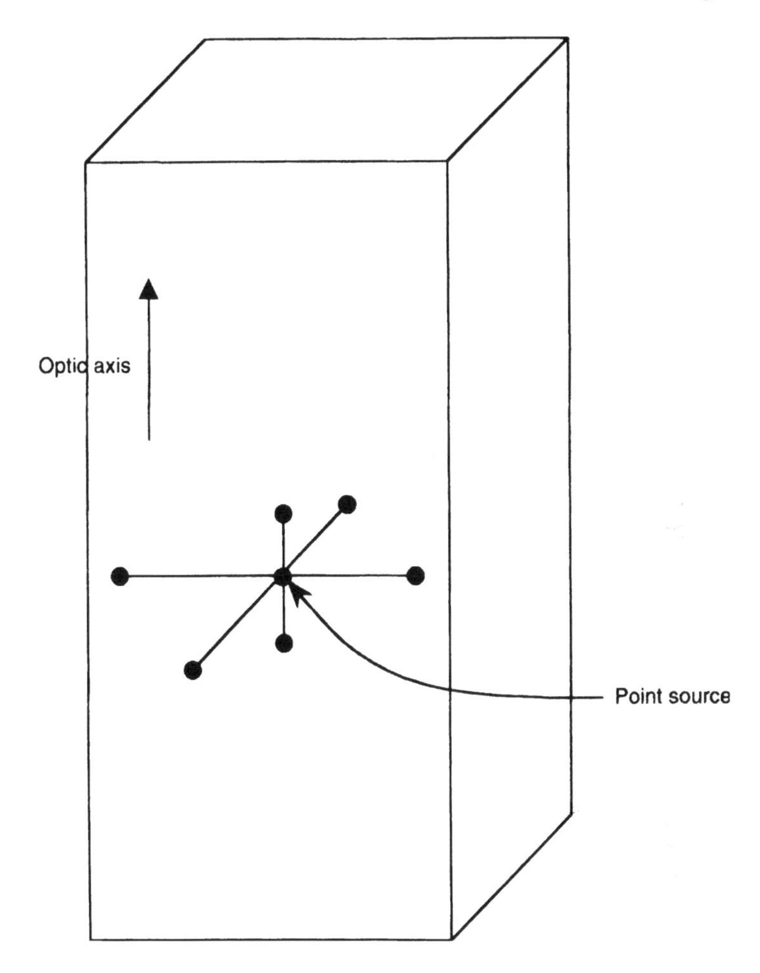

Figure 3.31 The lines show how far light from the point source travels in directions perpendicular to, and parallel to, the optic axis.

dicular to the optic axis vibrates only the long springs and, hence, experiences the "pure" extraordinary index.* In many crystals, the extraordinary index is smaller than the ordinary index; let us assume that is the case here. Thus, light moving perpendicular to the optic axis moves faster than light moving along the optic axis. The portions of the wavelet moving perpendicular to and along the optic axis are shown in Figure 3.31. To visualize the rest of the Huygens wavelet, just fill it in as in Figure 3.32. It turns out this is not an ellipsoid, a shape not unlike a gouda cheese.

*Extraordinary light propagating perpendicular to the optic axis vibrates only the long springs and, therefore, experiences only the "pure" extraordinary refractive index. Extraordinary light propagating at an oblique angle to the optic axis vibrates both long and short springs and, hence, experiences an extraordinary index somewhere between the "pure" extraordinary index and the ordinary index.

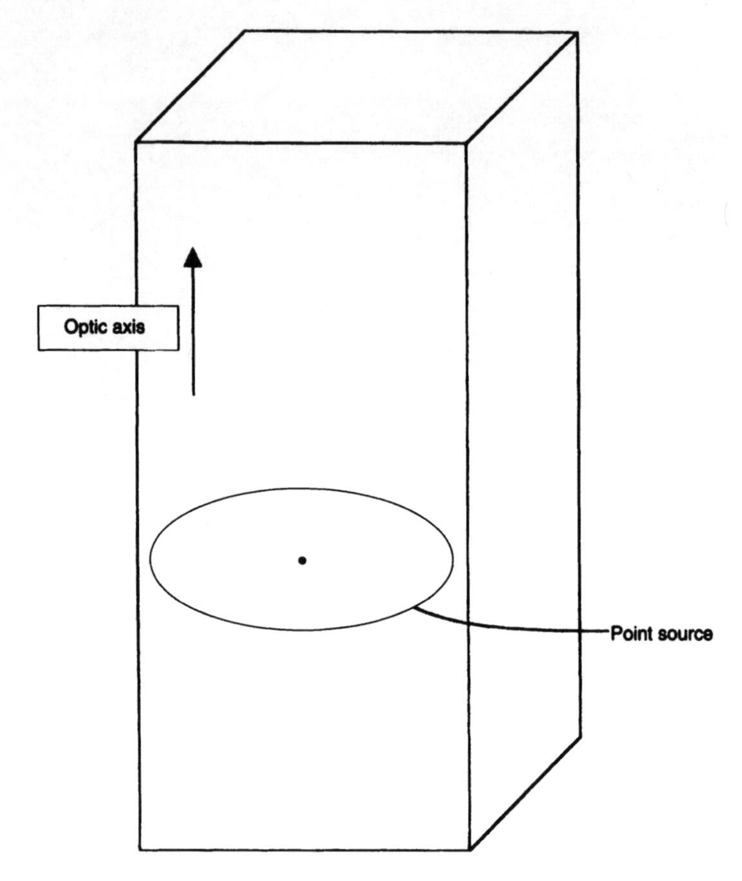

Figure 3.32 Visualizing the Gouda cheese shape of the extraordinary Huygens wavelet created by the point source.

What about the ordinary Huygens wavelet in a birefringent medium? We've already said that the ordinary light moves at a constant velocity, independent of propagation angle. (That is, the ordinary light interacts only with the short springs, no matter what direction it's traveling.) So the ordinary wavelet is simply a sphere.

Both Huygens wavelets are shown in Figure 3.33. These two surfaces can be visualized as a baseball inside a Gouda cheese. These surfaces represent the wave-

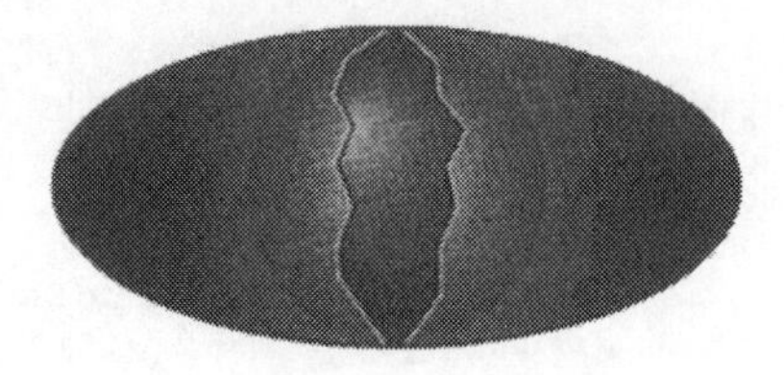

Figure 3.33 The ordinary and extraordinary Huygens wavelets in a birefringent material.

fronts produced by a point source inside a birefringent medium, just as a simple circle represents the wavefront produced by a pebble dropped into a pond. All the light polarized perpendicular to the optic axis goes into one of the wavefronts (the baseball), and the rest of the light goes into the other wavefront (the Gouda cheese).

The baseball in the Gouda cheese, together with Huygen's principle, will enable you to understand all birefringent phenomena. For example, consider a quarter-wave plate. The light propagates perpendicular to the optic axis, as shown in Figure 3.34. Just as in Figure 3.9, we have replaced the infinite number of point sources with two point sources, one at either edge of the beam. But now a side view of the baseball in the Gouda cheese replaces the spherical Huygen's wavelets of Figure 3.9. And now, by drawing tangents to the ordinary and extraordinary wavelets, we see that the incoming light is broken into two wavefronts in the birefringent medium. Of course, this is the same situation we dealt with earlier. In Figure 3.28, E_1 was the ordinary wave and E_2 was the extraordinary wave, which moved faster through the crystal.

Next, consider using the baseball/cheese model to explain propagation parallel to the optic axis. In Figure 3.35, you see that the two wavefronts travel at exactly the same velocity, so there is no distinction between the ordinary wavefront and the extraordinary wavefront.

Finally, let us consider the general case of propagation at an oblique angle to the optic axis, as shown in Figure 3.36. As in Figures 3.34 and 3.35, we construct the ordinary and extraordinary wavefronts by drawing tangents to their Huygens wavelets. But look at what is happening in this picture: the extraordinary wavefront is moving sideways, "walking off" from the ordinary wavefront. Thus, the single incoming beam of light will emerge as two parallel beams.

The effect we have just described—when light propagates at an oblique angle to the optical axis—is often called double refraction, for obvious reasons. The incoming light is refracted at two different angles.

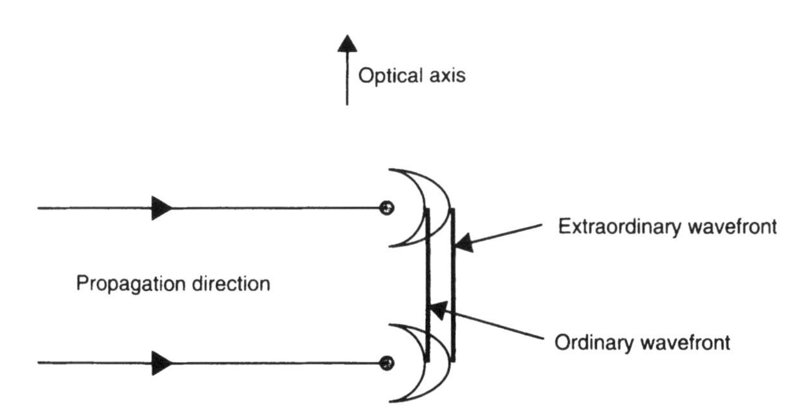

Figure 3.34 The extraordinary wavefront moves faster than the ordinary wavefront. The two crescent-moonlike shapes are side views of the baseball and Gouda cheese in Figure 3.33.

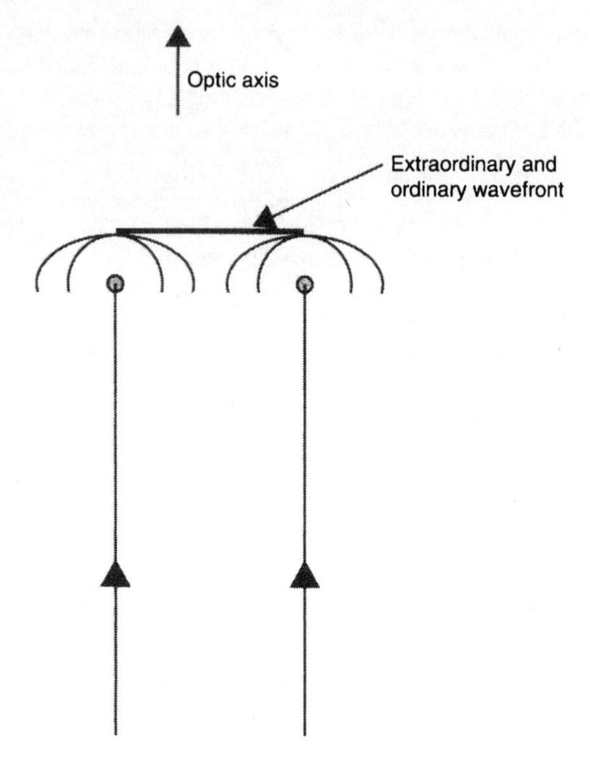

Figure 3.35 When light propagates parallel to the optic axis, the ordinary and extraordinary wavefronts move at the same velocity. The two semicircle-like shapes are again side views of the baseball and Gouda cheese, but in this case the top half is shown.

3.6 BREWSTER'S ANGLE

If you have ever glanced at your reflection in a plate-glass window, you know that light is reflected from the interface between glass and air. That's because any discontinuity in refractive index causes a partial (or sometimes total) reflection of light passing across the discontinuity. The reflectivity of a glass-air interface for light at normal incidence is about 4%. So when you glance at your reflection in that plate-glass window, you are seeing about 8% of the light incident on the window being reflected back to you—4% from each side of the glass.

The important point here is that the fraction of light reflected depends on the angle of incidence and the polarization of the light. Let us conduct a thought experiment to see how this works. If you set up a piece of glass and a power meter as shown in Figure 3.37, you can measure the reflectivity as you rotate the piece of glass (and the power meter). What will your results look like? That is shown in Figure 3.38, in which the reflectivities for both horizontal and vertical polarizations are given as a function of the angle of incidence. For the horizontal polarization (i.e., the polarization perpendicular to the plane of Figure 3.37), the reflectiv-

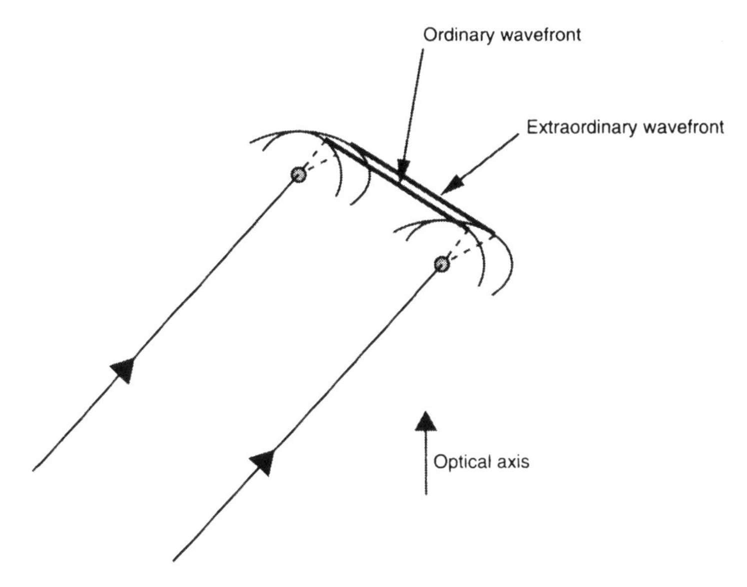

Figure 3.36 When light propagates at an oblique angle to the optic axis, the ordinary and extraordinary wavefronts travel at different speeds and in different directions. Here, the ordinary wavefront is moving straight ahead, while the ordinary wavefront is walking off at an angle shown by the dotted lines. You can convince yourself that the two beams are parallel when they emerge from the birefringent crystal by drawing the (spherical) Hugen's wavelets the beams produce in air. If the crystal is long enough, the two beams will not overlap at all.

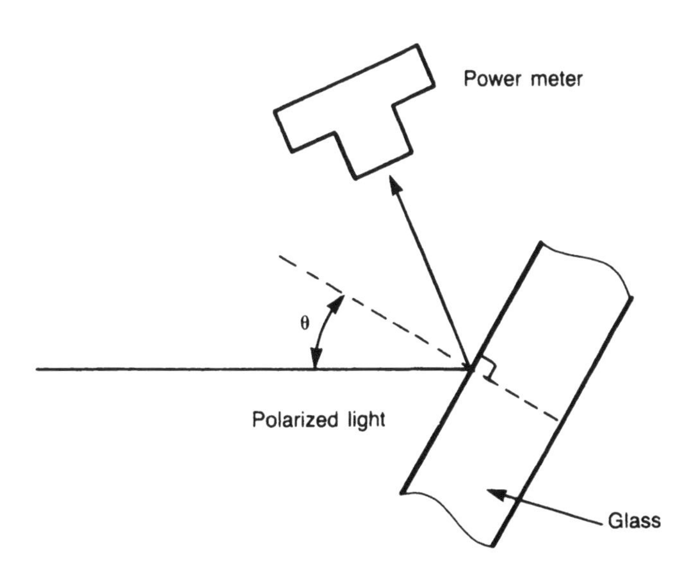

Figure 3.37 An experiment to measure the reflectivity of a piece of glass as a function of the angle of incidence.

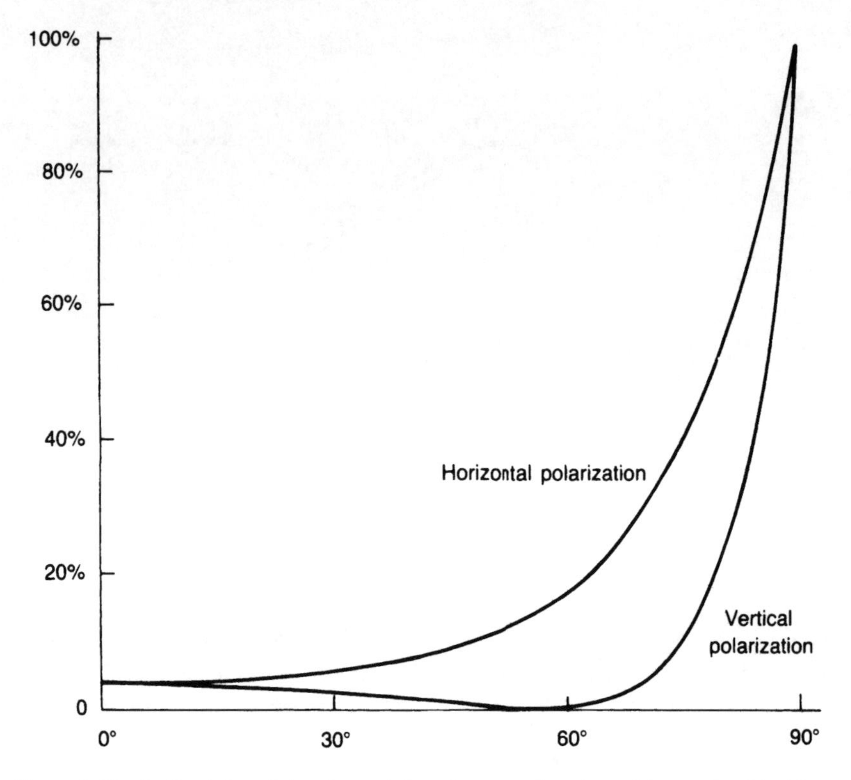

Figure 3.38 Reflectivity of the glass–air interface as a function of the angle of incidence.

ity increases gradually at first and then more rapidly as the angle of incidence approaches 90°.

For the vertical polarization, the reflectivity decreases to zero and then increases. The angle at which the reflectivity disappears is Brewster's angle, and it depends on the refractive indices of the two media.* A Brewster plate—a small piece of glass oriented at Brewster's angle—can be placed inside a laser to introduce a loss of about 30% (15% at each surface) to one polarization but no loss to the other polarization. This preferential treatment of one polarization is usually enough to make the laser lase in only that polarization.

3.7 BRIGHTNESS

When you say that one light source is brighter than another, you mean that the brighter source creates a greater intensity on the surface of your retina when you look at the source. The intensity on this surface depends on the intensity of the source and the extent to which the light spreads out after it leaves the source. The

*Brewster's angle is given by $\tan \theta_B = n_2/n_i$, in which n_2 is the refractive index of the medium on the right if the ray travels from left to right.

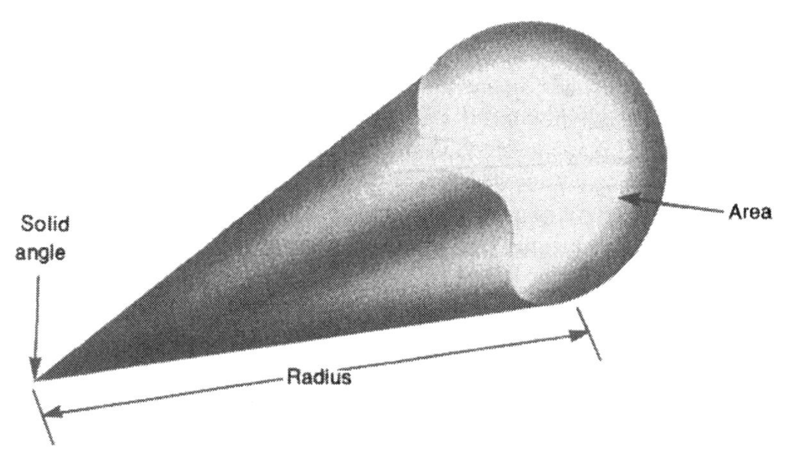

Figure 3.39 A solid angle.

faster light spreads out, the less reaches your eye. This spreading out of the light is called the *divergence* of the source, and it can be measured in terms of the solid angle formed by the light leaving the source.*

What is a solid angle? Well, you can think of a solid angle as the three-dimensional analogy to an ordinary plane angle. In general, a solid angle can have any irregular shape, like the one shown in Figure 3.39. Or it can be regular, like an ice-cream cone. The magnitude of a plane angle is measured in radians; the magnitude of a solid angle is measured in steradians. If you recall that the magnitude of a plane angle (in radians) is the ratio of the length of a circle's arc subtended by the angle to the radius of the circle, you can readily see how the magnitude of a solid angle (in steradians) is defined. It is the ratio of the area of a sphere's surface subtended by the angle to the square of the radius of the sphere. Why the square of the radius? Because the solid angle is a measure of how much the angle spreads out in three dimensions, and that quantity increases with the square of the radius. Since the square of the radius is divided into the area, steradians are dimensionless (as are radians).

In laser technology, the brightness of an optical source is defined as the source's intensity divided by the solid angle of its divergence:

$$B = \frac{P}{A\Omega}$$

in which P is the power of the source, A is its cross-sectional area, and Ω is divergence in steradians.

*Because the beams from commercial lasers are usually symmetrical, divergence is more conveniently measured in plane rather than solid angles. Thus, the divergence of most lasers is specified in radians rather than steradians.

Note that because steradians are dimensionless, the dimensions of brightness are watts per square centimeter, the same dimensions as for intensity.* But brightness is different from intensity because the intensity of a source does not depend on its divergence.

QUESTIONS

1. There are 2π radians in a circle. How many steradians are there in a sphere?
2. Calculate the brightness of a 1 W source with a 1 cm diameter that radiates into 10^{-6} sr. Repeat the calculation if the source radiates into 2×10^{-6} sr. Repeat the calculation if the source radiates into 10^{-6} sr but its diameter is increased to 2 cm.
3. Polarizing sunglasses cut the glare reflected from horizontal surfaces (e.g., roads) better than ordinary sunglasses. Use the information in Figure 3.38 to explain why this is so. Should the sunglasses reject horizontal or vertical polarization?
4. If you have two sets of polarizing sunglasses and hold them at right angles (so that one blocks horizontal polarization and the other blocks vertical), how much light can get through both lenses? Now place the lens of a third pair between the first two crossed lenses. If the third lens is oriented to pass light polarized at 45° to the vertical, what do you see when you look through all three lenses? (Hint: "Nothing" is the wrong answer. Explain why.)
5. The tip of the electric-field vector traces an ellipse (rather than a circle as in Figure 3.16) in elliptically polarized light. Explain how an elliptically polarized light can be resolved into two orthogonal components. What is the phase relation between the components?
6. Describe a fast and easy way, by observing the light reflected from a handheld microscope slide, to figure out whether a laser beam is linearly polarized and, if it is, whether it is vertically or horizontally polarized.

*The dimensions of brightness are often written $W/cm^2/sr$, in which W is watts and sr is steradian, even though a steradian is not a real dimension.

Chapter 4

Interference

The effects of optical interference can be observed with laser light or with ordinary (incoherent) light, but they can be explained only by postulating a wavelike nature of light. In this chapter, we explore what optical interference is and note several examples where it can be observed in everyday life. Then we examine two other examples that are especially important to laser technology.

The first of these is Young's double-slit interference experiment. Now, if you have ever taken an optics course, you have already studied this effect. It is probably covered in every introductory optics book that has ever been written because it is such a straightforward, easy-to-understand example of optical interference. But the principle embodied in Young's experiment is especially important in laser technology because it is exactly the same principle employed in acousto-optic (A-O) modulators—and A-O modulators are very important devices. Moreover, the principle of Young's experiment is also the underlying principle of holography. So you can see why it is important to understand what is going on in Young's double-slit experiment.

The second example is the Fabry–Perot interferometer. It is probably in just about every optics book ever written and for the same reason: it is an elegant demonstration of optical interference. But the Fabry–Perot is a very useful instrument in its own right, so it is doubly important to understand it. But the most important reason for studying the Fabry–Perot interferometer is that the physics that goes on inside a Fabry–Perot is quite similar to the physics that goes on inside a laser resonator. If you are to understand what goes on inside a laser, you have got to understand a Fabry–Perot first.

4.1 WHAT IS OPTICAL INTERFERENCE?

Two or more light waves interfere with each other when they're brought together in space to produce an electric field equal to the sum of the electric fields in the individual waves. The interference can be destructive if the waves are out of phase with each other or constructive if the waves are in phase. In Figure 4.1a, the two waves that are brought together are in phase, so a bright spot—brighter than either of the waves would produce by itself—is created on the screen. In Figure 4.1b the two waves are out of phase, so they cast no illumination on the screen.

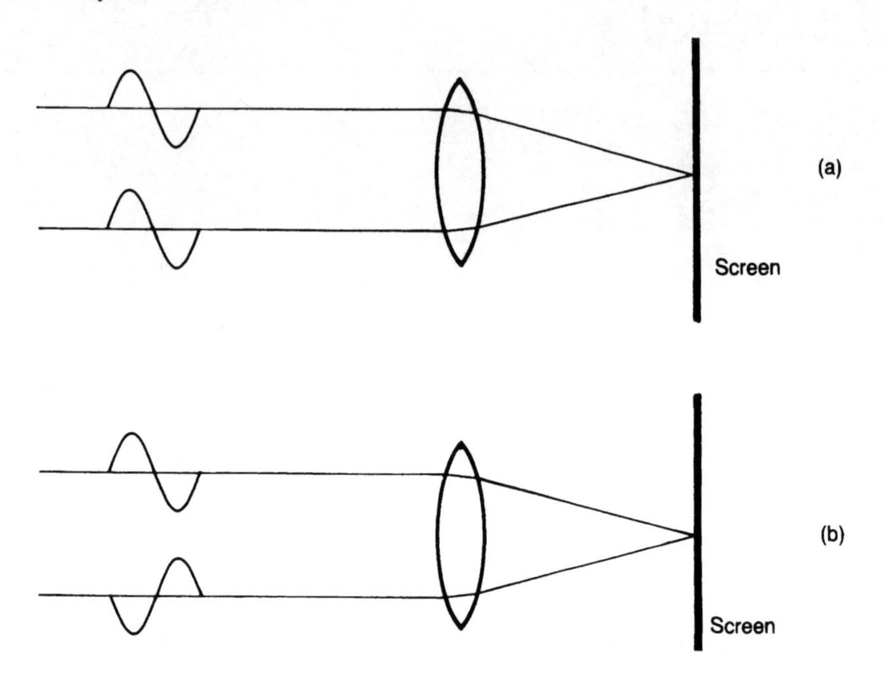

Figure 4.1 Constructive (a) and destructive (b) optical interference.

As an example of how optical interference can occur, consider two smooth microscope slides separated at one end by a human hair, as shown in Figure 4.2. An observer will see bright and dark bands if the slides are illuminated as shown because two waves will interfere at his or her retina. Figure 4.3 shows where those two waves come from: one from the upper slide and one from the bottom. (Let us ignore reflections from the other two surfaces of the slides; they would behave the same

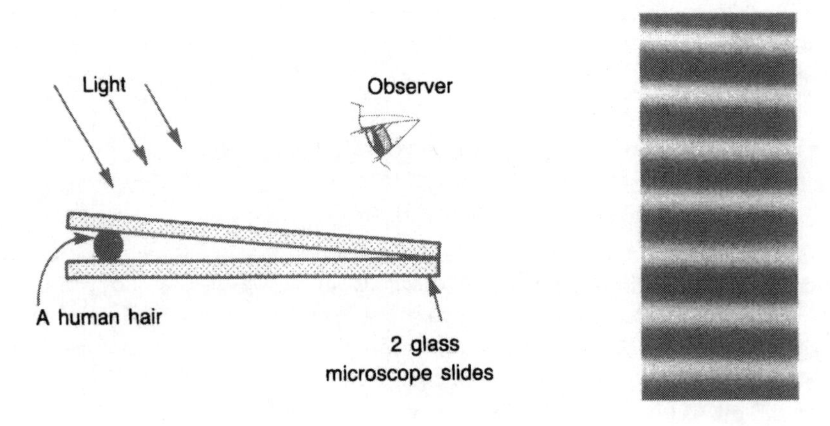

Figure 4.2 An example of optical interference. The observer sees evenly spaced bright and dark stripes, as shown on the right.

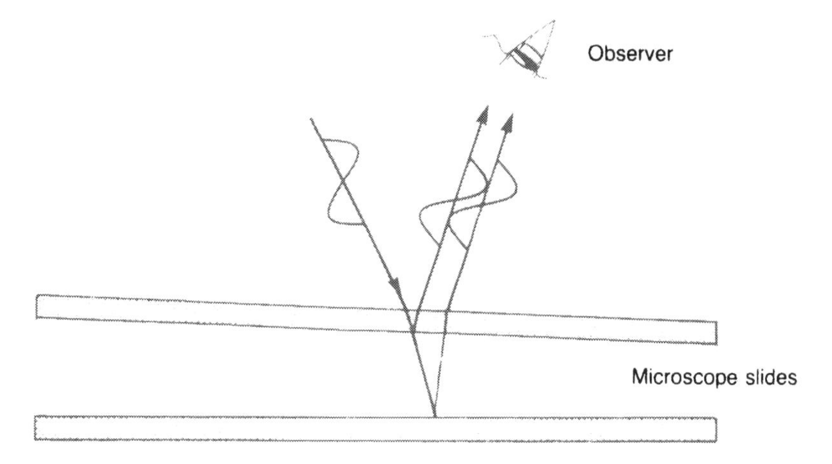

Figure 4.3 If the path difference for the two waves is an integral number of wavelengths, the observer sees constructive interference.

way as the ones we are considering.) These two waves are brought together at the observer's retina by the lens of the eye.

Now, these two waves started out being the same wave, but they were split apart by the partial reflection from the upper slide. The wave reflected from the bottom slide has a longer distance to travel before it reaches the observer's eye, and this path difference is important. If the path difference is an integral number of wavelengths, then constructive interference will occur at the observer's retina. On the other hand, if the path difference is an integral number of wavelengths plus one more half-wavelength, then the waves will be out of phase and will interfere destructively at the observer's retina.

Constructive interference is shown in Figure 4.3. But if the observer shifts her gaze slightly, so that she's looking at a different place, the path difference will also change and destructive interference can occur. Thus, as the observer shifts her gaze over the microscope slides, she sees a series of light and dark stripes, as shown in Figure 4.2.

4.2 EVERYDAY EXAMPLES OF OPTICAL INTERFERENCE

If you have ever noticed the colors reflected from an oil slick on a puddle of water or reflected from the side of a soap bubble drifting through the air, you have seen optical interference from a thin film. Take the oil slick as an example. It is a thin film of oil floating on top of the water, as shown in Figure 4.4. Sunlight striking the oil slick is reflected from the interface between oil and air and from the interface between water and oil. The two reflected waves are brought together at the observer's retina. Constructive interference will occur only if distances traveled by the two waves differ by an integral number of wavelengths.

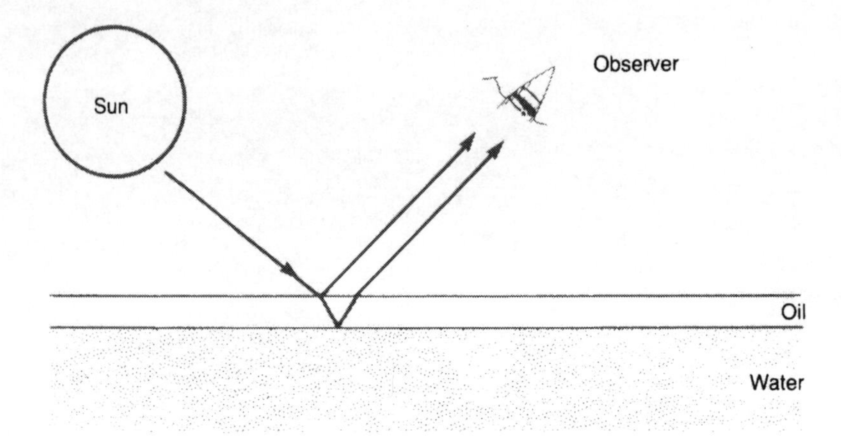

Figure 4.4 Optical interference from a thin film.

For some wavelengths (say, red), there will be constructive interference, whereas for other wavelengths the interference will be destructive. Thus, the observer sees red. When the observer shifts his or her gaze to another spot on the puddle, the path difference will change and a different color will be seen because there will now be an integral number of those wavelengths over the path difference. It is the same idea for the soap bubble, except that light is reflected from the inner and outer surfaces of the bubble.

When you look at a compact disc (CD), you will see rainbows of colors reflected from it. The surface of the CD is covered with millions and millions of little pits, and each of these pits reflects light to your eyes. The light from each pit follows a different path from the source to your eye, where it interferes destructively or constructively, depending on wavelength.

The CD acts like a diffraction grating, a device discussed further in Chapter 11. Letters and stickers made of inexpensive diffraction gratings are used to decorate automobiles. They are eye-catching because their color changes as the viewing angle changes. Sometimes, jewelry is made from diffraction gratings for the same reason.

4.3 YOUNG'S DOUBLE-SLIT EXPERIMENT

Suppose a screen with two closely spaced slits is illuminated from behind with monochromatic light, as shown in Figure 4.5. The pattern of light cast on the viewing screen is not what you might expect it would be. Instead of the two bright stripes shown in Figure 4.6, many bright stripes appear on the viewing screen, as shown in Figure 4.5. The brightest stripe is right at the center of the screen, midway between the two slits, where you'd expect the screen to be darkest.

This result, like any other interference effect, can be explained only by taking into account the wavelike nature of light. In Chapter 2 (Figure 2.6), we learned how

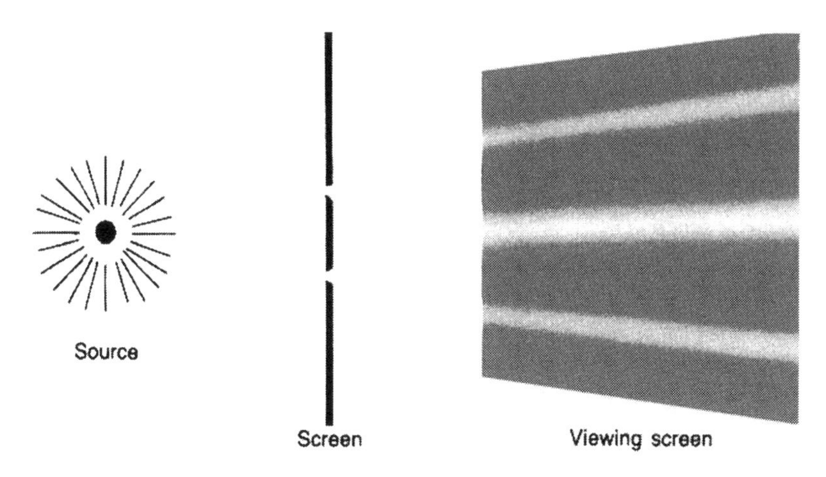

Figure 4.5 Young's double-slit experiment.

interference occurs between water waves from two sources, and the same logic will work with light waves. Figure 4.7 shows the light wavefronts incident on the slits and the new wavefronts that emanate from each of the slits. Constructive interference occurs wherever a wave crest from one slit coincides with a crest from the other slit. The heavy lines in Figure 4.7 show where constructive interference would cause a bright stripe to appear if a screen were placed at that location. The broken lines, on the other hand, show where the waves are exactly out of phase and would cast no illumination on a screen.

Figure 4.8 shows a different way of looking at the same effect. Here, an incoming wave passes through the slits. Figure 4.8 shows how that wave can arrive at two lo-

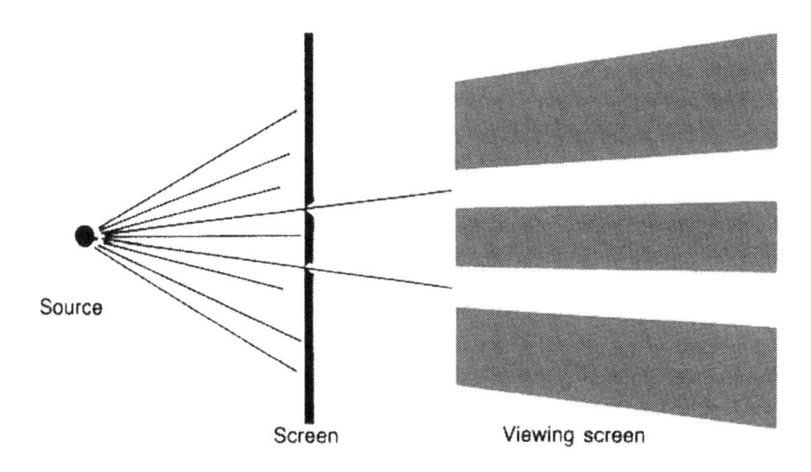

Figure 4.6 If light did not behave as a wave, only two bright stripes would be produced on the viewing screen.

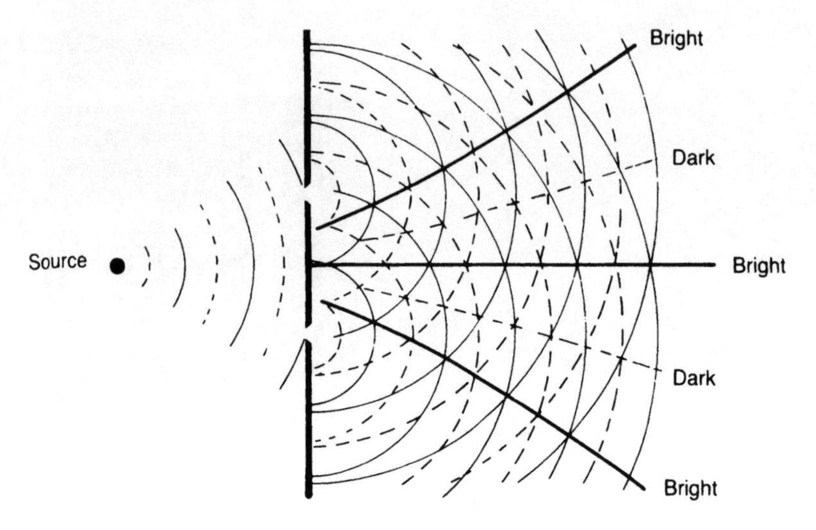

Figure 4.7 Alternating bright and dark bands are explained by the wavelike nature of light.

cations on the screen. At the center of the screen (P_1), the parts of the incoming wave that passed through the top and bottom slits have traveled the same distance and, hence, are in phase and interfere constructively to create a bright stripe. However, at a spot lower on the screen (P_2), the situation is different. Here the wave from the top slit has traveled farther than the other (it was on the outside of the curve) and has fallen behind. The two waves arrive at the screen out of phase, interfere destructively, and cast no light on the screen.

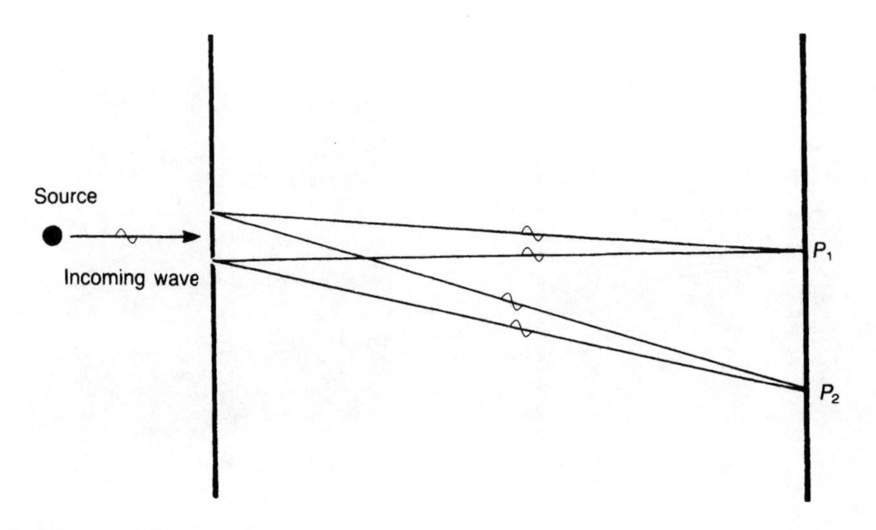

Figure 4.8 Screen illumination depends on the path difference traveled by light from each of the two slits.

How much light is cast on the screen between P_1 and P_2? As you move from P_2 to P_1, the screen will become increasingly brighter because the two waves will be arriving more closely in phase. Any other point on the screen will be brightly illuminated, dimly illuminated, or not illuminated at all, depending on the relative phases of the two waves when they arrive at that point. The screen will look as depicted in Figure 4.5.

Let us take a closer look at Figure 4.8 and derive a mathematical expression for the angles at which a bright stripe will be cast on the screen. If you assume that the distance from the slits to the viewing screen is much greater than the distance between the slits, then you can say that the paths taken by the two waves are approximately parallel. An enlargement of the part of the drawing showing the slits is presented in Figure 4.9. Here, θ is the angle that the (almost) parallel waves take when they leave the slits. The extra distance that the upper wave must travel to the screen is the distance labeled ℓ. If this distance is an integral number of wavelengths, then a bright stripe will appear at angle θ. But if you look at the triangle drawn in Figure 4.9, you see that $\ell = d \sin \theta$, in which d is the distance between the slits. Thus, the equation

$$n\lambda = d \sin \theta$$

defines the angles at which a bright stripe appears on the screen.

4.4 FABRY–PEROT INTERFEROMETER

A Fabry–Perot interferometer is a strange device. It takes things you think you understand—mirrors—and makes them behave in a way that does not seem to make sense.

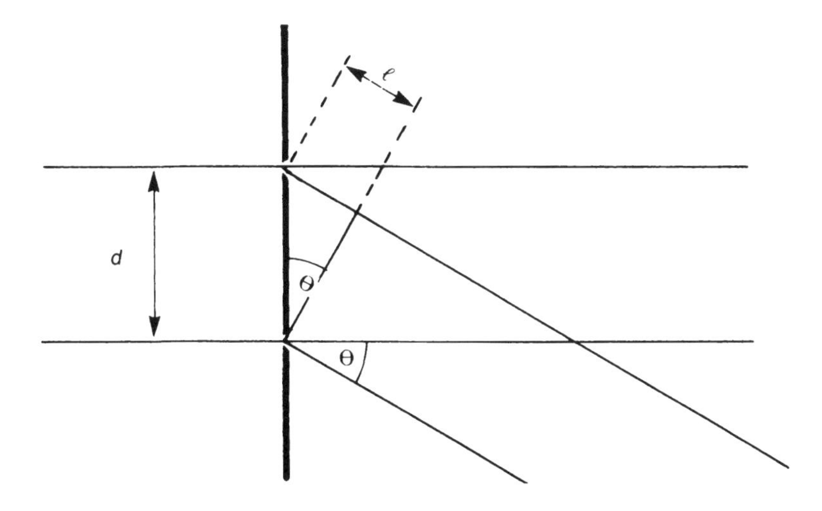

Figure 4.9 A closer look at Young's double-slit experiment.

Everybody knows what a mirror is, and the concept of a partially reflecting mirror is not very difficult to grasp. If you have a 95% reflecting mirror, it reflects 95% of the light that hits it and transmits the other 5%.

The part that does not seem to make sense happens when you take a second 95% mirror and place it directly behind the first. How much light would you expect to pass through both mirrors? If the first mirror M_1 reflects 95%, and the second mirror (M_2) reflects 95% of whatever gets through the first one, there should not be much light at all getting through the pair (5% of 5%, or 0.25% to be precise). But if you actually try this experiment and do it properly, you will find that 100% of the incident light is transmitted through the pair of mirrors!

The situation is illustrated in Figure 4.10. If 1 watt is incident from the left, 1 watt will be transmitted through the mirrors. But the stipulation about doing it properly in the previous paragraph is crucial. To understand what this means, let us look more closely at what goes on between the mirrors.

Figure 4.11 shows some of the light waves between the mirrors. The top wave is the 5% that first came through the left mirror. The next wave down is the 95% of the top wave that is reflected from the right mirror. And the bottom wave is the 95% of the middle wave that is reflected from the left mirror. Of course, we could go on indefinitely with this but, for simplicity, Figure 4.11 only shows the first three waves bouncing back and forth in the resonator. In reality, all these waves occupy the same volume in space; they're shown separated in Figure 4.11 for clarity. The arrowheads with each wave in Figure 4.11 show which direction the wave is moving.

The thing to look at carefully in Figure 4.11 is the phase of the waves moving from left to right inside the resonator. The phase is random; the light making its second crossing left to right has no particular phase relationship with the light making its first crossing. This is doing it improperly, and in this case the interferometer is nonresonant and very little light is transmitted.

For a Fabry–Perot to be resonant, the separation between its mirrors must be equal to an integral number of half-wavelengths of the incident light. Such an interferometer is shown in Figure 4.12. Note here that because the mirrors are separated by an integral number of half-wavelengths, the light is exactly in phase with itself after one round trip between the mirrors. Thus, all the waves traveling in one direction (say, left to right) are in phase with each other. And the waves moving right to left will likewise all be in phase.

In this case, all the individual waves between the mirrors add together and result in a substantial amount of power bouncing back and forth between the mirrors. For

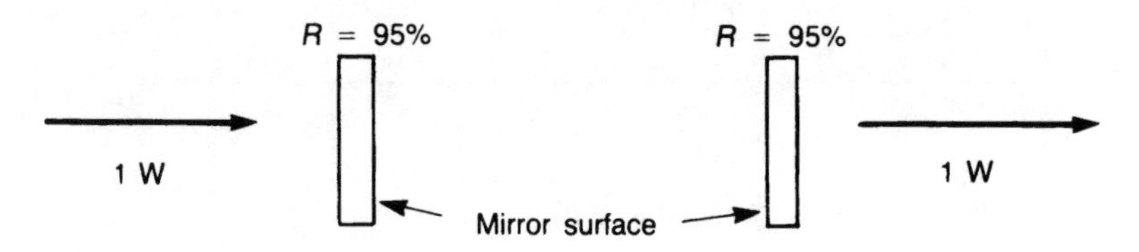

Figure 4.10 If a Fabry–Perot interferometer is resonant, it will transmit all the incident light, no matter how great the reflectivity of the individual mirrors.

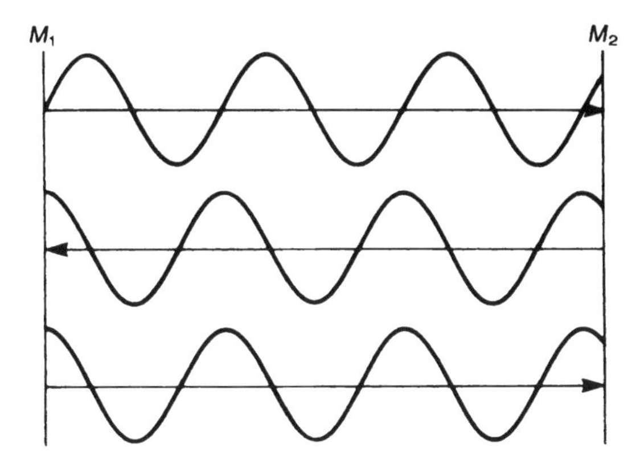

Figure 4.11 A nonresonant Fabry–Perot.

the interferometer shown in Figure 4.10, about 20 W will circulate between the mirrors, even though only 1 W is incident on the interferometer.

Those 20 watts are constantly reflecting off the second mirror, which transmits 5%. That is where the 1 W of transmitted light, shown in Figure 4.10, comes from.

Are you wondering about the light that travels right to left between the mirrors? It should be about 19 W (because one of the 20 W was transmitted through mirror M_2). What happens when this light strikes mirror M_1? M_1 is a 95% mirror, so is 5% of the 19 W transmitted through it?

The answer has to be an emphatic *no.* If 0.95 W (5% of 19 W) were transmitted, we would have a device on our hands that transmits 0.95 W to the left and 1 W to the right, even though it is receiving only 1 W. Thus, it would violate the law of conservation of energy.

But what happens to the 0.95 W that should have been transmitted through M_1? One way to understand what happens is to remember that there is 1 W incident on

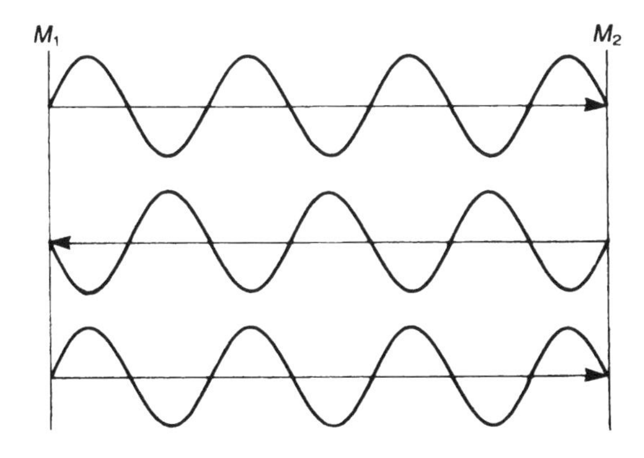

Figure 4.12 A resonant Fabry–Perot.

the interferometer. This light hits M_1 from the left, and if M_1 were acting like a proper 95% mirror, it would reflect 0.95 W back to the left. Now we have two "phantom" 0.95 W beams coming from M_1: one transmitted through M_1 and one reflected from M_1. These two beams are exactly out of phase with each other and cancel each other. And just as the energy magically disappeared from the two beams in Figure 4.1, the energy in the two phantom 0.95 W beams from M_1 disappears, then shows up again in the light circulating between the mirrors.

Thus, the input mirror M_1 does not act like a proper mirror when it is part of a Fabry–Perot interferometer because it does not reflect 95% of the incident light. The Fabry–Perot interferometer interacts with the electric and magnetic fields of the incident light as a single entity. It is misleading to try to figure out how one mirror by itself interacts with the light; you have to take the whole interferometer (both mirrors) into account.

Figure 4.13 summarizes the behavior of a Fabry–Perot interferometer. If the interferometer is resonant (i.e., if the spacing between its mirrors is equal to an integral number of half-wavelengths of the incident light), it will transmit the incident light, no matter how great the reflectivities of the individual mirrors. If the interferometer is nonresonant, it will reflect almost all the incident light (assuming that the mirrors are highly reflective).

A tunable source (e.g., a dye laser) illuminates the Fabry–Perot interferometer in Figure 4.14. As the wavelength of the source is changed, it will pass through several resonances with the interferometer, producing a series of transmission peaks as shown in the strip-chart recorder in Figure 4.14. Suppose that exactly 10,000 half-wavelengths of the source fit between the mirrors. Then the interferometer is resonant and it transmits. But transmission ceases when the wavelength emitted by the source decreases somewhat because an integral number of half-wavelengths will not fit between the mirrors. Eventually, though, the wavelength will become short enough so that 10,001 half-wavelengths fit between the mirrors. Then the interferometer will be resonant once again.

It is fairly simple to calculate the frequency difference between adjacent transmission peaks of a Fabry–Perot. The resonance requirement that the separation between the mirrors be equal to an integral number of half-wavelengths can be expressed mathematically as

$$n \times \frac{\lambda}{2} = \ell$$

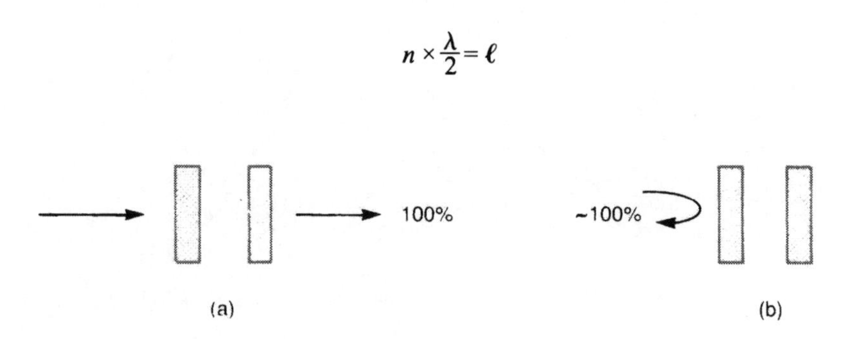

Figure 4.13 Almost all the incident light is transmitted through a resonant Fabry–Perot (a) and almost all the incident light is reflected from a nonresonant Fabry–Perot (b).

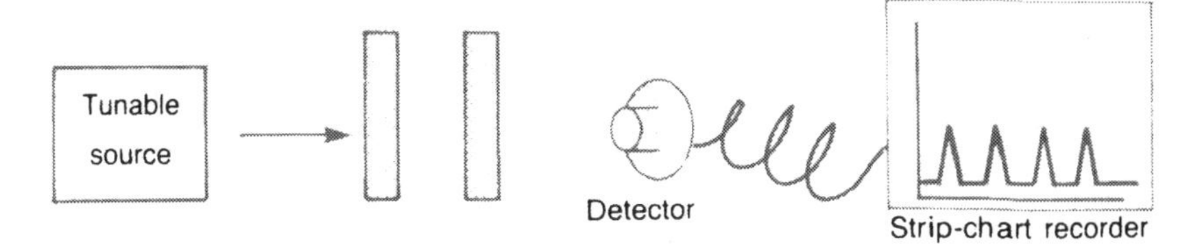

Figure 4.14 If light from a tunable source is incident on a Fabry–Perot, the interferometer will transmit whenever the incoming wavelength satisfies the resonance condition.

in which λ is the wavelength of the light, ℓ is the mirror separation, and n is the integral number of half-wavelengths between the mirrors.

This equation can be solved for the wavelength of the resoannt light:

$$\lambda = \frac{2\ell}{n}$$

Since $f = c/\lambda$ (see Chapter 1), the frequency of light at the nth resonance (i.e., the resonance where there are exactly n half-wavelengths between the mirrors) is given by

$$f_n = n\frac{c}{2\ell}$$

And it follows that the frequency of light at the $(n + 1)$th resonance is simply

$$f_{n+1} = (n+1)\frac{c}{2\ell}$$

Remember that we are finding the frequency difference between adjacent transmission peaks. To do that, we subtract the frequency of one from the frequency of its neighbor and find

$$\Delta f = \frac{c}{2\ell}$$

The frequency separation of adjacent transmission peaks depends only on the spacing between the mirrors.

QUESTIONS

1. In a Young's double-slit experiment, calculate the distance from the central bright line on the viewing screen to the next bright line if the screen is 1 m from the slits and the slits are 0.1 mm apart and illuminated with a 632.8 nm helium–neon (HeNe) laser.

2. Suppose a Fabry–Perot interferometer is constructed with mirrors that are highly reflective at both 488 and 532 nm. Calculate the frequency separation between adjacent transmission peaks at each wavelength if the mirror separation is 1 cm. Are you surprised by the result? What would the frequency separation between adjacent transmission peaks be if the interferometer were used with a HeNe laser at 632.8 nm?
3. If a Fabry–Perot interferometer whose mirrors are separated by 1 cm is illuminated with a Nd:YAG laser whose nominal wavelength is 1.064 μm, what is the precise wavelength nearest to 1.064 μm that is transmitted? How many half-waves are there between the mirrors? Suppose the wavelength illuminating the interferometer is reduced until one more half-wave will fit between the mirrors. What is this new wavelength? How much does it differ from the previous wavelength? What is the frequency difference between the light transmitted in these two cases?
4. Complete the description of the Fabry–Perot begun in Figure 4.12. Draw the waves transmitted through the left mirror and show that they interfere destructively with the wave reflected from that mirror. To make this work you have to know that the phase change of a light wave reflected from an internal surface is 180° different from the phase change of a light wave reflected from an external surface. (Strictly speaking, Figures 4.11 and 4.12 are incorrect because there is a 180° phase change on reflection from an external surface.)

Chapter 5

Laser Light

The previous chapters discussed some of the characteristics and behavior of light in general: the wavelike and particle-like nature of light, how it propagates through a dielectric medium, the polarization of light, and the phenomenon of optical interference. All these considerations apply to laser light as well. But laser light has some unique characteristics that do not appear in the light from other sources.

For example, laser light has far greater purity of color than the light from other sources. That is, all the light produced by a laser is almost exactly the same color, or monochromatic.

Another unique characteristic of laser light is its high degree of directionality. All the light waves produced by a laser leave the laser traveling in very nearly the same direction. One result of this directionality is that a laser beam can be focused to a very small spot, greatly increasing its intensity.

These characteristics of monochromaticity and directionality, together with the phase consistency of laser light, are combined into a single descriptive term: coherence. Coherence is what makes laser light different from the light produced by any other source.

5.1 MONOCHROMATICITY

Because a glass prism is dispersive, it separates white light into its component colors (Figure 5.1a). The bandwidth of white light is as wide as the whole visible spectrum, about 300 nm. If light that is nominally red—maybe white light that is passed through a fairly good red filter—falls on the prism, it is separated into its component wavelengths, too. In this case, however, the bandwidth is far narrower, perhaps only 10 or 20 nm. The prism will produce a narrower band of colors, ranging from dark red to light red, as shown in Figure 5.1b. But the prism will have no discernible effect on the red laser light in Figure 5.1c because its bandwidth is vanishingly small compared with that of the red light from the filter in Figure 5.1b. The bandwidth of an He–Ne laser is typically somewhat less than 1 nm, and it can be reduced far below that amount by techniques described in Chapter 10.

Note, however, that even a laser cannot be perfectly monochromatic. The light produced by a laser must have some nonzero bandwidth, even though that width is very slight by most standards. Why? Because a perfectly monochromatic optical

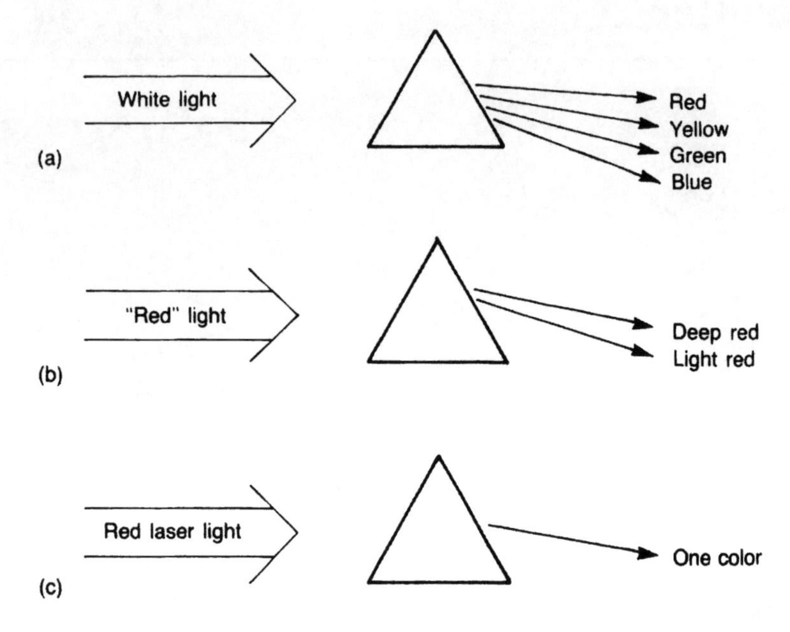

Figure 5.1 A prism can be used to understand the concept of monochromaticity.

source would violate the uncertainty principle, a foundation of modern physics. This principle holds that if you know (with no uncertainty) the wavelength of a source, you can know absolutely nothing about how long it has been on or how long it will stay on. That is, you have to suppose that it has been on forever and that it will remain on forever—clearly an impossible situation.

5.2 DIRECTIONALITY

Everyone has seen the publicity spotlights at circuses and used-car dealerships. Their beams of light penetrate the heavens, apparently diverging little as they disappear into the night sky. But while these spotlights produce beams that do not expand much over hundreds of yards, a laser beam the same size would propagate hundreds of miles without expanding very much.

The divergences of lasers are typically measured in milliradians. This very small divergence results from the requirement that light must make many round-trips of the laser resonator before it emerges through the partially transmitting mirror. Only rays that are closely aligned with the resonator's center line can make the required number of round-trips, and these aligned rays diverge only slightly when they emerge (Figure 5.2).

But just as it is impossible for a laser to be perfectly monochromatic, it is impossible for a laser (or anything else) to produce a nondiverging beam of light. Although the divergence of a laser beam can be very small when compared with that of light from other sources, there will always be some divergence. This is a basic prop-

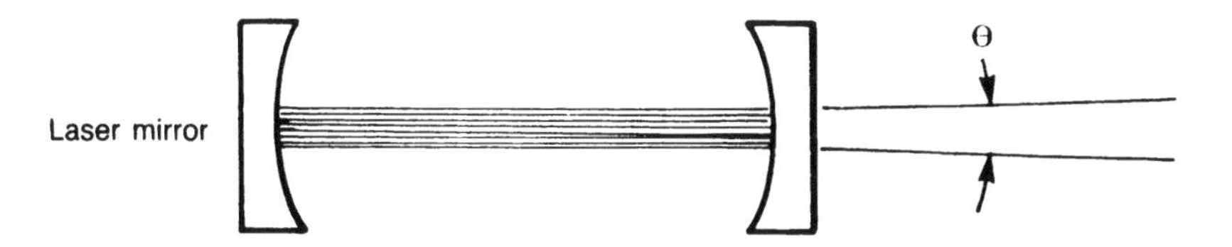

Figure 5.2 Because the light in a laser makes many round-trips between the mirrors, it emerges with small divergence.

erty of light, called diffraction, which was explained in 1678 by Dutch physicist Christian Huygens.

Now let us use Huygens' principle to explain the phenomenon of diffraction. Figure 5.3 shows a plane wave incident on an aperture and some of the points that might produce secondary wavelets. These secondary wavelets spread out as shown, but some of them are blocked by the edges of the aperture. A surface of tangency to the remaining wavelets tends to wrap around at the edges as shown, so the light diverges after it has passed through the aperture.

If the aperture is much larger than the wavelength of the light passing through it, the divergence will be small. A small aperture, on the other hand, produces a large divergence. Mathematically, the full-angle divergence (in radians) is given by

$$\theta = 2.44\ \lambda/D$$

in which D is the diameter of the aperture.

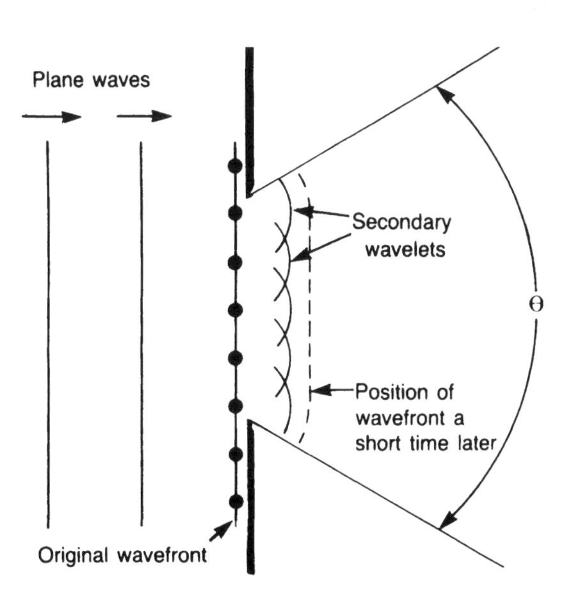

Figure 5.3 Divergence of an aperture plane wave, according to Huygens' principle.

However, Figure 5.3 does not tell the whole story because it doesn't take interference into account. If you let the light that has passed through the aperture in Figure 5.3 fall onto a screen, you will not see a single bright spot. What you will see is a diffraction pattern, as shown in Figure 5.4. If you recall Young's double-slit experiment in Chapter 4, it is easy to understand where this pattern comes from. In Young's experiment, the light intensity at any point on the screen depends on how the waves from each slit add together at that point. For example, the point will be dark if the waves add up out of phase with each other. In plane-wave diffraction, illustrated in Figures 5.4 and 5.5, the intensity at any point on the screen depends on how all the secondary wavelets add together at that point. The result will be a central bright disk surrounded by light rings, or the diffraction pattern in Figure 5.4. The equation above specifies the angle subtended by the central disk, which contains 84% of the light that passes through the aperture.

It is important to understand the significance of diffraction. For example, you cannot reduce the divergence of light to an arbitrarily small angle by placing an aperture a great distance from a point source. Figure 5.5 shows that the light that passes through the aperture will have a greater divergence than the geometrical divergence of the incident light if the aperture is small enough.

The divergence of a laser beam can be very small, even smaller than the divergence of plane waves diffracted through an aperture. Chapter 9 discusses transverse modes and the many shapes that a laser beam can have, but for the moment let us be concerned only with Gaussian beams, which have intensity profiles as shown in Figure 5.6. Mathematically, the intensity never vanishes completely, but conventionally the "edge" of the beam is taken to be the place where the intensity has dropped to about 14% (or $1/e^2$) of its maximum value.

The divergence of a Gaussian beam (in radians) is given by the following equation:

$$\theta = 1.27\ \lambda/D$$

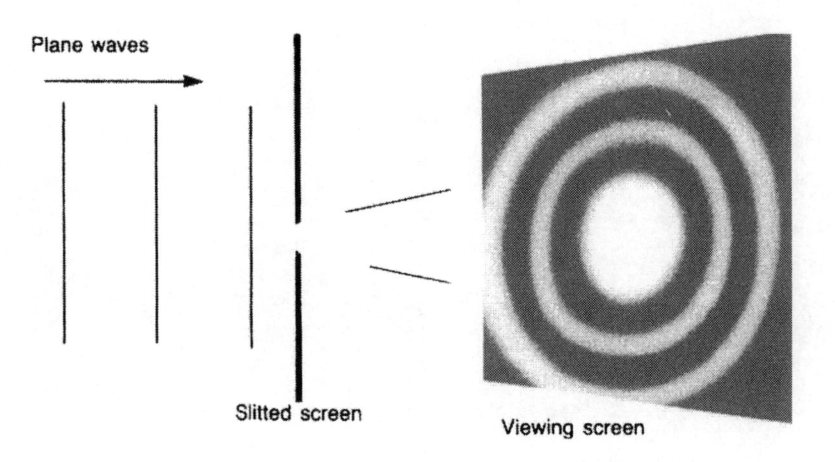

Figure 5.4 Diffraction of plane waves through an aperture.

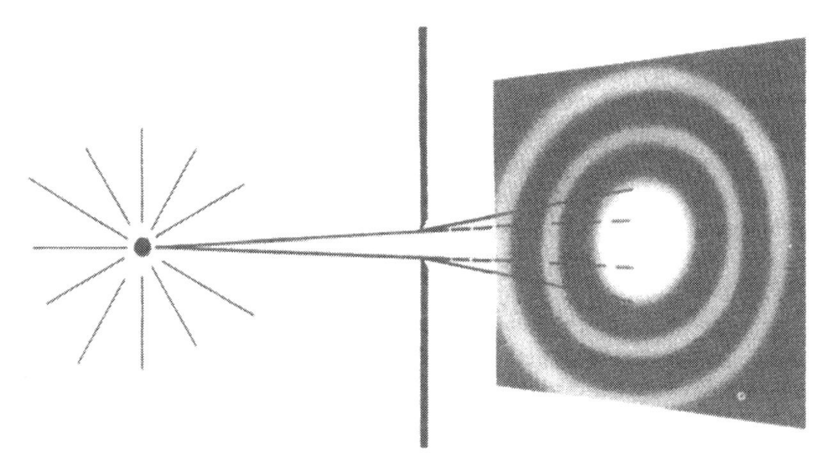

Figure 5.5 Diffraction makes the actual divergence of light greater than the geometrical divergence.

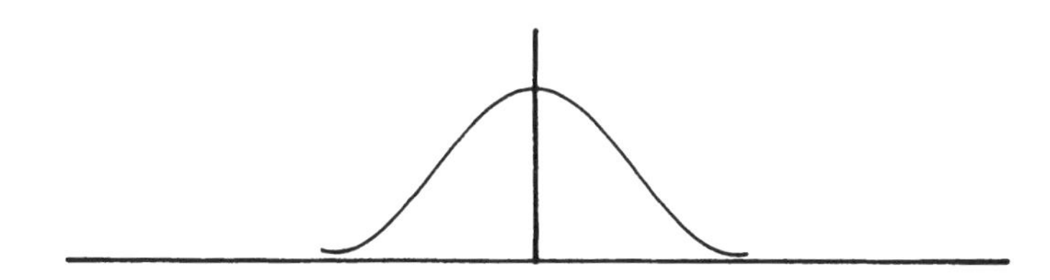

Figure 5.6 Intensity profile of a Gaussian beam; it is brightest at the center and dimmer toward the edges.

in which D is the diameter of the beam from edge to edge at its smallest point, or at its "waist."* Note that this is just about half the divergence of a plane wave passing through an aperture of diameter D. Moreover, a Gaussian beam does not produce diffraction rings like the light that is diffracted from a plane wave through an aperture. As it propagates through space, a Gaussian beam just expands; it does not change shape.

However, do not conclude that a Gaussian beam with diameter D can pass through an aperture of diameter D without ill effect. Remember that there is light in the Gaussian beam outside its nominal diameter, and this light is blocked when the beam passes through the aperture. In fact, the beam will diverge in an angle greater than the ideal $1.27\ \lambda/D$ radians, and faint diffraction rings will appear around the main beam. But these negative effects will decrease rapidly and become almost negligible as the aperture is increased to two or three times the beam diameter.

*The word waist has by convention come to mean the radius of the beam at its smallest point, not the diameter. If you know the waist of a beam, you must multiply it by 2 before using this equation.

One benefit of a laser's small divergence is that the beam can be focused to a smaller spot than the more-divergent light from a conventional source. To see why this is true and to derive an expression for the diameter of the focused beam, it is necessary to understand two simple laws of optics. These laws, which are illustrated in Figure 5.7, state that (1) all parallel light rays that pass through a lens are focused to the same point and (2) any ray that passes through the center of a lens is undeviated by the lens.

Figure 5.8 shows a diverging beam focused to a small spot by a lens.* To calculate the diameter S of the focused spot, we will add two more rays, c and d, which are parallel to b and a, respectively, and which pass through the center of the lens (Figure 5.9). Because rays c and d pass through the center of the lens, they are not bent, and because ray b is parallel to ray c, it will be focused to the same point on the screen as ray c. The same logic holds for rays a and d. The parallelism ensures that the angle between rays c and d is equal to the angle between rays a and b. But the angle θ, in radians, is given by $\theta = S/f$ (this can be deduced from Figure 5.9), in which f is the focal length of the lens. Thus, the diameter of the focused spot is $S = \theta f$. You can conclude that the smaller the divergence of the beam, the smaller the focused spot. And the smaller the focused spot, the greater the intensity of the light.

5.3 COHERENCE

Two waves in a laser beam are shown in Figure 5.10. These waves illustrate the unique characteristics of laser light. They have very nearly (1) the same wavelength, (2) the same direction, and (3) the same phase. Together, these three properties make the light coherent, and this coherence is the property of laser light that distinguishes it from all other types of light.

All the things that can be done only with laser light can be done because laser light is coherent. Monochromatic laser light can be used to probe the structure of atoms or to control complex chemical reactions, because it is coherent. Highly directional laser light can transport energy over large distances or focus that energy to very high intensities, because it is coherent. Phase-consistent laser light can produce realistic three-dimensional holograms or create ultrashort pulses of light whose duration is only a few optical cycles—because it's coherent. The coherent light from a laser is indeed a different breed of light from that emitted by any other source.

Furthermore, the light from a laser exhibits both spatial coherence and temporal coherence. Spatial coherence means that light at the top of the beam is coherent with light at the bottom of the beam. The farther you can move across the beam and still find coherent light, the greater the spatial coherence. Temporal coherence, on the other hand, comes about because two waves in a laser beam remain coherent for a long time as they move past a given point. That is, they stay in phase with each other for many wavelengths. If you think about that for a minute, you will see that the more monochromatic a laser is, the greater its temporal coherence.

*Figure 5.8 shows the beam about the same size as the lens for clarity. In fact, the lens should be at least twice the diameter of the beam, or else it will act as an aperture and introduce diffraction effects.

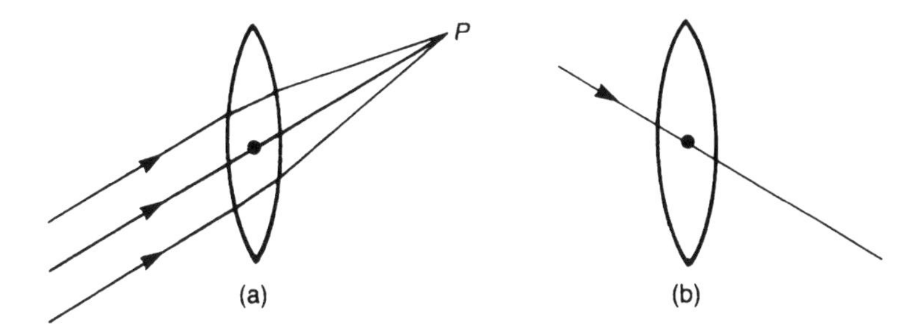

Figure 5.7 Two lens rules: (a) All parallel rays are focused to a common point; (b) any ray through the center of the lens is undeviated.

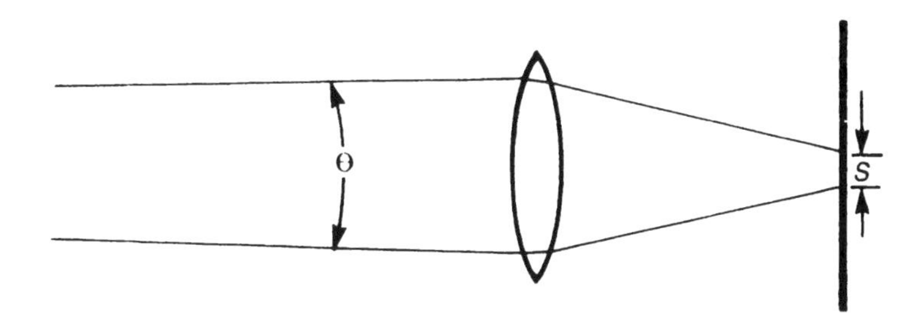

Figure 5.8 A beam of divergence θ is focused onto a screen.

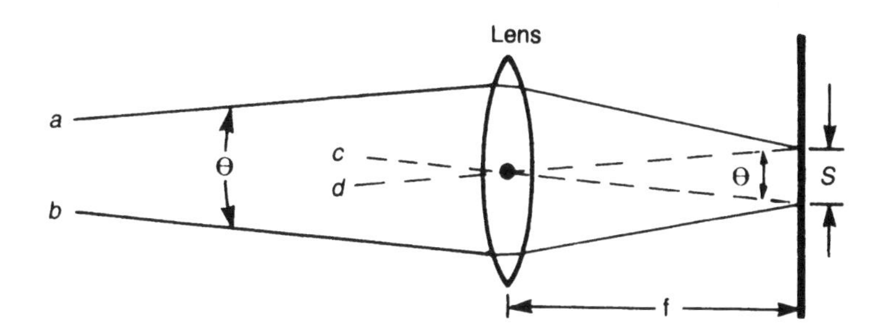

Figure 5.9 A construction used for calculating the diameter of the focused beam.

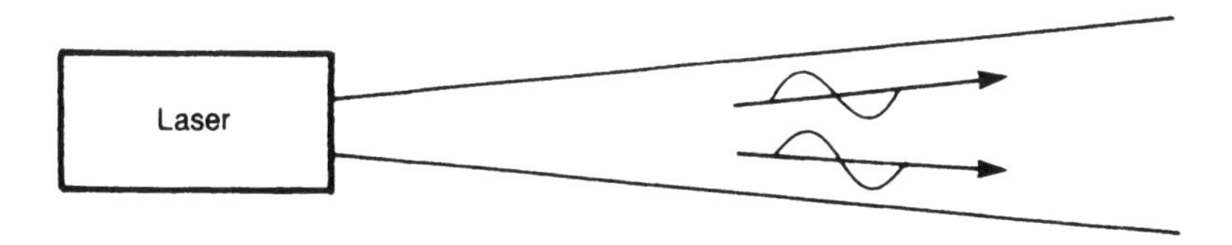

Figure 5.10 The waves of light from a laser are coherent; they all have very nearly (1) the same wavelength, (2) the same direction, and (3) the same plane.

QUESTIONS

1. Use Huygens' principle to explain the refraction (i.e., the bending) of light at an interface between air and glass. (Hint: see Figure 3.1.)
2. What is the divergence of an He–Ne beam whose waist (radius) is 1 mm?
3. U.S. astronauts have placed a laser retroreflector on the surface of the moon. If an Nd:YAG laser on earth is pointed at the moon, calculate the size of the beam when it reaches the moon. (The wavelength of Nd:YAG is 1.06 μm, and you may assume that the beam has a 1 mm diameter at the laser and is not affected by the earth's atmosphere. The moon is roughly 250,000 miles from earth.) How would you modify the laser to make more of its light hit the retroreflector?
4. What phenomenon of wave propagation lets you hear around corners?
5. Explain why temporal coherence means the same thing as monochromaticity.
6. Calculate the intensity (laser power/beam area) that would be created at your retina if you stared into the bore of a 0.5 mW He–Ne laser. Assume (1) that the beam is significantly smaller than your pupil and has 0.5 mrad divergence, (2) that your retina is 2 cm from the lens of your eye, and (3) that the lens has a 2 cm focal length. Compare this intensity with the intensity of the beam outside your eye.
7. One night, a careless technician pointed an argon-ion laser at a police helicopter flying overhead and temporarily blinded the pilot. (This is a true story. The pilot was able to land the helicopter safely and the technician was subsequently arrested.) If the 5 W laser were operating at a wavelength of 514 nm and a divergence of 5 mrad and if the helicopter were 100 m directly overhead, what would the intensity (laser power/beam area) be at the pilot's retina? Assume his night-accustomed pupil was 7 mm in diameter and that the retina of his eye was 2 cm from the lens, which had a focal length of 2 cm. For the purpose of calculation, assume that the beam propagated like a Gaussian beam but had uniform intensity across its diameter. (Hint: the light hitting the pilot's eye diffracted through his pupil like a plane wave.)

Chapter 6

Atoms, Molecules, and Energy Levels

Chapters 6 through 10 are in many ways the most fundamental in the book, because they explain the basic mechanisms that make a laser work. Chapters 1 through 5 dealt with light itself, and little has been said so far about lasers. But now that the behavior of light has been explained, it is time to look inside the laser and see what happens there.

We will see that two things are needed to make a laser work: laser gain and a resonator. Chapters 6 and 7 explain how energy can be stored in atoms or molecules to create laser gain, and Chapters 8 and 9 explain how mirrors can provide the feedback necessary to create a resonator. You will notice many similarities between a laser resonator and the Fabry–Perot interferometer that we discussed in Chapter 4.

In Chapter 2, we saw that quantum mechanics is an explanation of nature that allows light to behave both as a wave and as a particle. But there are further implications of quantum mechanics—specifically, how it predicts that energy is stored in atoms and molecules. The surprising and far-reaching conclusion is that energy can be added to or taken from an atom or molecule only in discrete amounts. That is, the energy stored in an atom or molecule is quantized. This means that whereas you might be able to add 1.27 or 1.31 eV of energy to a particular atom, you cannot add 1.26 or 1.28 or 1.30 eV.* This is a perplexing situation, like having a bucket that will hold 1.27 or 1.31 cups of water but not 1.26 or 1.28 or 1.30 cups. That certainly makes no sense.

In this chapter, we will see how the requirement that energy be quantized affects the behavior of atoms and molecules, and in Chapter 7 we will see that this behavior leads directly to laser action.

6.1 ATOMIC ENERGY LEVELS

Recall the basic structure of an atom. As shown in Figure 6.1, it is a positively charged nucleus surrounded by a cloud of negative electrons, and each of these electrons moves in its own orbit around the nucleus. When energy is absorbed by the

*An electron-volt (eV) is a unit of energy.

Introduction to Laser Technology, Fourth edition. C. Breck Hitz, J. Ewing, and J. Hecht

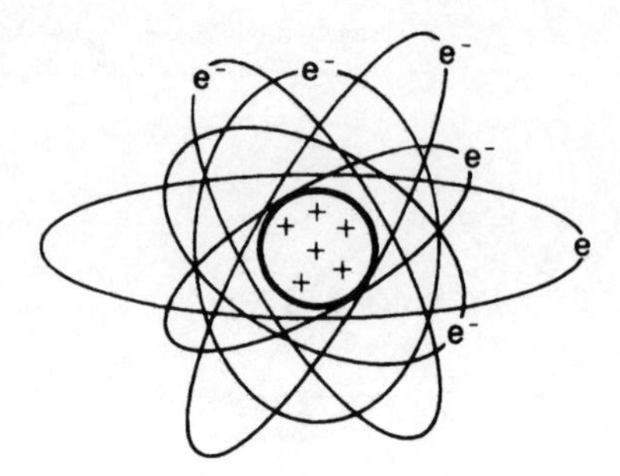

Figure 6.1 The positively charged nucleus of an atom is surrounded by an orbiting cloud of negative electrons.

atom, the energy goes to the electrons. They move faster, or in different orbits. The crucial point is that only certain orbits are possible for a given electron, so the atom can absorb only certain amounts of energy. And once the atom has absorbed some energy, it can lose energy only in specified amounts because the electron can return only to allowed lower-energy orbits.

The behavior of an atom can be shown schematically with an energy-level diagram like the one in Figure 6.2. Here, the allowed energies for the atom are represented by different levels in the diagram. An atom in the ground state has energy E_0, whereas an atom in the first excited state has energy E_1, and so on. The atom loses energy, $E = E_1 - E_0$, when it moves from level 1 to level 0. But an atom in level 1 cannot lose any other amount of energy; it must either keep all its energy or lose an amount equal to $E_1 - E_0$ all at once.

On the other hand, an atom in the ground state (level 0) can absorb only certain allowed amounts of energy. For example, the ground-state atom of Figure 6.2 could absorb $E_1 - E_0$ and move to the first excited state (level 1), or it could absorb $E_2 - E_0$ and move to the second excited state, and so on. But the atom cannot absorb an amount of energy less than $E_1 - E_0$, nor can it absorb an amount of energy between $E_2 - E_0$ and $E_1 - E_0$.

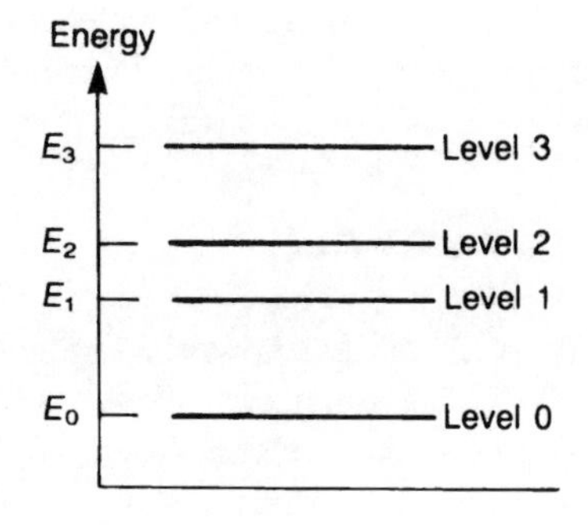

Figure 6.2 The allowed energy levels for an atom correspond to different orbital configurations of its electrons.

This seemingly bizarre behavior on an atomic scale, the quantization of energy, is one of the fundamental results of quantum mechanics. As we discovered in Chapter 2, quantum mechanics explains that nature behaves differently on an atomic scale than it does on a people-sized scale. The theory seems bizarre to us because our intuition is based on our experience with nature on a people-sized scale. But the validity of quantum mechanics has been proved in many experiments that make the atomic-scale behavior of nature show up in the real, people-sized world.

One of the ways an atom can gain energy is to absorb a photon. But the atom must absorb a whole photon because partial absorption is not allowed. That means that the energy of the photon must correspond exactly to the energy difference between two levels of the atom. For example, the ground-state atom in Figure 6.2 could absorb a photon of energy $E_1 - E_0$ and move to the first excited state.

Because the energy of a photon is $E = hc/\lambda$, there is a restriction on the wavelength of light that can be absorbed by a given molecule. For the atom in Figure 6.2, light of wavelength $\lambda = hc/(E_0 - E_1)$ will be absorbed and boost the atom to its first excited state. But light whose wavelength does not correspond to an energy-level difference in the atom will not be absorbed. Red glass is red because it contains atoms (or molecules) that absorb photons of blue light but cannot absorb red light. Ordinary glass has no atoms (or molecules) in energy levels that are separated from other levels by the amount of energy in visible photons.

6.2 SPONTANEOUS EMISSION AND STIMULATED EMISSION

There are several ways an atom in an excited state can lose its energy. The energy can be transferred to other atoms, it can come off in the form of heat, or it can be emitted as light. If it is emitted as light, the wavelength of the emitted light will correspond to the energy lost by the atom. There are two mechanisms by which the light can be emitted: spontaneous emission and stimulated emission.

The absorption and subsequent spontaneous emission of a photon is shown in Figure 6.3. First, a photon whose energy is exactly right to boost the atom from its

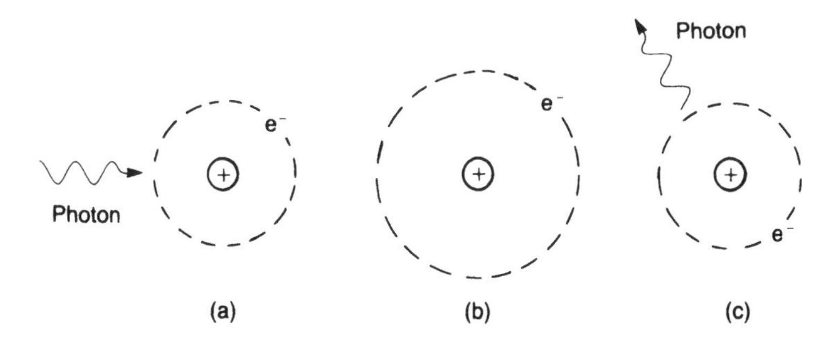

Figure 6.3 An atom absorbs a photon (a), which excites it for a period of time (b). Later, the photon is spontaneously emitted (c).

ground state to its first excited state is absorbed. The average atom will stay in this excited state for a period of time known as the *spontaneous lifetime,* which is characteristic of the particular transition. Many atomic transitions have spontaneous lifetimes of nanoseconds or microseconds, although much longer and shorter lifetimes are known. Eventually, the atom will spontaneously emit the photon and return to its ground state. As indicated in Figure 6.3, the photon is emitted in a random direction.

The process of stimulated emission is shown in Figure 6.4. A second photon—one with exactly the same energy as the absorbed photon—interacts with the excited atom and stimulates it to emit a photon. Interestingly, this emission can take place long before the spontaneous lifetime has elapsed. The second photon is not absorbed by the atom, but its mere presence causes the atom to emit a photon. As indicated in Figure 6.4, the light is emitted in the direction defined by the stimulating photon, so both photons leave traveling in the same direction. Of course, since the stimulating photon has the same energy as the emitted photon, the emitted light has the same wavelength as the stimulating light. The polarization of the emitted light is also the same as that of the stimulating light. Moreover, the emitted light has the same phase as the stimulating light: the peaks and valleys of the electromagnetic waves are all lined up with each other.

As we will discover later, stimulated emission is crucial to laser action. Indeed, the word laser is an acronym whose third and fourth letters stand for "stimulated emission." And, as we will learn in Chapter 7, the light emitted by a laser is coherent because its waves are all traveling in the same direction, all have the same wavelength, and are all in phase with each other.

6.3 MOLECULAR ENERGY LEVELS

As you know, a molecule is composed of two or more atoms. Figure 6.5 shows a simple molecule of only two atoms in which some electrons stay with their original nuclei and one electron is shared by both nuclei. Because a molecule is more complex than an atom, it has more types of energy levels than an atom does. In fact, three types of energy levels are possible in a molecule: electronic, vibrational, and rotational.

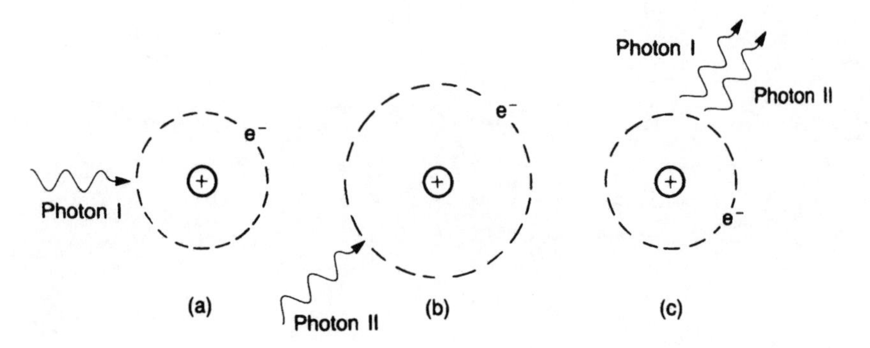

Figure 6.4 A second photon can stimulate the atom to emit in a time shorter than the spontaneous lifetime.

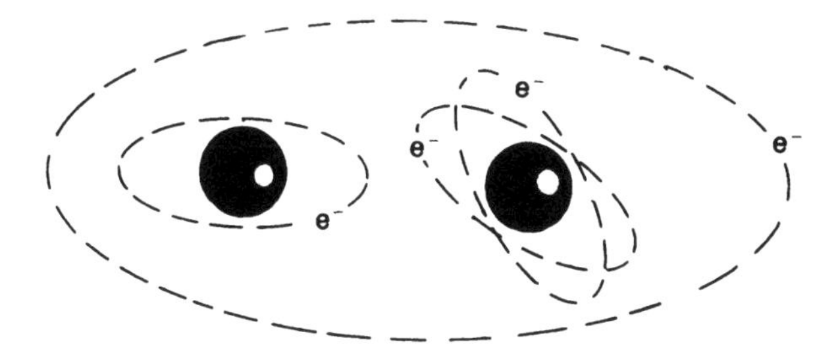

Figure 6.5 In a molecule, some electrons stay with their original nucleus, whereas shared or completely transferred to the other nucleus. (Not all the molecule's electrons are shown here).

A molecule can have electronic energy levels that are exactly analogous to the electronic energy levels of an atom. That is, the molecule's electrons move to more-energetic orbits when the molecule absorbs energy. Of course, these levels are quantized, and each transition has its own spontaneous lifetime.

But a molecule can also vibrate, which is something an atom cannot do. If you think of the force holding the molecule together as a spring, you can visualize how the nuclei can vibrate back and forth, as shown in Figure 6.6. Thus, a molecule can absorb energy by absorbing a photon, for example—and the absorbed energy can turn into vibrational energy in the molecule.

The vibrational energy of a molecule is quantized, just as its electron energy is. That means that a molecule cannot absorb just any amount of energy; it can only absorb enough energy to move it from one allowed energy level to another. So you can draw an energy-level diagram for a molecule's vibrational levels just as you can draw a diagram for its electronic levels.

Finally, a molecule can absorb energy and start rotating about its axis, as shown in Figure 6.7. An atom cannot store energy this way because, unlike a molecule, an

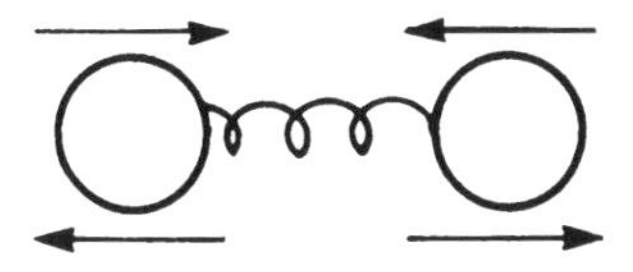

Figure 6.6 A molecule can store energy by vibrating.

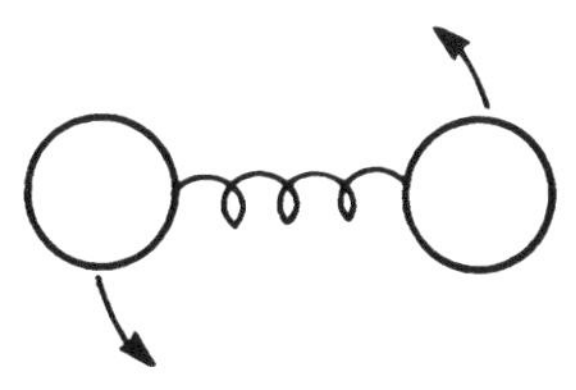

Figure 6.7 A molecule can store energy by rotating.

Table 6.1 Types of energy levels and their transition wavelengths

Level	Atoms	Molecules	Approximate λ of most transitions
Electronic	Yes	Yes	Visible or ultraviolet
Vibrational	No	Yes	Near infrared
Rotational	No	Yes	Far infrared

atom's mass is completely symmetric. The rotational energy a molecule can possess is quantized so that a molecule can absorb or emit only the exact amounts of energy corresponding to a transition between allowed levels.

How do these three types of energy levels compare with each other? In general, transitions between electronic levels are the most energetic, and transitions between rotational levels are the least energetic. This conclusion is implicit in Table 6.1, which summarizes the three types of energy levels and the wavelengths of transitions for each. Recall that the energy in a photon is inversely proportional to its wavelength and you will see that electronic transitions in general absorb or emit a greater amount of energy than vibrational or rotational transitions, and vibrational transitions in general involve more energy than rotational transitions. (There are, of course, exceptions to these rules.)

An energy-level diagram for a hypothetical molecule shows all three types of levels, as depicted in Figure 6.8. The closest-spaced levels in this simplified diagram are the rotational ones, whereas the electronic levels are furthest apart.*

Because a molecule has so many different energy levels, its spectrum can be much more detailed than an atom's. For example, Figure 6.8 shows only two electronic levels, and an atom with only two electronic levels would have only one emission or absorption line. But the hypothetical molecule in Figure 6.8 could emit and absorb light at dozens of different wavelengths, corresponding to transitions involving each of many of the pairs of levels shown. In discussing the carbon dioxide laser in Chapter 17, we will learn that not every transition you could imagine is possible; selection rules arising from the conservation laws prohibit many transitions.

6.4 SOME SUBTLE REFINEMENTS

The foregoing explanation of how energy is stored in atoms and molecules has deliberately trod roughly on some of the niceties of quantum theory in order to create a simple and somewhat intuitive model. And although this simple model is adequate to understand most of the principles of laser technology, you should know that it is an approximation that is not quite correct from a theorist's point of view. In this section, we discuss several of these subtleties of quantum mechanics.

*The energy-level diagram for most real molecules would be more complex than shown in Figure 6.8, with many overlapping levels. For example, the higher rotational levels of the ground vibrational state might be more energetic than the ground rotational level of the first excited vibrational state.

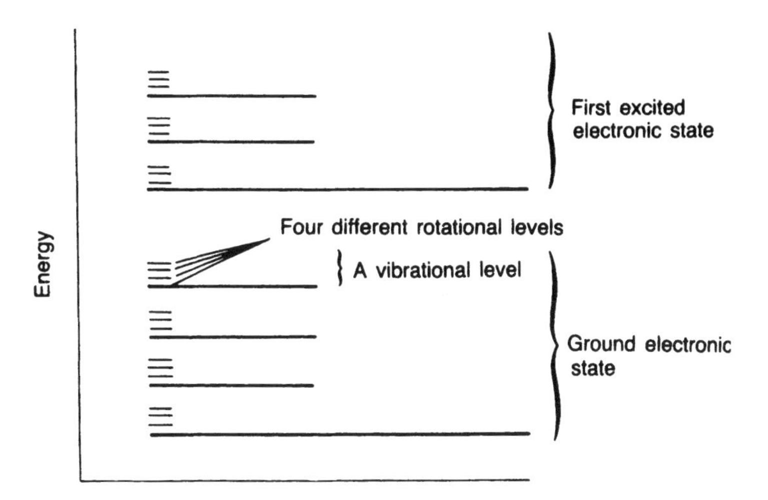

Figure 6.8 Electronic, vibrational, and rotational levels for a hypothetical molecule.

The picture of an atom composed of marble-like electrons orbiting around a nucleus is not consistent with modern quantum theory. The uncertainty principle states that the electrons are more like a negatively charged cloud surrounding the nucleus, not distinct particles. When the atom absorbs energy, the shape of this cloud changes to accommodate the extra energy.

In introducing the concept of spontaneous decay, we explained that an excited atom will stay excited for a period of time known as its *spontaneous lifetime.* The concept of spontaneous lifetime is valid for a collection of atoms, in the sense that they will decay from the excited state with an average time constant equal to the spontaneous lifetime. But one particular atom can stay in an excited state for a longer or shorter time than the spontaneous lifetime. You can think of an average atom staying excited for its spontaneous lifetime in the same sense that you think of the average American family having 1.8 children.

On an even more subtle level, the concept of an atom's being in one energy level and then moving abruptly to another energy level as it absorbs or emits a photon is incorrect. According to quantum theory, a single atom exists in a number of different energy levels simultaneously. When you perform a measurement of the atom's energy, you force the atom to have the amount of energy you measure at the instant you perform the measurement. But, in general, the only correct description of an atom's energy is one of the probability of making different measurements: "If I measure the energy in this atom, there is a 7% chance I'll find 2 eV, a 20% chance I will find 1.4 eV, and a 73% chance I will find 1 eV." (Remember, quantum mechanics is not supposed to make perfect sense on an intuitive level until you have spent several years watching how it behaves, by studying for a graduate degree in quantum physics, for example.)

How does the atom just described interact with light? That is, how does rigorous quantum mechanics explain optical absorption and stimulated emission? Quantum mechanics says that if the photon energy of the light (given by $E = hc/\lambda$) is equal to the difference between the possible measured energies of the atom, the light will cause the atom's energy-measurement probabilities to change. Specifically, the probabilities of finding the atom in either of the two energy levels separated by the light's photon energy tend to become equal.

Suppose the atom we have described is irradiated with light whose photon energy is 1 eV (i.e., light whose wavelength is about 1.2 μm). Initially, the atom has 7% probability of having 2 eV and 73% probability of having 1 eV. But if you made the measurement after the atom had been exposed to the 1.2 μm light for a while, you would find that the probabilities were becoming equal. If you repeated the experiment 100 times (exposing the atom to the light for the same amount of time each instance), you might find that the atom had 2 eV 30 times, 1 eV 50 times, and 1.4 eV in the remaining 20 measurements. In other words, you would find that the probability of the atom's having 2 eV had increased from 7% to 30% , and that its probability of having 1 eV had decreased from 73% to 50%, as shown in Figure 6.9.*

If you left the atom of Figure 6.9 exposed to the light long enough, its probabilities would reach new equilibrium values. If the light were intense enough, the new equilibrium values would be equal (i.e., a 40% probability of finding 2 eV and a 40% probability of finding 1 eV). In this case, the transition is saturated because the probabilities do not change, no matter how much more light you pump into the atom.

In the situation depicted in Figure 6.9, the atom ends up with more energy than it started with. Energy has been transferred from the light to the atom, so Figure 6.9 represents the quantum mechanical version of optical absorption.

Figure 6.10 shows the quantum mechanical explanation of stimulated emission. If the atom from Figure 6.9b is immediately exposed to light whose photon energy is 0.6 eV, the probabilities for the first and second excited states will change, and that change will tend to make the probabilities equal. Thus, the probability of finding the atom in the second excited state might decrease from 30% to 28%, and the probability of finding it in the first excited state might increase from 20% to 22%. In this case, energy has been transferred from the atom to the light, so the light has been amplified by stimulated emission.

If the probabilities of Figure 6.9a are the equilibrium values for the atom under a particular set of conditions, then the situation depicted in Figure 6.10a is a nonequilibrium condition, and the probabilities will automatically drift back to those of Figure 6.9a. The amount of time it takes them to drift back is determined by the spontaneous lifetime of the level. The energy given off by the atom as it drifts back may be in the form of spontaneous (optical) emission, it may be heat, or it may be collisionally transferred to the other atoms.

The picture of an atom absorbing or emitting a photon and instantaneously changing energy levels as it does so is not wrong. In fact, it is a very useful model. But be aware that the model is a simplification, and that the true quantum mechanical explanation is far more complex.

*If you had irradiated the atom with light whose photon energy did not correspond to any transitions in the atom—say, light whose photon energy was 0.7 eV—nothing would happen. The light wouldn't interact with the atom, and the probabilities would remain unchanged.

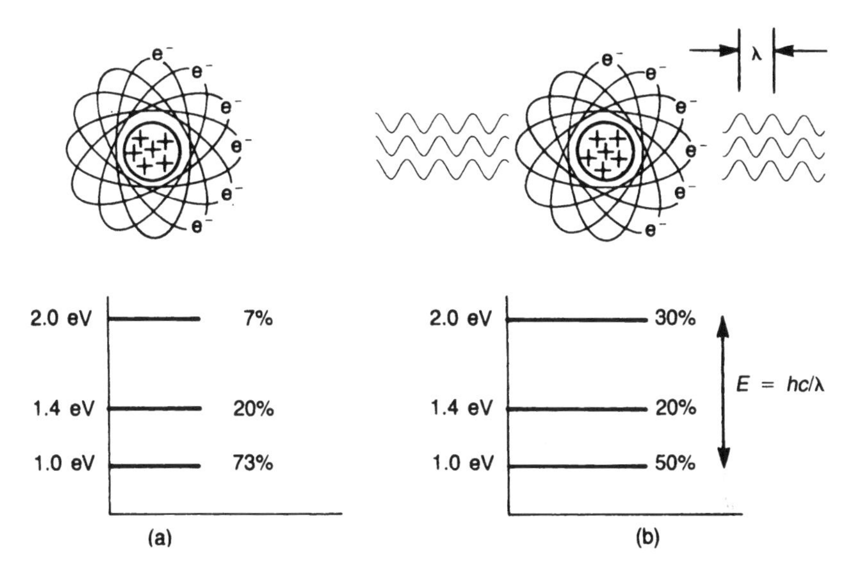

Figure 6.9 An atom might initially have 73% probability of being in the ground state, 20% probability of being in the first excited state, and 7% probability of being in the second exited state (a). When the atom is exposed to light whose photon energy ($E - hc/\lambda$) corresponds to the energy difference between the ground state and second excited state, the probabilities of finding the atom in those states start to change (b).

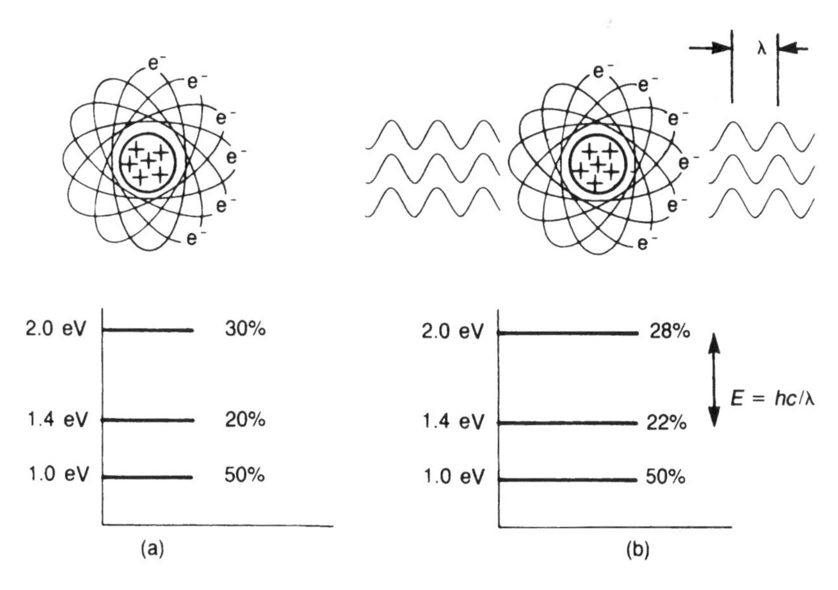

Figure 6.10 If the atom in Figure 6.9b is exposed to light whose photon energy corresponds to the energy difference between the first and second excited states, the probabilities of finding the atom in either of these two states tend to become equal. In (b), the atom is shown shortly after the exposure has begun—before the probabilities have become equal.

QUESTIONS

1. The 1.2 μm light in Figure 6.9b has a photon energy of about 1 eV. Calculate the photon energy in joules, and from your result calculate the approximate conversion for converting electron-volts to joules. What is the wavelength of the light in Figure 6.10b?
2. Suppose that the ground state of the hypothetical molecule in Figure 6.8 lies 3.1×10^{-19} J below the lowest vibrational-rotational level of the first excited electronic state. Calculate the wavelength of light associated with the following transitions if each change in vibrational level requires 4×10^{-20} J and each change in rotational level requires 5×10^{-21} J.
   (a) From the first excited rotational level of the ground vibrational level of the first excited electronic state to the ground state.
   (b) From the ground rotational level of the first excited vibrational level of the second excited electronic state to the ground state.
   (c) From the second excited rotational level of the ground vibrational level of the first excited electronic level to the first excited rotational level of the first excited vibrational level of the ground electronic level.

Chapter 7

Energy Distributions and Laser Action

In Chapter 6, we learned how energy is stored in atoms and molecules. It is stored in discrete amounts, and an atom can be thought of as making a transition from one energy level to another as it absorbs or emits energy.

However, in Chapter 6, we limited our attention to one atom or molecule at a time. In this chapter, we examine the behavior of a collection of atoms and look at how the energy in a collection of atoms is divided among the individual atoms. In other words, we find the answer to these questions: If a jar holds 100 atoms of the same element, how many of these atoms are in the ground state? How many are in the first excited state? And so on.

And we discover that for an unusual type of energy distribution among a collection of atoms (or molecules), it is possible for light to be amplified as it passes through the collection. This amplification is the basis of laser action, and understanding it is absolutely crucial to understanding the operation of a laser.

This chapter concludes with an examination of the two energy-level schemes of common lasers. Some of the mechanisms for pumping energy into common lasers are also explained.

7.1 BOLTZMANN DISTRIBUTION

Suppose you have a sealed jar that contains 100 atoms of some element, as shown in Figure 7.1. (Of course, a realistic-sized jar normally holds more like 10^{16} or 10^{20} atoms, but it is easier to work with 10^2.) In addition to the 100 atoms, the jar contains some energy. The first question you might ask yourself is, "How does the energy manifest itself?" That is, "What form does the energy take? How does it show up?"

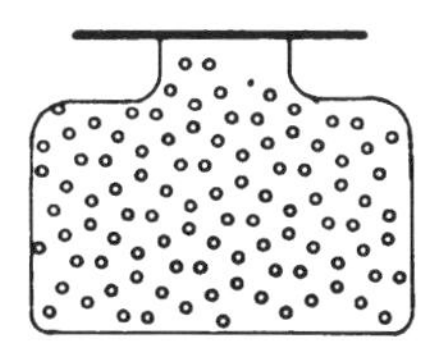

Figure 7.1 A jar containing exactly 100 atoms of an element.

Introduction to Laser Technology, Fourth edition. C. Breck Hitz, J. Ewing, and J. Hecht

Let us limit our discussion to the thermal energy in your jar. The more thermal energy in the jar, the higher the temperature will be. There are two ways that thermal energy in a collection of atoms manifests itself. Some of the thermal energy in your jar will show up in the motion of the atoms themselves: they carom around inside the jar, bouncing off the walls and each other. And some of the energy shows up in the electronic energy levels of the atoms. Thermal energy will boost some atoms to the first excited state, some to the second excited state, and so forth. But how many get to each excited state?

Boltzmann's law is one of the fundamental laws of thermodynamics, and it dictates the population of each energy level if the atoms in the jar are in thermal equilibrium. Figure 7.2 shows schematically the prediction of Boltzmann's law.* Here, the length of the bar representing each level is proportional to the population of that level. You can draw an important conclusion immediately: no energy level will ever have a greater population than that of any level beneath it. Level E_3 is populated by more atoms than levels E_4, E_5, or E_6, but, on the other hand, it has fewer atoms than levels E_2 or E_1.

Remember that the distribution shown in Figure 7.2 is an equilibrium distribution. That means that it is the normal way the atoms in the jar behave. It is possible to create an abnormal distribution for a short time (more about that later). But as long as the temperature of the jar does not change, the atoms will eventually return to the distribution in Figure 7.2.

Now you know that the picture of a jar containing so many atoms in a certain energy level is not quite accurate because an atom in general does not really exist in only one energy level. The truly correct way to describe the jar in Figure 7.2 is to say that each atom has a 25% probability of being in the ground state if you measured it, a 23% probability of being in the first excited state, and so on through the higher excited states. But that means that if you take the trouble to measure all 100 atoms, you will probably find 25 in the ground state and so on. Thus, in that sense it is acceptable to say that the jar contains 25 atoms in the ground state, 23 in the first excited state, and so forth.

What happens to the distribution in the jar if you add energy to it? Suppose you put the jar in a furnace. Its temperature rises and the distribution changes from that of Figure 7.2 to that of Figure 7.3. There are fewer atoms in the low-lying energy levels, and some of the previously empty upper levels are now populated. But the important conclusion you drew earlier is still valid: no energy level will ever have a population greater than that of any level beneath it. Even in the limit of the highest imaginable positive temperature, that conclusion will be valid for any collection of atoms (or molecules) in thermal equilibrium.

*Mathematically, Boltzmann's law predicts that the population of any energy level is related to the population of the ground level by the following equation:

$$N_i = N_0 \exp - (E_i/k_B T)$$

in which N_i is the population of the ith level whose energy above ground level is E_1; N_0 is the population of the ground level; k_B is Boltzmann's constant, 1.38×10^{-23} J/K; and T is the temperature. The results of this complicated-looking equation turn out to be fairly simple, as shown in Figures 7.2–7.4.

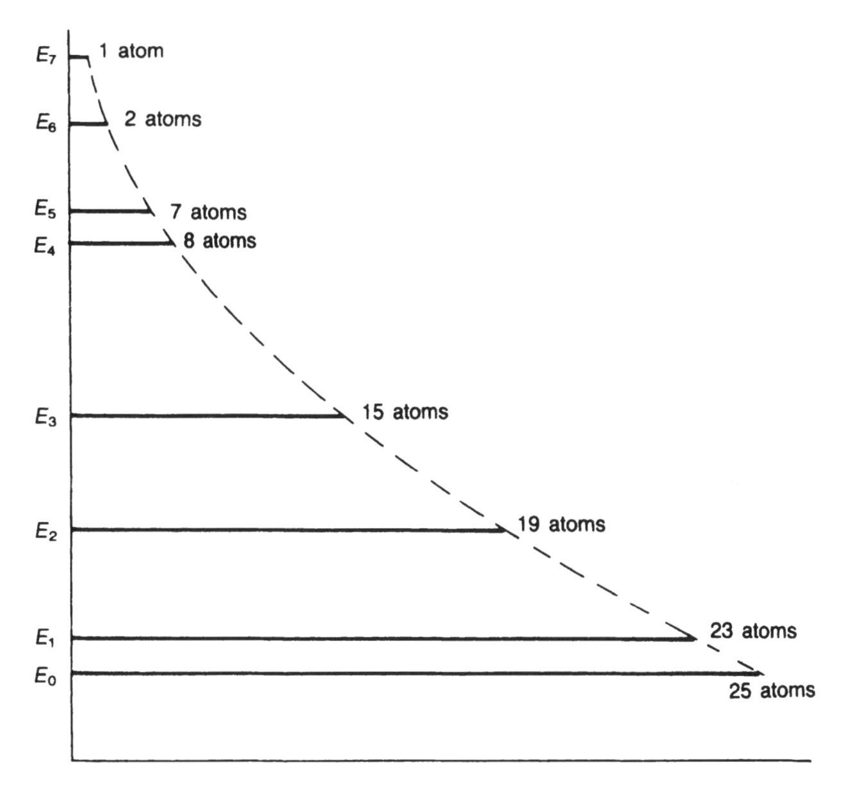

Figure 7.2 If the atoms in the jar are in thermal equilibrium. Boltzmann's law predicts that they will be distributed as shown here, with increasing populations in the lower energy levels.

Next, let us ask, What happens if you put your jar in a freezer? Thermal energy is removed from the jar, and you will wind up with a distribution that looks like the one in Figure 7.4. There are no atoms in the higher levels, and the lower levels are very highly populated. What would happen if the jar were chilled all the way to absolute zero? In that case, all 100 atoms would be in the very bottom energy level, E_0. Of course, you will have noticed that the previous conclusion about populations still holds in Figure 7.4.

The same Boltzmann's law logic holds if your jar contains 100 molecules instead of 100 atoms, but the situation is a little more complicated. Recall that, in addition to electronic energy levels, molecules have vibrational levels and rotational levels. Altogether, there are four forms that thermal energy can take in a jar of molecules: as translational energy when the molecules bounce around inside the jar, as rotational energy in the molecules, as vibrational energy in the molecules, and as electronic energy in the molecules. Raw energy flows back and forth among these four modes, keeping them in equilibrium with each other, Moreover, Boltzmann's law governs the distribution of energy among the molecules. That is, the diagrams in Figures 7.2–7.4 could equally well represent the populations of rotational energy levels of a collection of molecules at three different temperatures.

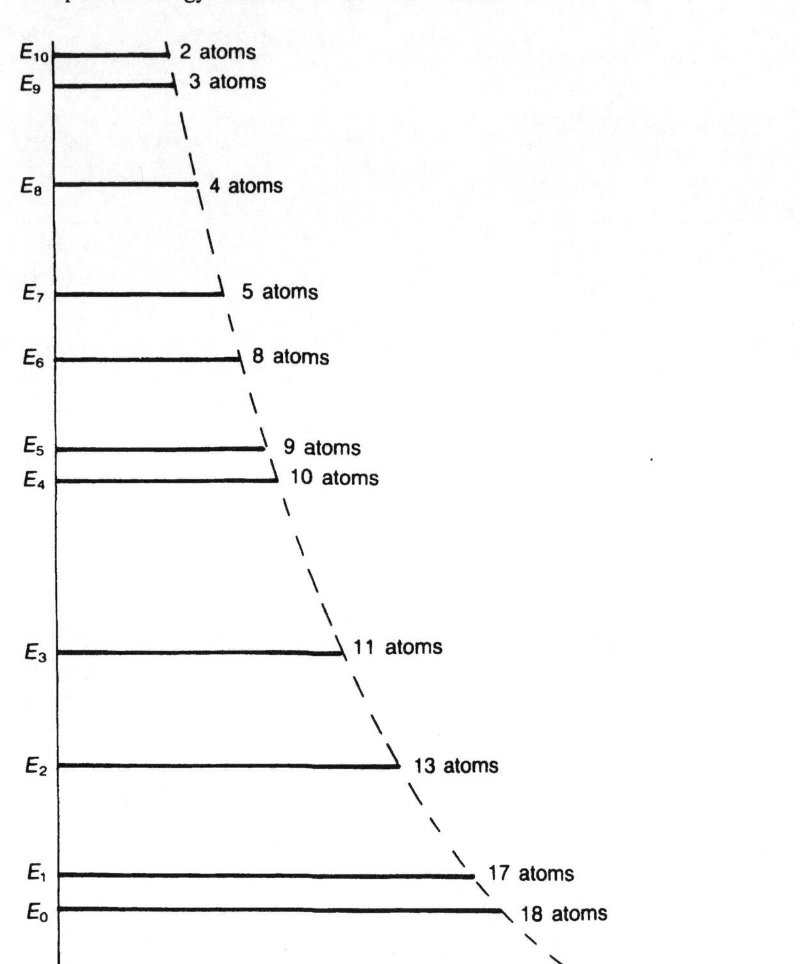

Figure 7.3 If the jar is heated, the atoms will redistribute themselves among energy levels as shown, but there will still always be increasing populations in the lower energy level.

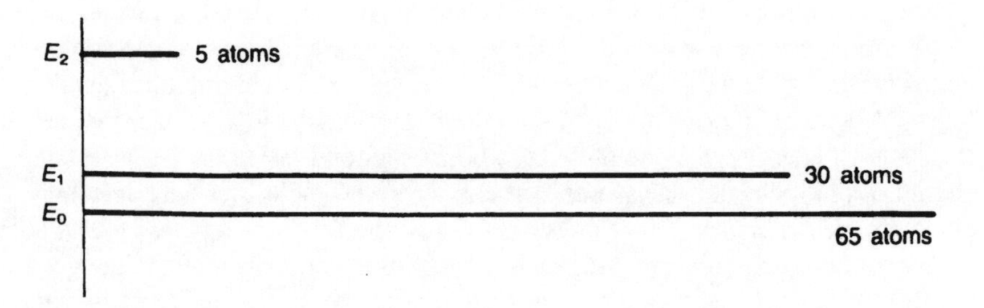

Figure 7.4 When the jar is chilled, a new equilibrium distribution results, with less total energy.

7.2 POPULATION INVERSION

As long as a collection of atoms is in thermal equilibrium, energy will be partitioned among them according to Boltzmann's law. But it is possible to create a collection of atoms that is not in thermal equilibrium. The collection will not stay in that non-equilibrium condition long, but for a short period you will have a collection of atoms that violates the conclusion you drew in the previous section. For example, suppose you somehow plucked seven of the E_0 atoms out of the jar whose distribution is diagramed in Figure 7.2. Then, for an instant at least, you would have a distribution like that of Figure 7.5. This distribution shows a *population inversion* between the E_1 and E_0 levels because the equilibrium populations are inverted: there are more atoms in level E_1 than in level E_0.

And there is more than one way to create a population inversion. Suppose you added energy to the jar, but instead of adding random thermal energy, you added energy in very precise amounts. For example, you might think of setting your jar down in front of a gun shooting out a beam of electrons, each electron having the same velocity (and therefore the same energy) as all the other electrons. When one of the electrons collides with one of the 100 atoms in the jar, it could transfer its energy to the

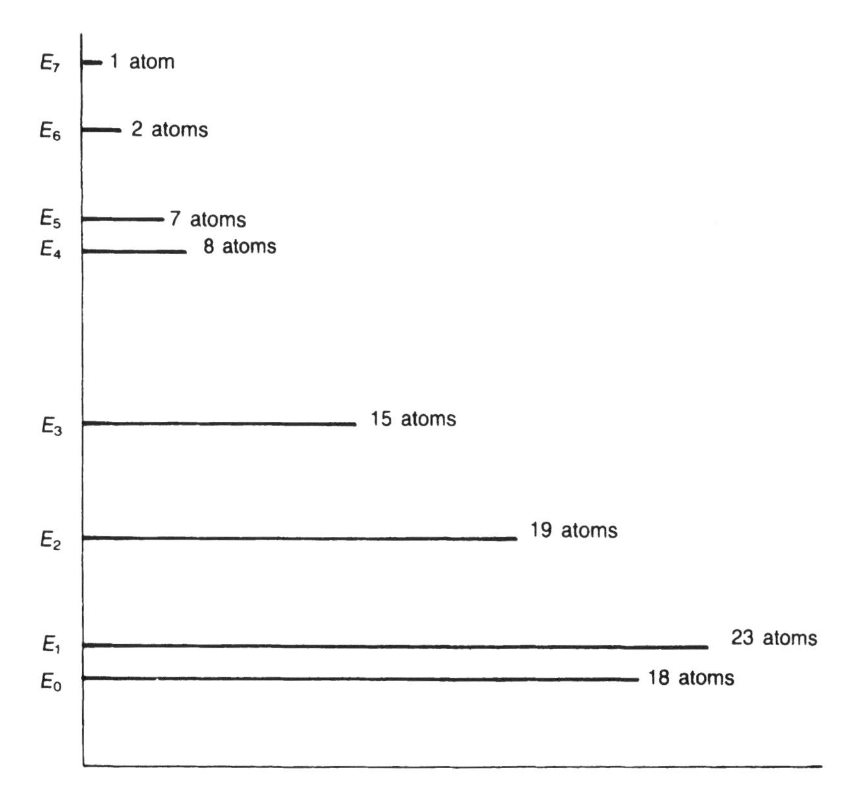

Figure 7.5 A population inversion could be created by plucking some E_0 atoms from the jar described in Figure 7.2.

atom. Now, suppose the energy of the electrons coming from the gun were exactly equal to the energy difference between the E_2 and E_3 levels. If one of these electrons collided with an atom in the E_2 level, it could excite that atom to the E_3 level. If the gun shot out electrons fast enough, it could pump atoms up to E_3 faster than they could decay spontaneously from that level. Hence, you could create a population inversion between the E_3 and E_2 levels, as shown in Figure 7.6.

The population inversions in Figures 7.5 and 7.6 are nonequilibrium distributions, ones that will not last very long. If the electron gun in the previous example were turned off, the atoms in the jar would quickly revert to the distribution of Figure 7.2. But there is nothing terribly unnatural about a nonequilibrium situation. You do not need to understand the intricacies of atomic energy levels to understand a nonequilibrium situation. Figure 7.7a is another example of a nonequilibrium situation. The system in Figure 7.7 is shown in its equilibrium condition in Figure 7.7b. To create the nonequilibrium situation again, energy is input into the system from an external source, as shown in Figure 7.7c. Do you understand how the waterwheel and sun of this example are analogous to the jar of atoms and the electron gun of the previous example?

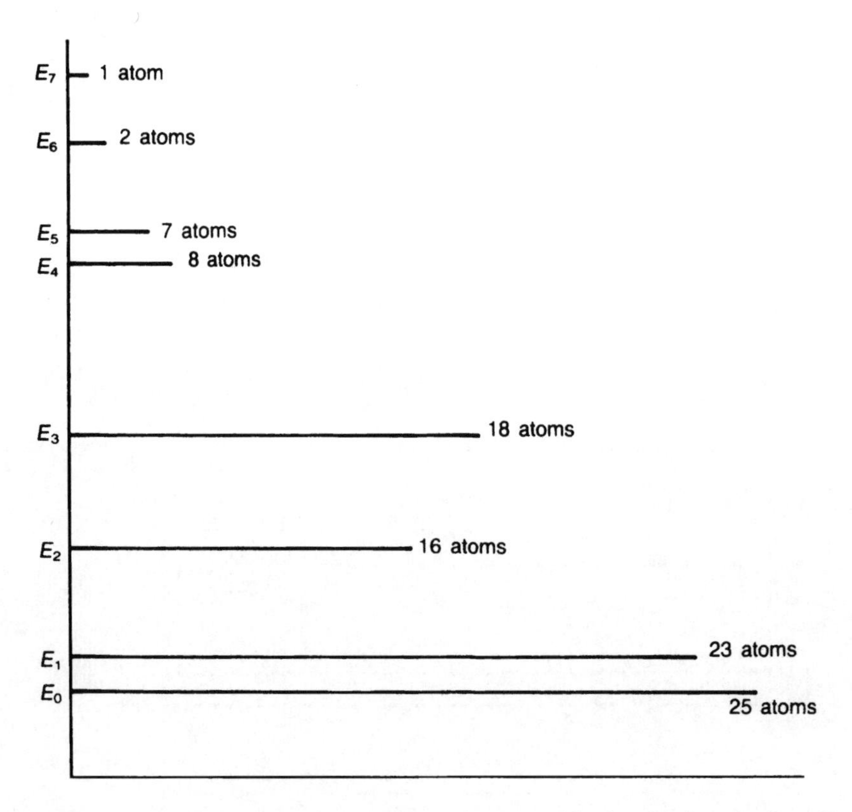

Figure 7.6 A population inversion could also be created by bombarding the jar with a monoenergetic electron beam.

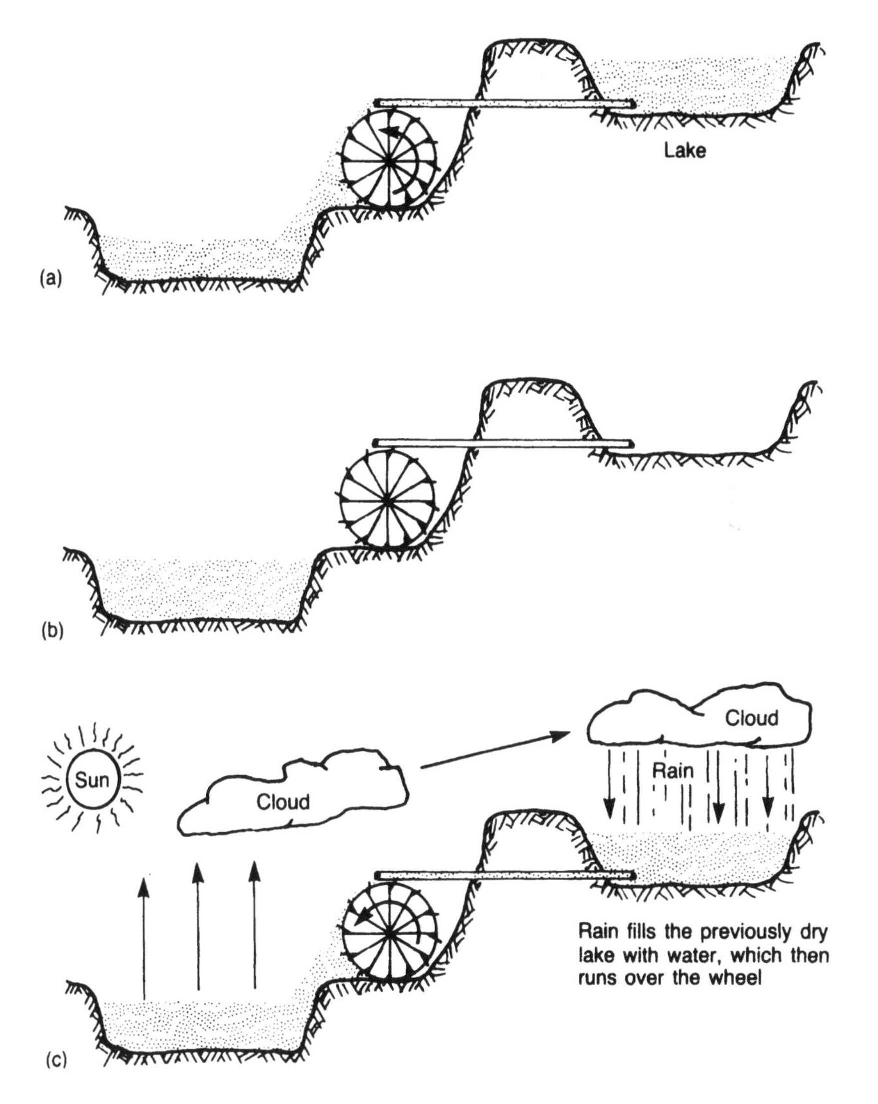

Figure 7.7 A system in a nonequilibrium condition (a) and an equilibrium condition (b). Energy from an external source can create a nonequilibrium condition (c).

7.3 L.A.S.E.R.

In Chapter 6 we discussed stimulated emission, and in this chapter we've discussed population inversions. These are the two concepts necessary to understand the fundamental principle of the laser. As you know, the letters in the word laser stand for light amplification by stimulated emission of radiation, or L.A.S.E.R.

Let's return to the hypothetical jarful of atoms we have been discussing. For a moment let us take all but one of the atoms out of the jar. Suppose that the remaining atom is in the ground energy level, E_0. Also, suppose that a photon comes along whose energy exactly corresponds to the energy difference between the ground level and, say, the second excited state, that is,

$$hc/\lambda = E_2 - E_0$$

in which λ is the wavelength of the photon. What happens when the photon interacts with the atom? Because its energy is exactly correct, the photon can be absorbed by the atom and the atom will be boosted to the E_2 level. Figure 7.8a shows before-and-after drawings of the situation.

Now go back and start with the atom already in the E_2 level. What happens this time when the photon interacts with the atom? Because its energy is exactly correct, the photon can stimulate the excited atom to emit a photon. The atom ends up in the E_0 level, and the two photons depart in the same direction and in phase with each other, as explained in the previous chapter. This sequence of events is shown in Figure 7.8b.

The next thing to do is to put all 100 atoms back into the jar. Let's say the jar is chilled to absolute zero and all 100 atoms are in the ground state. Suppose three photons come along, each having the correct wavelength, as shown in Figure 7.9a. The odds are that all three will be absorbed, leaving 97 atoms in the E_0 ground state and three atoms in the E_2 excited state.

Next, let us start with 50 of the atoms in the E_2 excited state and 50 in the ground state. (This is not an equilibrium distribution, so assume that everything happens quickly compared to the time it takes the atoms to revert to their equilibrium distribution.) When the three photons come along, each photon has a 50–50 chance of interacting with an excited atom and a 50–50 chance of interacting with a ground-state

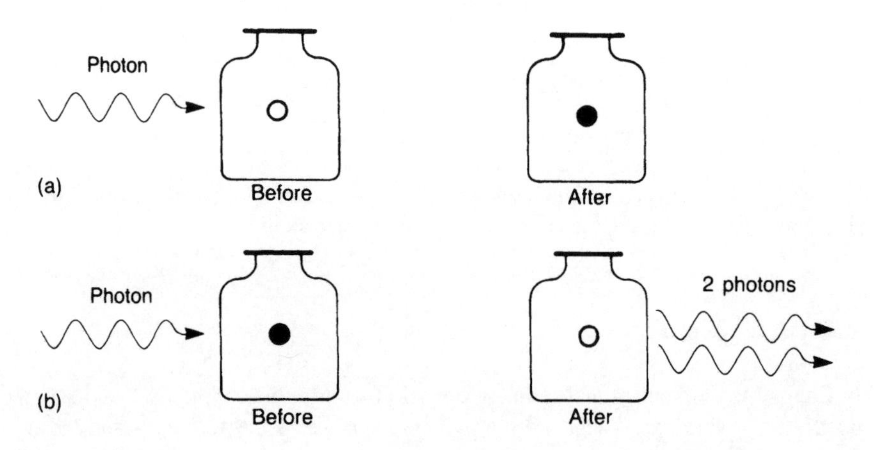

Figure 7.8 (a) An atom in the ground state (unshaded) is boosted to an excited state (shaded) when it absorbs a photon; (b) when the atom starts in an excited state, the incident photon can stimulate it to emit.

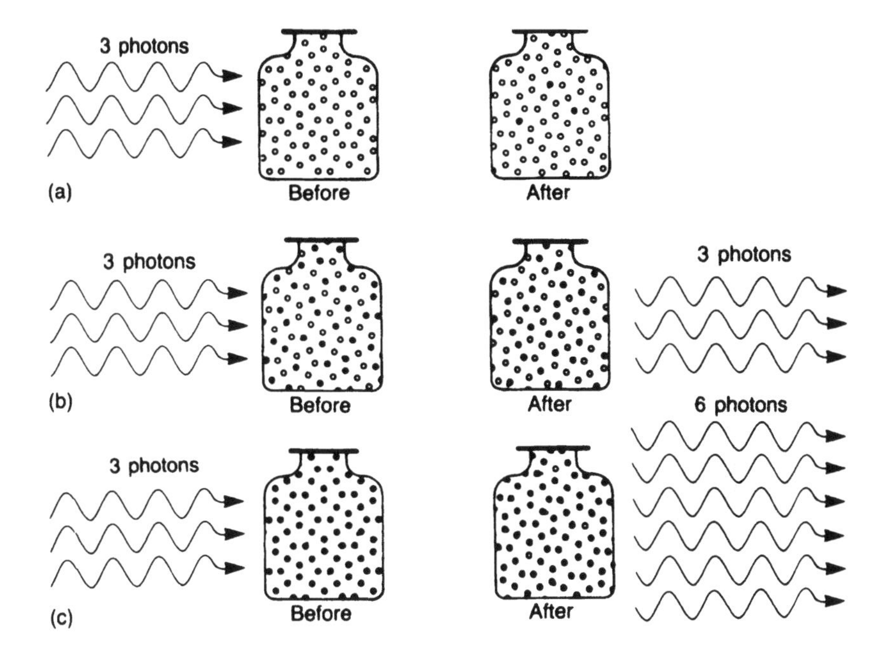

Figure 7.9 (a) Three photons are absorbed, exciting three atoms from their ground states; (b) if half of the atoms are excited initially, each photon will have a 50–50 chance of being absorbed or of stimulating emission; (c) if all the atoms are excited initially, each incident photon probably will stimulate emission from one of the atoms.

atom. If the photon interacts with an excited atom, it will stimulate the atom to emit; if it interacts with a ground-state atom, the photon will be absorbed. Thus, for every photon that is created by stimulated emission, another photon disappears by absorption. The number of photons departing the jar in the "after" picture in Figure 7.9b is equal to the number of incident photons in the "before" picture. There is no way of telling one photon from another of the same frequency and polarization, but if there were you would find that the three photons leaving the jar were not necessarily the same three that entered.

Finally, let us start with all the atoms in the E_2 excited state. When the three photons come along, the odds are that every one of them will stimulate an atom to emit, and the number of photons leaving the jar (maybe six) will be greater than the number that entered.

That is all there is to it. That is Light Amplification by Stimulated Emission of Radiation (L.A.S.E.R.). The light is amplified—three in, six out—when stimulated emission adds photons (radiation) to what is already there. It does not work without a population inversion, as you can see in Figure 7.9b. Of course, the 100% population inversion in Figure 7.9c is not required. Any population inversion—even 51 excited atoms in the jar—will provide some amplification. But the bigger the population inversion, the bigger the amplification.

Figure 7.9c is based on the somewhat simplistic assumption that each photon interacts with only a single atom. But if the atoms are big enough in cross section, the photons will interact with atoms soon after they enter the jar, and the stimulated photons produced by those interactions could stimulate other atoms to emit. (On the other hand, if the atoms are small enough in cross section, one or more of the photons might pass through the jar without interacting with any of the atoms.) Thus, there may or may not be exactly six photons emerging from the jar. The point is that more emerge than enter.

7.4 THREE-LEVEL AND FOUR-LEVEL LASERS

In the example in the previous section, we investigated a population inversion between the ground level and the second excited state. In real lasers, usually three or four energy levels are involved in the process of creating a population inversion and then lasing.

In a three-level system, shown in Figure 7.10a, essentially all the atoms start in the ground state. An external energy source excites them to a pump level from which they spontaneously decay quickly to the upper laser level. The energy released by this decay is usually heat rather than light. In most lasers, the upper laser level has a long spontaneous lifetime, so the atoms tend to accumulate there, creating a population inversion between that level and the ground state. When lasing takes place, the atoms return to the ground state, each emitting a photon.

A four-level laser is different from a three-level laser in that it has a distinct lower laser level, as shown in Figure 7.10b. Often, essentially all the atoms start in the ground state, and some are pumped into the pump level. They decay quickly to the upper level, which usually has a long lifetime. (Because it has a long lifetime, it is called a metastable level.) But now when lasing takes place, the atoms fall to the

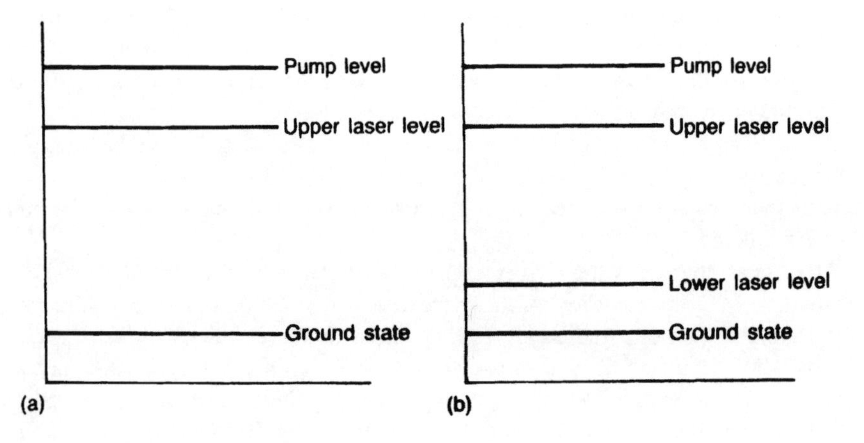

Figure 7.10 An energy-level diagram for a three-level (a) and a four-level laser system (b).

lower laser level rather than to the ground state. Once the atoms have undergone the stimulated transition to the lower laser level, they decay spontaneously to the ground state. The energy released in this decay is usually heat.

In which type of system do you think it would be easier to create a population inversion? That is, in which system do you have to pump up more atoms to create the inversion? If you answer that question correctly, you will be surprised to learn that the first laser was chromium-doped ruby—a three-level laser.

With the exception of ruby, most common lasers are four-level systems. In some lasers the pump level is not a single level; it is a collection of several levels that would be more correctly designated as a pump band. Nonetheless, all these lasers function essentially like the simplified model in Figure 7.10b.

There is a variation of the four-level scheme that entails pumping directly to the upper laser level rather than to a pump level. In this case, there are only three levels (the ground state, the upper laser level, and the lower laser level). Because there is a distinct lower laser level, such a laser has more in common with a normal four-level system than a three-level system. Some metal-vapor lasers operate on this scheme. The advantage of this system is that it is more efficient. A normal four-level (or three-level) laser loses energy when it spontaneously decays from the pump level to the upper laser level; this energy loss is absent if the laser can be pumped directly to the upper laser level.

There is another variation of the four-level system that should be mentioned here. In a sense, these lasers are midway between the three- and four-level systems and are sometimes referred to as "quasi-three-level" lasers. They are actually four-level systems, but the lower laser level is so close to the ground level that it is significantly populated at normal operating temperatures. (Recall Boltzmann's law from the beginning of this chapter.) In a normal four-level system like the one illustrated in Figure 7.10b, the lower laser level is sufficiently distant from the ground level that it has no significant thermal population at room temperature. But in quasi-three-level systems, the lower laser level is so low that it has a significant thermal population. Nonetheless, as we will see in Chapter 15, these lasers can be chilled to depopulate the lower laser level, and then they can be quite useful.

7.5 PUMPING MECHANISMS

How can the energy be input to a collection of atoms or molecules to create a population inversion? You know the energy cannot be put in thermally; that would just heat the collection, not create a population inversion. Earlier in this chapter, we explored two approaches. The population inversion in Figure 7.5 was created by plucking some atoms out of the bottom energy level so that it had less population than the level above it. The very first laser-like device ever built, the ammonia maser (microwave amplification by stimulated emission of radiation), created a population inversion that way. Ammonia molecules were forced through a filter that selectively blocked molecules in the lower of two energy levels so that the molecules that emerged exhibited a population inversion. That technique has not proven successful with modern lasers.

However, the second technique, electrical pumping, is practical for lasers. If the laser medium is placed in an electron beam, the electrons can create a population inversion by transferring their energy to the atoms when they collide. Several types of high-power gas lasers are pumped this way. More common is another technique of electrical pumping, the direct discharge. As shown in Figure 7.11, an electric discharge is created in a tube containing the gaseous laser medium, similar to the discharge in a fluorescent lamp. A population inversion is created in the ions or atoms of the discharge when they absorb energy from the current. Helium–neon lasers and most other common gas lasers are pumped by an electrical discharge.

Some gas lasers, particularly carbon dioxide lasers, are sometimes pumped with a radio-frequency (rf) energy. A sealed, electrodeless tube of the gas is excited by an intense rf field, creating a population inversion.

Another variation of electrical pumping creates a population inversion in semiconductor diode lasers. When a current passes through the interface between two different types of semiconductors, it creates mobile charge carriers. If enough of these carriers are created, they can produce a population inversion. We will explain the pumping mechanism of diode lasers more fully in Chapter 14.

In other common lasers, such as chromium-doped ruby and neodymium-doped YAG, the atoms that lase are embedded in a solid material instead of being in gaseous form. These lasers cannot easily be pumped by an electrical current or an electron beam. Instead, they are optically pumped, as shown in Figure 7.12. The laser material is bombarded with photons whose energy corresponds to the energy difference between the ground level and the pump levels. The atoms absorb energy from the pump photons and are excited to their pump bands.

Chemical energy sometimes can create a population inversion. In a chemical laser, two or more materials react, liberating energy and forming a new material. But the new material—it can be an element or a compound—is created with a built-in population inversion because many of its atoms or molecules have been excited by the chemical energy. Chemical lasers are not very common but because they are

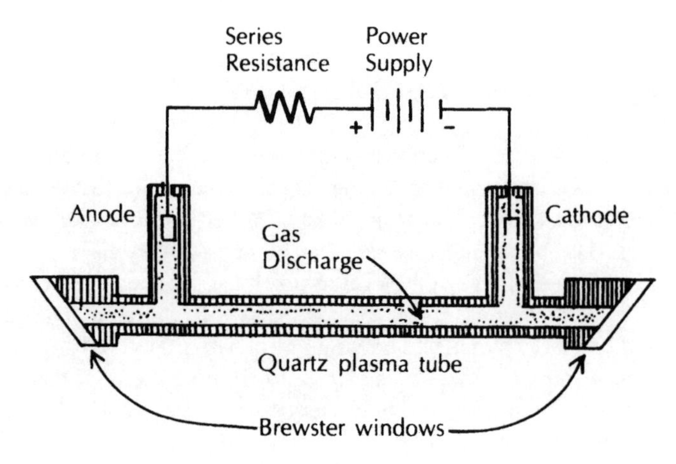

Figure 7.11 Electrical discharge pumping can create a population inversion.

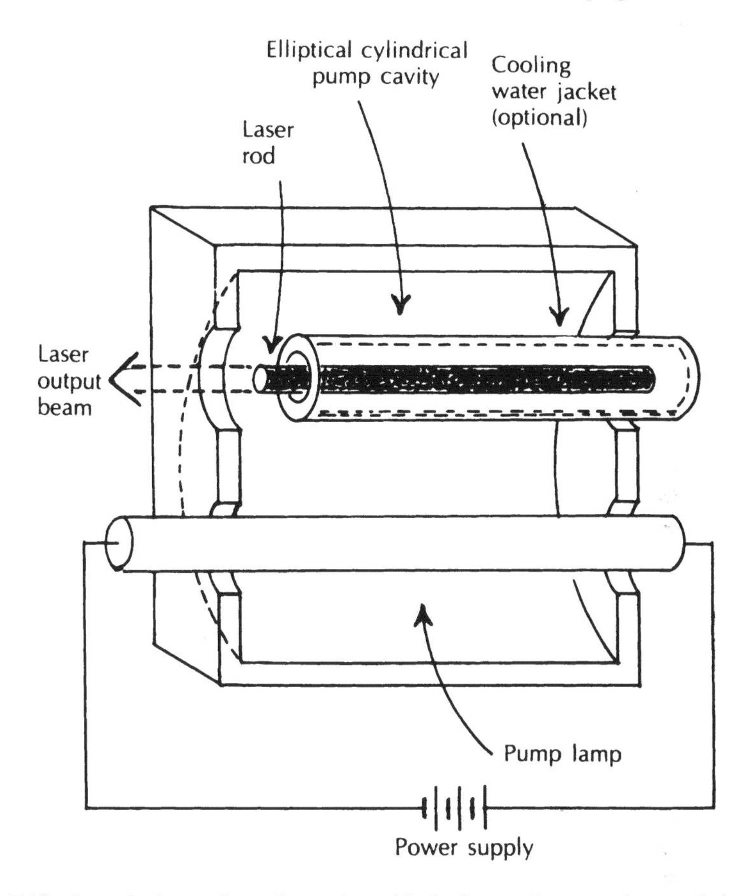

Figure 7.12 In optical pumping, a lamp alongside the laser rod creates the population inversion. The elliptical pump cavity focuses rays from the lamp into the rod.

capable of extremely high powers, they have been developed in the past by the military for weapons applications.

An exotic pumping mechanism should be mentioned here also, for the sake of completeness. Lasers have been pumped by nuclear particles, usually from nuclear bombs. The energy in the particles creates a population inversion in the laser medium, and a pulse of laser output is obtained before the shock wave and other energy from the bomb destroys the laser. Such lasers may someday have military applications.

Finally, free-electron lasers are pumped by high-energy electrons. The output of a free-electron laser is produced from electrons that are free rather than bound to an atom or molecule. Thus, the physics of a free-electron laser is much different from that of other lasers. Free electrons have no fixed energy levels like bound electrons, so the laser's wavelength can be tuned. Although free-electron lasers are promising for future applications, they are highly experimental and have been operated in only a few laboratories worldwide.

QUESTIONS

1. Why is it impossible to have a continuous-wave, optically pumped two-level laser?
2. Name several other examples of systems not in thermal equilibrium. For each of your examples, what is the approximate "spontaneous lifetime," that is, the time it takes the system to relax into thermal equilibrium?

Chapter 8

Laser Resonators

This chapter introduces the concept of the laser resonator, the crucial device that provides the feedback necessary to make a laser work. In Chapter 7 we found that there must be a population inversion for stimulated emission to occur. In this chapter we will find that the resonator is necessary if the stimulated emission is going to produce a significant amount of laser light.

It turns out that a resonator is not absolutely necessary to make a laser work but, as you will see in the first section of this chapter, a laser without a resonator usually just is not very practical.

Then we will discuss the concept of circulating power, the light that literally circulates back and forth between the mirrors of a laser resonator. We will learn how that power experiences both gain and loss as it circulates, and we will come to the important conclusion that, in a steady-state laser, the gain must equal the loss. We will also examine the concept of gain saturation, which allows a laser to satisfy this requirement.

Finally, we will discuss an important class of resonators called unstable resonators, which can sometimes produce more output than the more conventional stable resonators.

Incidentally, there are two terms you will come across that are used interchangeably in laser technology: resonator and cavity. The second word is a holdover from microwave technology because a microwave oscillator is a completely enclosed cavity. Because the word cavity can mean several things in laser technology, we will not use it to mean resonator.

8.1 WHY A RESONATOR?

Let us think about what kind of a laser you might have with just a population inversion and nothing else—specifically, no resonator. Suppose that you created a big population inversion in a laser rod, and suppose that one of the atoms at the far end of the rod spontaneously emitted a photon along the axis of the rod. The photon might stimulate another atom to emit a second photon, and, since we are assuming a big population inversion, one of these two photons might even find a third excited atom and create another stimulated photon. Thus, screaming out the near end of the laser rod come three photons.

Introduction to Laser Technology, Fourth edition. C. Breck Hitz, J. Ewing, and J. Hecht

Three photons. Now, the energy in a visible photon is about 10^{-19} J, whereas a real laser might produce 1 J. Obviously, we have a way to go from the three photons we have generated so far.

What next?

Well, we might try adding more laser rods, as shown in Figure 8.1. If each incoming photon in the second rod results in three output photons, we will have nine photons coming out of the second rod. A third rod will result in 27 photons.

But we want roughly 10^{19} photons to produce 1 J of output, and it should be obvious that it is going to take a long time to get there. Is there a better way?

The better way, of course, is to use mirrors, as shown in Figure 8.2. The photons are reflected back and forth for many passes through the rod, stimulating more and more emission on each pass. As indicated in Figure 8.2, the mirrors can be gently curved so they tend to keep the light concentrated inside the rod. In a practical laser, one of the mirrors is 100% reflective, but the other mirror transmits part of the light hitting it. This transmitted light is the output beam from the laser. The transmission of the output mirror varies from one type of laser to another but is usually somewhere between 1% and 50%.

For the sake of completeness, we should mention here that some lasers have such enormous gain that 10^{19} photons can be produced in a single pass. These superradiant lasers do not need resonators. Nitrogen lasers sometimes operate in this fashion.

8.2 CIRCULATING POWER

If the photons in Figure 8.2 bounce back and forth between the mirrors for long enough, the laser will reach a steady-state condition and a relatively constant power will circulate between the mirrors. This circulating power is not absolutely constant, as indicated in Figure 8.3. Part of it is lost when it hits the output mirror, and this lost power is replaced when the light passes through the gain medium.

Figure 8.3 shows how the circulating power varies inside the resonator. You can follow it around, starting, for example, at point A and moving to the right. The circulating power drops at the output mirrors because part of the light is transmitted through the mirror. The remaining light travels through the gain medium, where it is partially replenished. There is a small loss at the back mirror because no mirror is a perfect reflector. Then the light returns for a second pass through the gain medium, where it is fully restored to its previous level at point A.

Have you realized as you have read this that the power in the output beam is determined by the amount of circulating power and the transmission of the output mirror? Suppose that there are 50 W of circulating power inside the laser and the output

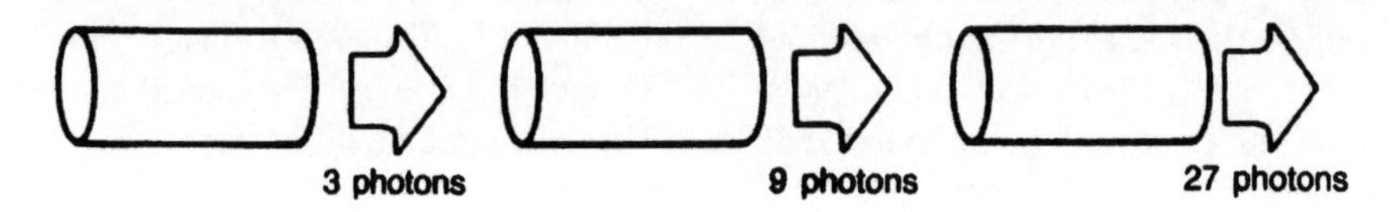

Figure 8.1 How to make a laser without a resonator.

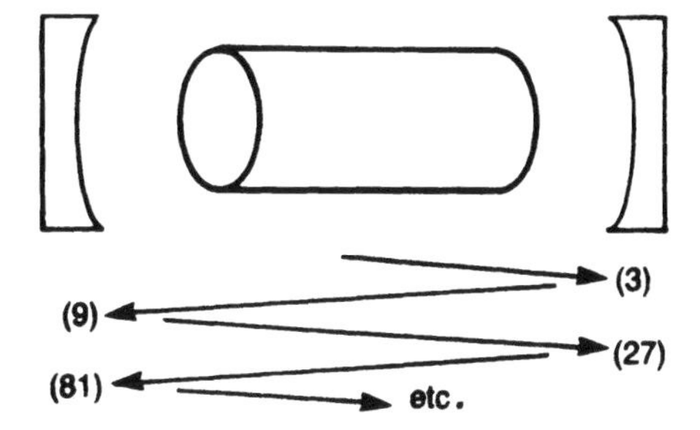

Figure 8.2 How to make a laser with a resonator.

mirror has 2% transmission. Then the output beam is simply 2% of the circulating power that is incident on the mirror, or 1 W. Mathematically, the concept is expressed this way:

$$P_{out} = \tau P_{circ}$$

in which τ is the transmission of the output mirror and P_{out} and P_{circ} are the output and circulating powers, respectively.

You can also use this concept to calculate the circulating power inside a resonator. Although a direct measurement of the circulating power would be difficult to make, it is easy to measure the output power and then simply divide by the mirror transmission to find what the circulating power really is.

The previous equation says that the output power is proportional to the mirror transmission. So does that mean you can always increase the output power by increasing the mirror transmission? No, because as the transmission of the output mirror is increased, the circulating power will decrease. What happens to the output power depends on whether the circulating power decreases faster than the mirror

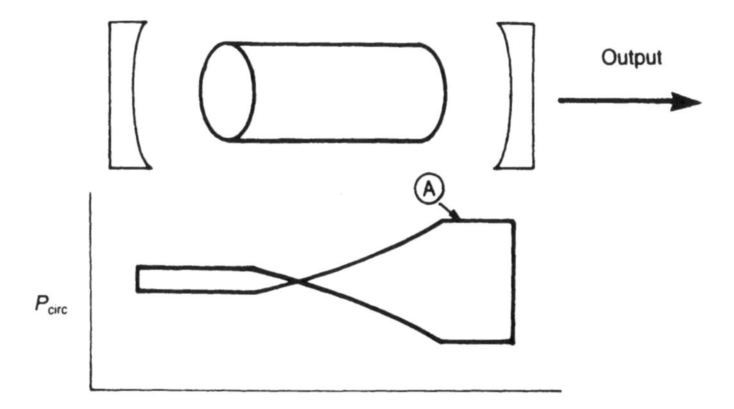

Figure 8.3 A schematic representation of the circulating power inside a laser resonator. The lower drawing shows the power in the resonator.

transmission increases, and this depends on the particular laser you're looking at. In fact, for any laser there will be an optimum value for the transmission of the output mirror that will produce the maximum possible output power.

8.3 GAIN AND LOSS

The loop in the diagram in Figure 8.3 is closed. That is, the circulating power is restored to precisely its initial value after a round-trip through the laser resonator. This is true for any steady-state or continuous-wave laser. In most pulsed lasers, the situation is different because energy moves so quickly from the population inversion to circulating power to output power that it never has time to reach an equilibrium. But for now let us limit our attention to continuous-wave lasers in which the circulating power can settle down to a steady-state behavior like that shown in Figure 8.3.

If the circulating power is restored to its original value after a round-trip of the resonator, the round-trip gain must be equal to the round-trip loss. If the round-trip gain is less than the round-trip loss, the laser will not lase. On the other hand, if the round-trip gain is greater than the round-trip loss, the gain will saturate until it is reduced to the same value as the round-trip loss. We will discuss this saturation phenomenon in more detail later in this section.

Note that we have just identified a second requirement for lasing. In Chapter 7 we found that there must be a population inversion in the gain medium. In this chapter, we see that merely having a population inversion is not enough; the population inversion must be large enough so that the round-trip gain is at least as large as the round-trip loss. In the parlance of laser technology, the gain that is just barely sufficient for lasing is called the *threshold gain.*

What are the causes of loss inside a laser resonator? Obviously, the transmission of the output mirror is one source of loss, but some circulating power is also lost at every optical surface in the resonator because there is no such thing as a perfect surface. Some light will be scattered and reflected no matter how well the surface is polished. As we have said, some light is lost at the imperfect rear mirror. Other light is lost by scattering as it propagates through refractive-index inhomogeneities in the gain medium. Some loss by diffraction occurs because of the finite aperture of the laser beam inside the resonator. Altogether, these resonator losses add up to a few percent per round-trip in most continuous-wave resonators.

Now, if you think about it for a moment, it seems as if we have gotten ourselves into a contradictory situation. The round-trip loss depends solely on the passive qualities of the resonator (e.g., its mirror transmission), and the round-trip gain must be equal to this round-trip loss. Do you see the problem? The problem is that we determined what the round-trip gain is without knowing anything about how hard the laser is being pumped. In other words, we are apparently saying that the gain is independent of how hard the laser is pumped. Can that be right?

It turns out that it is. To understand why, you must understand that there are two kinds of laser gain: saturated gain and unsaturated gain. The difference between the two is diagramed in Figure 8.4. Unsaturated gain is sometimes referred to as small-signal gain because it is the gain observed with only a very small input signal. In Figure 8.4a, 100 photons are amplified to 102 photons by the laser rod. The unsatu-

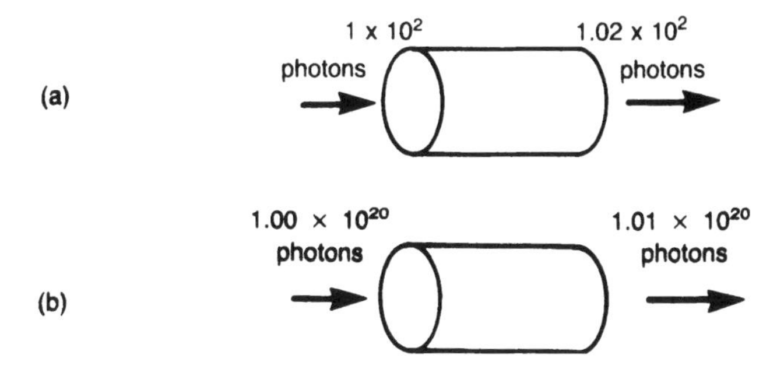

Figure 8.4 Unsaturated gain (a) and saturated gain (b).

rated gain is therefore 2%. But in Figure 8.4b the same laser rod provides a gain of only 1% when 10^{20} photons pass through it. What happened? So many photons passed through the rod, stimulating emission as they went, that the population inversion was significantly depleted by stimulated emission. Hence, the gain was reduced—saturated—from its unsaturated value in Figure 8.4a.

In the unsaturated case in Figure 8.4a, atoms pumped to the excited state are not stimulated to emit. These atoms can get out of the excited state only by spontaneous emission or by collisional deexcitation processes.* But in the saturated case in Figure 8.4b, about 10^{18} atoms leave the excited state by stimulated emission. The absence of these 10^{18} atoms in the excited state accounts for the gain difference between the two cases.

Now how does this resolve the seeming contradiction we previously mentioned? Let us start by thinking about a continuous-wave laser sitting an a table, lasing away with a 3% round-trip loss and a 3% round-trip gain. What happens when you turn up the pump power? Well, instantaneously the population inversion gets bigger and so does the gain. But this larger gain produces more circulating power, and the increased circulating power causes the gain to saturate more than it was before. The gain quickly decreases back to 3%, but the circulating power—and the output power—is now greater than it was before you turned up the gain. On the other hand, if you turn down the pump power, the reverse process takes place and you wind up with less circulating power but the same 3% gain.

Therefore, the actual gain in a laser resonator is in fact independent of the pump power. It is the circulating power and, hence, the output power, that varies with the pump power.

8.4 ANOTHER PERSPECTIVE ON SATURATION

There is another way to understand gain saturation. The result will be the same as in the previous section: the round-trip gain in a resonator is exactly equal to the round-trip loss, independent of pump power.

*Two atoms are stimulated to emit the two extra photons produced in Figure 8.4a, but the remaining 10^{19} atoms must get out of the excited state some other way.

Figure 8.5 shows the circulating power and the gain as a function of pump power. Think about what these plots show. At zero pump power, there is no gain and no circulating power. As the pump power increases, it pumps atoms into the upper laser level. In a true four-level system, you have a population inversion (and gain) as soon as you have any population in the upper level. But you do not have any circulating power yet because the gain is less than the intracavity loss (a in Figure 8.5).

As the pump power continues to increase, it finally reaches the point where the gain is as large as the loss. This point is called laser threshold (P_{th} in Figure 8.5). Now the laser begins to lase. Photons start bouncing back and forth between the mirrors. But as they bounce back and forth, they pass through the gain medium and saturate the population inversion. As the pump power increases, it pumps more atoms into the upper level, but these atoms are removed just as quickly by the increased circulating power. The net result is that the population inversion remains constant. The gain is exactly equal to the loss, independent of pump power.

But the circulating power (and hence the output power) increases with increasing pump power above threshold. In summary, Figure 8.5 shows that, below threshold, there is no output and gain increases with pump power. Above threshold, gain remains constant as pump power increases, while laser output increases.

8.5 RELAXATION OSCILLATIONS

The smooth output of many lasers, especially solid-state lasers, is marred by so-called relaxation oscillations. A laser's energy can oscillate between the population

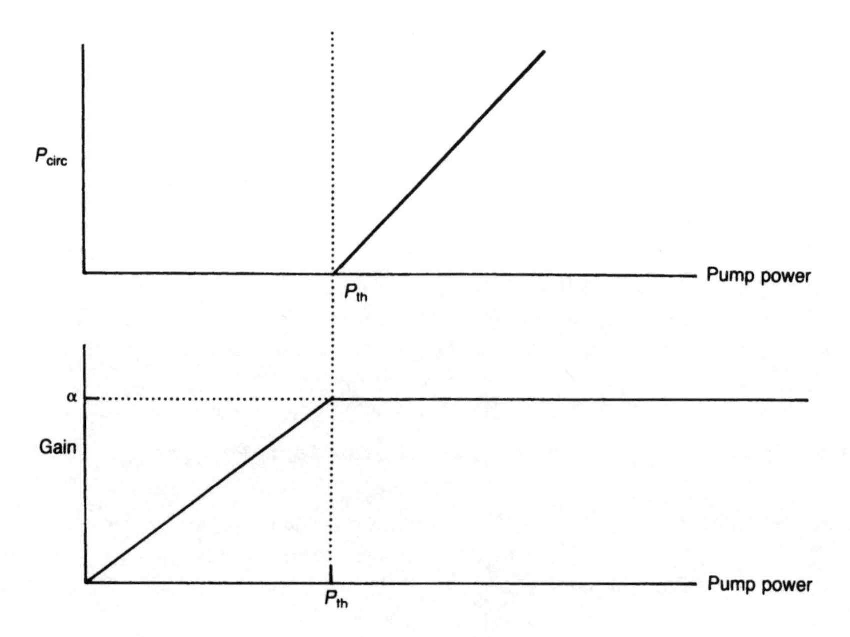

Figure 8.5 Circulating power and intracavity gain as a function of pump power. Laser threshold occurs when gain is equal to the total cavity loss (α).

inversion and the optical fields, much as a pendulum's energy oscillates between kinetic energy and potential energy. Since the laser output is directly proportional to the circulating power at any instant, this oscillation inside the laser shows up as a fluctuation on the output beam.

Gravity is the mechanism that drives a pendulum's oscillation, and stimulated emission drives a laser's. Figure 8.6 shows the circulating power and gain of a laser as a function of time. These plots begin when the laser has just reached threshold: the gain is equal to the loss, and the circulating power has just become greater than zero. The pump power continues to pump atoms into the upper level and, for an instant, there are not yet enough photons bouncing between the mirrors to saturate the gain. The gain increases briefly above its equilibrium value, but the circulating power increases rapidly, and soon it is great enough to saturate the gain.

Look carefully at what's happening in Figure 8.6. The gain saturates back down to its equilibrium value, but there are still all those photons bouncing between the mirrors. (The circulating power is at a maximum.) These photons continue to deplete the population inversion, driving the gain below its equilibrium value. Now the gain is less than the loss, so the circulating power decreases. As the circulation power diminishes, it saturates the gain less and the gain begins to increase. And now the whole process starts to repeat itself: the gain increases briefly above its equilibrium value, creating excessive photons, which saturate the gain back below its equilibrium value, and so forth.

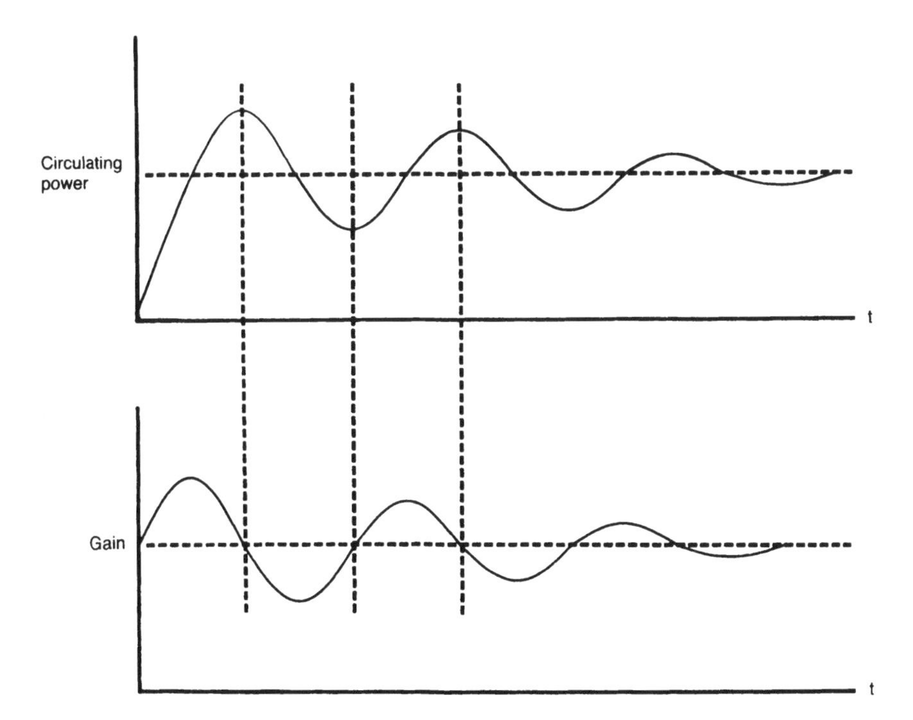

Figure 8.6 Relaxation oscillations occur when energy flows back and forth between the population inversion and the circulating power.

These relaxation oscillations occur at a resonant frequency in the range of several hundred kilohertz. Their modulation depth varies from a few percent to 100%. As shown in Figure 8.6, they are damped out and normally disappear quickly. The problem is that it is very easy to excite a new set of relaxation oscillations in a laser. Any mechanical vibration at the resonant frequency will set the laser off on another spasm of oscillation. In common lasers, bothersome relation oscillations can be induced merely by tapping the table on which the laser sits. In water-cooled lasers, the slight vibration caused by flowing water will keep the laser oscillating continuously unless special precautions are taken in the laser's mechanical design.

8.6 OSCILLATOR-AMPLIFIERS

When the power produced by a pulsed laser oscillator is insufficient for a particular application, an amplifier is sometimes added to the oscillator. This configuration, shown in Figure 8.7, is called an oscillator-amplifier, or a master–oscillator/power amplifier (MOPA).

The intense pulse from the oscillator passes through the amplifier only once, but that single pass is sufficient to deplete most (or all) of the population inversion in the amplifier.

There is an advantage to using an oscillator–amplifier configuration, rather than simply using a larger oscillator. A large oscillator will usually produce a lower-quality beam than a small oscillator because thermal distortions and other optical problems are more severe in a large oscillator. But the output of the oscillator–amplifier in Figure 8.7 will have all the desirable spatial, temporal, and spectral characteristics of the oscillator, while also having much higher power than the oscillator alone could produce.

8.7 UNSTABLE RESONATORS

Most lasers have stable resonators in which the curvatures of the mirrors keep the light concentrated near the axis of the resonator. If you trace the path of a ray of light

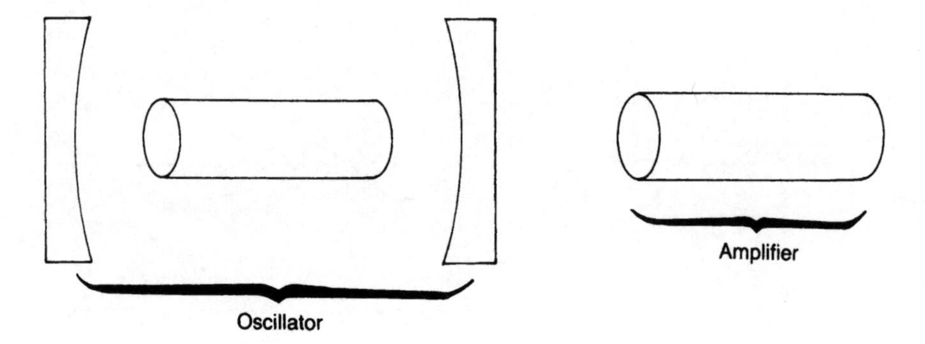

Figure 8.7 Oscillator-amplifier configuration for high-power, pulsed lasers.

between the mirrors of a stable resonator, you will find that the ray is eventually reflected back toward the resonator axis by the mirrors, as shown in Figure 8.8. The only way light can escape from the resonator is to go through one of the mirrors.

In the unstable resonator diagramed in Figure 8.9, the light rays continue to move away from the resonator axis until eventually they miss the small convex mirror altogether. The output beam from this resonator will have a doughnut-like shape with a hole in the middle caused by the shadow of the small mirror. (There are clever ways to design unstable resonators that avoid the hole-in-the-center beam.) The advantage of unstable resonators is that they usually produce a larger beam volume inside the gain medium so that the beam can interact with more of the population inversion and thereby produce more output power. Because the light passes through an unstable resonator only several times before emerging, these resonators are usually used only with high-power, pulsed lasers.

Incidentally, do not be confused about the use of the word stable here. A stable resonator is, by definition, one in which a ray is trapped between the mirrors by their curvature. The word stable implies nothing about a resonator's sensitivity to misalignment nor about an absence of fluctuations in its output power.

8.8 LASER MIRRORS

Laser mirrors are different from bathroom mirrors. As you probably know, an ordinary bathroom mirror is fabricated by depositing a metallic surface on the back of a piece of glass. Although metal mirrors are sometimes used in very high-power lasers, most laser mirrors are dielectric mirrors. They are made by depositing alternate layers of high- and low-index materials on a substrate. A fraction of the incident light is reflected from each interface, and the total reflectivity depends on the relative phase of these reflections.

A high-reflectivity mirror, for example, might have several layers of transparent material deposited on a glass substrate. Each layer would be exactly the right thickness so that the light reflected from its front surface would be precisely in phase with the light reflected from its rear surface. It is possible to fabricate dielectric mirrors like this whose reflectivity is greater than 99.9999%.

The thicknesses of the layers of materials deposited on a laser mirror must be carefully controlled. If there is an error on only a tiny fraction of a wavelength, the desired interference effect will be lost. The necessary control is achieved by fabricating laser mirrors by vacuum deposition. The bare mirror substrate, together with

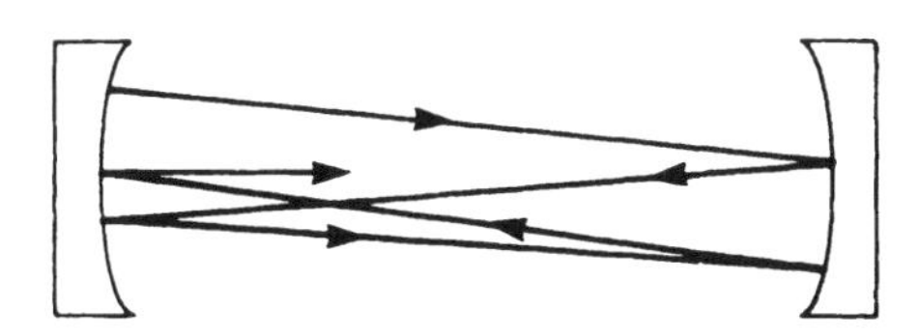

Figure 8.8 A ray is always reflected back toward the center by the curved mirrors of a stable resonator.

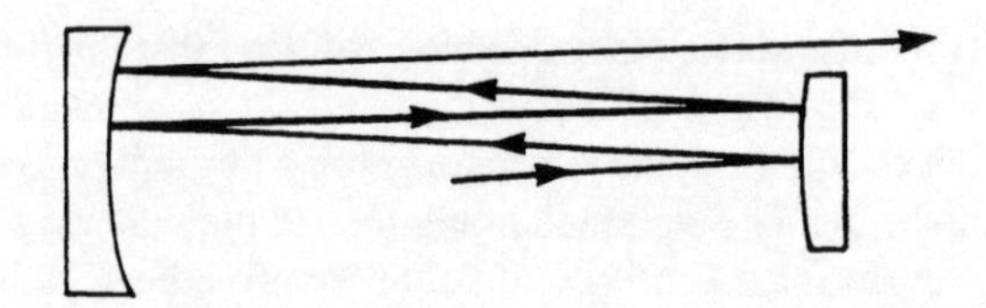

Figure 8.9 In this unstable resonator, a ray will eventually be reflected past one of the mirror.

a pot of the material that will form a layer on the mirror, is placed in a chamber. The chamber is evacuated, and then the material in the pot is heated until it evaporates. It condenses on the cool substrate, in a layer whose thickness can be easily controlled. This process is repeated for each layer on the mirror.

The same approach can be used to produce partially reflecting mirrors or to produce antireflection coatings. To produce an antireflection coating, the thicknesses of the layers are adjusted so that the reflection from the front of each layer is precisely out of phase with the reflection from the back of the surface.

QUESTIONS

1. Suppose you wanted to make a laser that could produce 1 J (10^{19} photons) without mirrors, as indicated in Figure 8.1. How many laser rods would you need?
2. Calculate the circulating power inside a continuous-wave Nd:YAG laser that produces a 10 W output beam and has a 6% transmissive output mirror. Calculate the circulating power inside an He–Ne resonator that has a 1% mirror and produces 2 mW.
3. Consider the laser shown in the following diagram.

   Calculate the threshold round-trip gain if the output mirror has 3% transmission, each optical surface (including the mirrors) has 0.25% scatter, and the back mirror has 0.5% transmission.
4. In an Nd:YAG laser, the unsaturated gain is related to the saturated gain by the

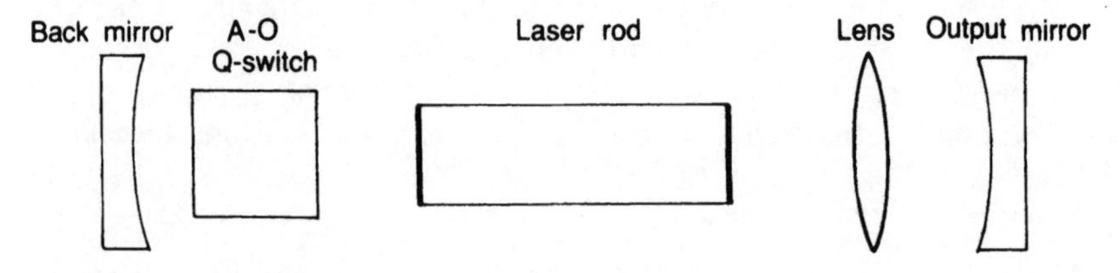

following equation:

$$g = \frac{g_0}{1 + \beta P_c}$$

in which g is the saturated gain; g_0 is the unsaturated gain; P_c is the circulating power; and β is the saturation parameter, an experimentally determined quantity. From this equation and the equation $P_{out} = \tau P_{circ}$, derive an equation expressing the output power from the laser in terms of unsaturated gain (g_0), saturated gain (g), saturation parameter (β), and mirror transmission (τ). Assume that the mirror transmission is the only significant loss to the circulating power.

Chapter 9

Resonator Modes

In Chapter 8, we learned why a resonator is necessary and how the optical power circulates back and forth between the mirrors of a laser resonator. In this chapter, we take a close look at that circulating power, and, in particular, we examine its spatial distribution within the resonator.

We will find that studying the transverse spatial distribution of energy within the resonator leads to an understanding of how a laser beam propagates through space outside the resonator. This chapter introduces some equations that can be used to calculate how the characteristics of a beam change as it propagates. We will also learn how to select the proper resonator mirrors to produce a laser beam of a given size and divergence. Because some resonator configurations are inherently unstable, we will learn how to determine whether a resonator is stable. Finally, we will see how the longitudinal spatial distribution of energy within a resonator affects the output beam from the laser.

9.1 SPATIAL ENERGY DISTRIBUTIONS

It is recorded, in this book if nowhere else, that an early philosopher in the field of photonics tried to capture light in a Greek vase. Standing outdoors on a sunny afternoon, he turned the open mouth of the vase toward the sun and let light flood into the container. Quickly, he slipped a lid over the mouth of the vase and hurried into a darkened cave, where he carefully removed the lid to let the light escape. But he was disappointed, for the interior of the vase was pitch black every time he performed the experiment.

Of course, you know that the experiment was doomed to failure. But do you really know why? What happened to the light that had been in the vase? Where did it go? And how long did it take to go wherever it went?

At one instant the vase was full of photons, bouncing around off the sides of the vase, and the next instant the vase was capped and all the photons had disappeared. Where did they go? The photons were absorbed by the walls of the vase. Even a polished white Greek vase absorbs maybe 10% of the light incident on it. So after the light has bounced around 100 or so times inside the vase, it is practically all gone. How long does that take? Well, if you figure it is a big Greek vase 1 ft in diameter, and if you remember that the speed of light is roughly 1 ft/ns, you will see that the light will be gone after about 100 ns. So the philosopher actually had the right idea; he just did not move fast enough.

Introduction to Laser Technology, Fourth edition. C. Breck Hitz, J. Ewing, and J. Hecht

In the ensuing centuries, modern science has improved on the Greek vase as a light-storing device. A laser resonator is, in fact, nothing more than a modern light storage device. Admittedly, it is designed with a figurative hole in it because the output mirror allows part of the stored energy to "leak" out. But the gain medium replaces the energy as fast as it is lost through mirror transmission and the other losses discussed in Chapter 8.

When you talk about resonator modes, you are talking about the *spatial distribution* of stored light energy between the laser mirrors. It turns out that energy is not stored uniformly in a resonator, like water in a glass. Instead, the energy exists in clumps, somewhat like cotton balls stored in a jar. The resonator mode is determined by the spatial arrangement of these clumps of light energy.

There are two kinds of modes: transverse and longitudinal. To visualize the transverse mode of a laser, imagine that the resonator is cut in half along a plane transverse to the laser axis, as shown in Figure 9.1a. If you then examined the distribution of energy along this plane, you would see the shape of the transverse laser mode. On the other hand, if you sliced the resonator as shown in Figure 9.1b, you would see the shape of the longitudinal laser mode.

This concept is analogous to mapping the spatial distribution of cotton balls in a jar, as shown in Figure 9.2. You can imagine drawing a map corresponding to the distribution of cotton along the plane of the figure. However, in a laser resonator the energy map does not change if the plane moves. That is, if the "cut here" plane of Figure 9.1a moves to the left or right (or up or down in Figure 9.1b), the shape of the energy distribution you would see in that plane does not change (although its size might).

9.2 TRANSVERSE RESONATOR MODES

It is not necessary to cut the resonator in half, as in Figure 9.1a, to see the shape of the transverse modes of a resonator. All you need to do is look at the shape of the

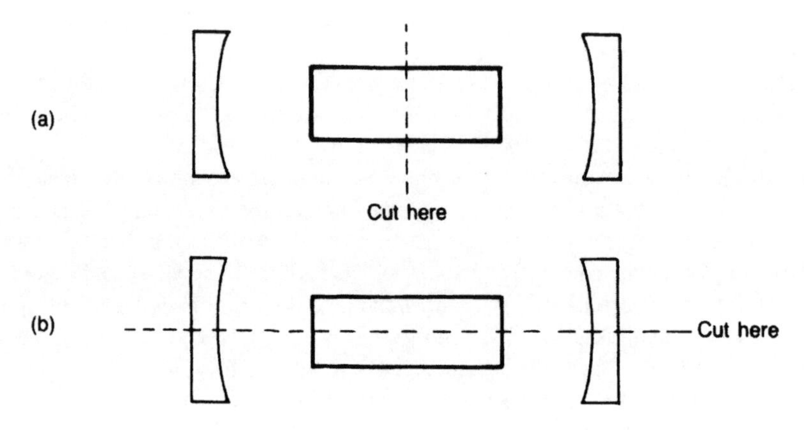

Figure 9.1 How to visualize transverse laser modes (a) and longitudinal laser modes (b).

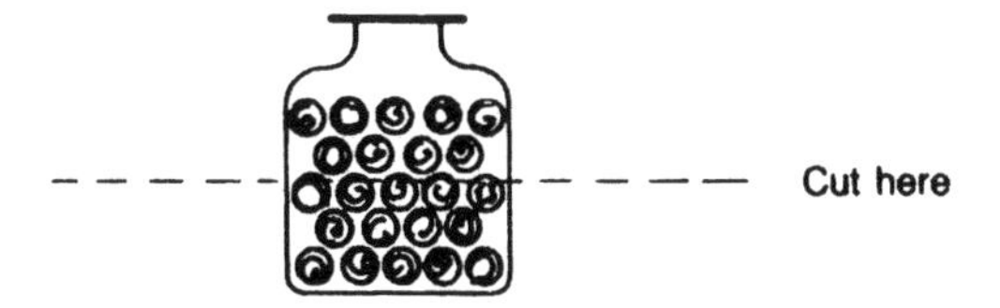

Figure 9.2 How to visualize horizontal cotton-ball storage modes.

output beam because the pattern inside the resonator moves out through the mirror and becomes the shape of the beam. The beam can have a number of profiles, as shown in Figure 9.3.

Theoretically, dozens of transverse modes can oscillate simultaneously in a resonator and each can have a different frequency, but, in practice, only several (or sometimes only one) oscillate. The mode shapes in Figure 9.3 were created by forcing a resonator to oscillate in only one mode at a time.

Note that each mode has a different designation in Figure 9.3. If you remember that the number of dark stripes in the pattern corresponds to the subscript, you will always be able to name the mode properly. Incidentally, there is no accepted system for deciding which subscript comes first; one person's TEM_{41} mode is another's TEM_{14}, and both are correct.*

As shown in Figure 9.4, high-order modes are larger than low-order modes. For many laser applications, it is important that the laser oscillate only in the TEM_{00} mode. How can you prevent a laser from oscillating in its higher-order modes?

The answer has to do with the relative sizes of the different modes. The TEM_{00} mode is smaller in diameter than any other transverse mode. Thus, if you place an aperture of the proper size (as shown in Figure 9.4) inside the resonator, only the TEM_{00} mode will fit through it. Higher-order modes will be extinguished because the loss imposed on them by the aperture will be greater than the gain provided by the active medium. Some TEM_{00} lasers come equipped with apertures like the one shown in Figure 9.4, whereas in others the small diameter of the active medium acts as an effective aperture.

In Figure 9.4, the TEM_{11} mode occupies a larger volume in the gain medium than the TEM_{00} mode does. The TEM_{11} mode can, therefore, interact with more of the population inversion and extract more power from the laser. For this reason, lasers oscillating in high-order modes usually produce more power than otherwise similar lasers limited to TEM_{00} oscillation. However, the advantages of the TEM_{00} mode often outweigh the cost of reduced power.

9.3 GAUSSIAN-BEAM PROPAGATION

The TEM_{00} mode is so important that there are several names for it in laser technology, all meaning the same thing. The TEM_{00} mode is called the *Gaussian mode,* the

*TEM stands for *transverse electromagnetic,* a name that derives from the way the electric and magnetic fields behave at the resonator's boundary.

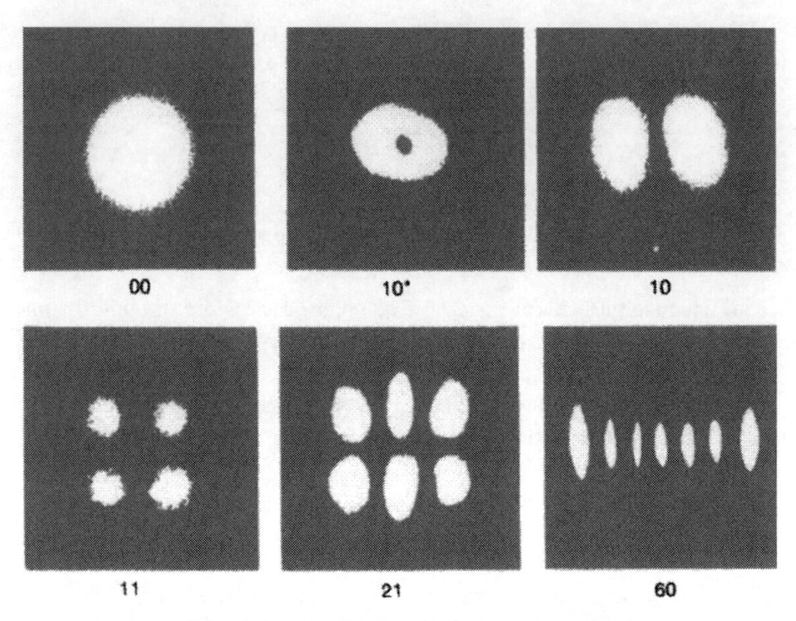

Figure 9.3 The shapes of transverse laser modes.

fundamental mode, or, sometimes, the *diffraction-limited* mode. No matter what it is called, it is a very important mode, and this section describes how the light produced by a Gaussian-mode laser propagates through space.

But before you can understand how a Gaussian beam propagates, you must understand two parameters that characterize the beam. One is easy to understand; the other is a little more subtle. The easy one is the beam radius, the radius of the spot the beam would produce on a screen. It is a somewhat arbitrary parameter because a Gaussian beam does not have sharp edges. The intensity profile of a Gaussian beam is given by the following equation:

$$I = I_0 e^{-2x^2/w^2}$$

in which I_0 is the intensity at the center, x is the distance from the center, and w is the beam radius. The intensity profile is pictured in Figure 9.5. The "edge" of this beam is defined to be the point where its intensity is down to $1/e^2$ (about 13%) of its

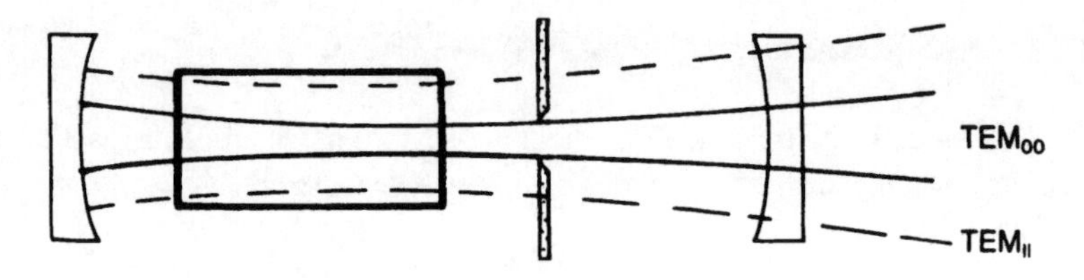

Figure 9.4 An aperture in the resonator can force it to oscillate only in the TEM_{00} mode.

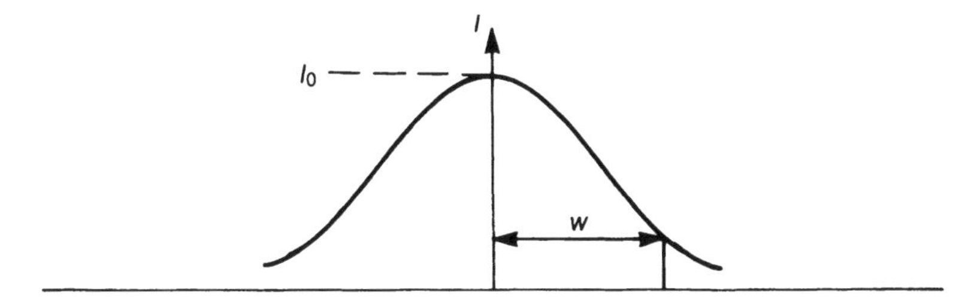

Figure 9.5 The intensity profile of a Gaussian beam.

intensity at the center. To the eye, the place where the intensity has dimmed to $1/e^2$ of its maximum value looks like the edge of the beam.

The other parameter that characterizes a Gaussian beam—the more subtle parameter—is the radius of curvature of the beam's wavefront. Remember that one of the characteristics of coherent light is that all the waves are in phase with each other. If you were to construct a surface that intersected all the points of common phase in a Gaussian beam, that surface would be spherical. The idea is shown graphically in Figure 9.6. The dotted spherical surface passes through the trough of each wave in the beam.

Both the beam radius and the radius of curvature change as the beam propagates. You can think of a Gaussian beam having the appearance shown in Figure 9.7. (Of course, you cannot really see the radii of curvature in the beam, but you can imagine them.) The light could be traveling either direction in this drawing—left to right or right to left. The beam radii and the radii of curvature would be the same in either case. The radius of curvature is infinite at the beam waist, drops sharply as you move away from the waist, and then begins increasing again as you move farther from the waist. At long distances from the waist, the radius of curvature is equal to the distance from the waist. The beam radius increases steadily, of course, as you move away from the waist.

The following equations allow you to calculate both the beam radius (w) and the radius of curvature (R) at any distance (z) from the waist if you know the beam radius at the waist (w_0) and the laser wavelength (λ):

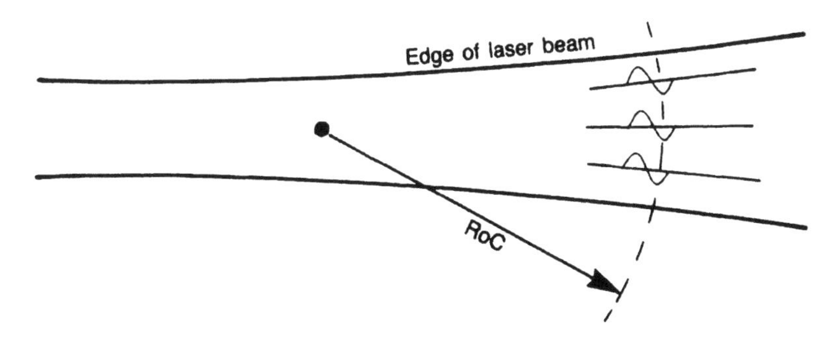

Figure 9.6 In a Gaussian beam, the surface of constant phase (dotted line) is spherical.

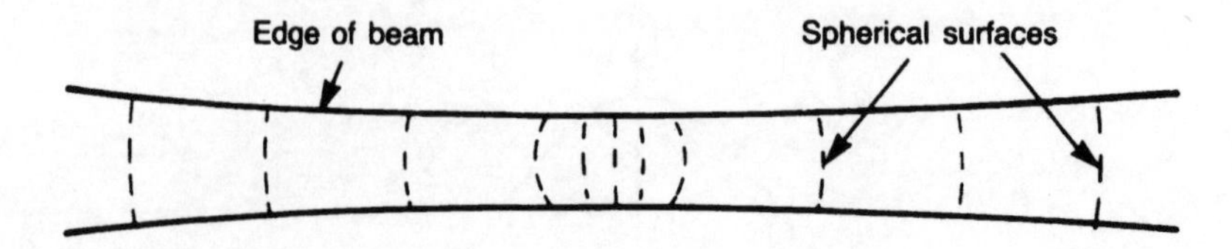

Figure 9.7 Both the beam radius and the wavefront radius of curvature change as a Gaussian beam propagates through space.

$$w = w_0\left[1+\left(\frac{\lambda z}{\pi w_0^2}\right)^2\right]^{1/2}$$

$$R = z\left[1+\left(\frac{\pi w_0^2}{\lambda z}\right)^2\right]$$

The first equation gives the beam radius, and the second gives the wavefront radius of curvature. Let us look at some examples in which these equations would be useful.

Suppose you wanted to do a laser-ranging experiment to measure the distance from the earth to the moon. You would aim a pulse of laser light at the retroreflector left on the moon by the astronauts. By very carefully measuring how long it took the pulse to travel to the moon and back, you would be able to figure out the distance with an accuracy of feet or even inches. But how wide is the beam by the time it gets to the moon? If it is too wide, only a tiny fraction of the light will be reflected back by the retroreflector, and the amount that returns to earth could be too small to detect.

Let us suppose that we are using an Nd:YAG laser whose wavelength is 1.06 μm and whose beam waist is 0.5 mm radius. The approximate earth–moon distance is 239,000 mi. We must invoke the first of the two equations:

$$w = w_0\left[1+\left(\frac{\lambda z}{\pi w_0^2}\right)^2\right]^{1/2}$$

Let us write down the known quantities and express them all in the same dimension (meters):

$$\lambda = 1.06 \times 10^{-6}\ \text{m}$$

$$w_0 = 5 \times 10^{-6}\ \text{m}$$

$$z = 239{,}000\ \text{mi} = 3.84 \times 10^{8}\ \text{m}$$

Next, substitute these values into the equation:

$$w = (5\times10^{-4}\,\text{m})\left\{1+\left[\frac{(1.06\times10^{-4}\,\text{m})(3.84\times10^{8}\,\text{m})}{\pi(5\times10^{-4}\,\text{m})}\right]^2\right\}^{1/2}$$

Finally, do the arithmetic:

$$w = (5 \times 10^{-4}\text{ m})[1 + (5.2 \times 10^{8})^2]^{1/2}$$
$$\approx (5 \times 10^{-4}\text{ m})(5.2 \times 10^{8}\text{ m}) = 2.5 \times 10^{5}\text{ m}$$

This beam is too wide. The retroreflector is only about a meter in diameter, so only a minuscule fraction of the light will be reflected back toward earth. What could you do to decrease the size of the beam on the moon?

Take a look at the equation again. There are three parameters: λ, z, and w_0. The laser wavelength is fixed, and you cannot move the moon any closer to the earth to make the problem easier. But you can adjust the radius of the laser beam, and that is the solution. The larger the waist of a Gaussian beam, the smaller its divergence will be. If you expand the beam with a telescope before it leaves earth, you can greatly reduce its divergence.

There are many different Gaussian beams, one for each size of waist. If you know the size of the beam at its waist, you can calculate the beam size and its radius of curvature at any point in space. The characteristics of a Gaussian beam are completely determined from its waist size (assuming, of course, that the wavelength does not change). By using a lens, or a system of lenses or focusing mirrors, it is possible to convert one Gaussian beam into another. In Figure 9.8, a small-waisted divergent Gaussian beam is changed into a less-divergent, larger-waisted beam by a lens.

You can imagine numerous applications for the first of the two equations we have discussed. A manufacturer making laser surveying equipment would want to know how big the beam would be a few hundred yards from the laser so that it could make the detectors the proper size. An engineer designing a laser scanner for a grocery store would need to know the size of the beam on the window where the groceries are scanned. But where would you use the second equation? What good does it do to know the radius of curvature of a Gaussian beam?

For a Gaussian beam to exist in a resonator, its wavefronts must fit exactly into the curvature of the mirrors. The Gaussian beam from Figure 9.7 is shown again in

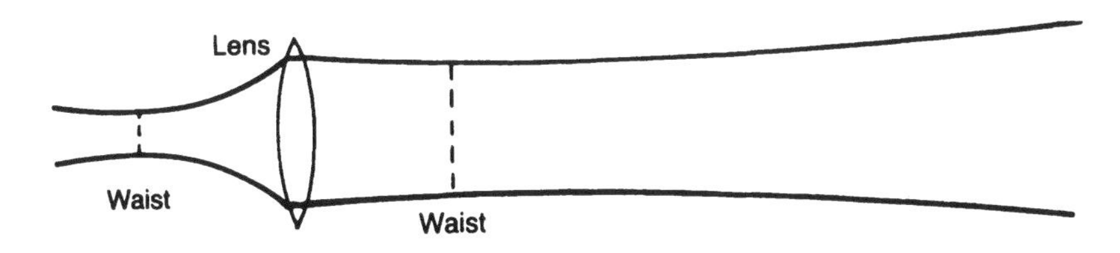

Figure 9.8 A lens converts on Gaussian beam into another.

Figure 9.9, together with some of the resonators that would support this particular beam. For example, the portion of the beam between E and H could oscillate in a resonator composed of a flat mirror and a mirror whose curvature matched the wavefront at H. Or two curved mirrors could match the beam's wavefronts at B and H. It is even possible to have a stable resonator configuration with a convex mirror, as shown in the bottom drawing of Figure 9.9.

It is easy to understand why the wavefront of a Gaussian beam must fit exactly into the curvature of a resonator mirror. In a Gaussian beam (or any wave) energy flows perpendicular to the wavefront. If you want to see the direction of energy flow in a wave, simply visualize little arrows all along the wavefront, perpendicular to the wavefront where they meet it. If the mirror curvature exactly fits the wavefront, all the energy in the wave is exactly reflected back on itself and the resonator is stable.

Now do you see the value of the second equation? It tells you what mirrors you must use to produce a given Gaussian beam in a resonator. Let us look at an example. Suppose you wanted to design an argon laser ($\lambda = 514.5$ nm) whose beam had a 0.5 mm diameter right at the center of the laser and whose mirrors were 1 m apart. Which mirrors would you use?

You would want to calculate the wavefront radius of curvature at the points where the mirrors are to go, then obtain mirrors with the same curvature. Begin by writing the equation:

$$R = z\left[1 + \left(\frac{\pi w_0^2}{\lambda z}\right)^2\right]$$

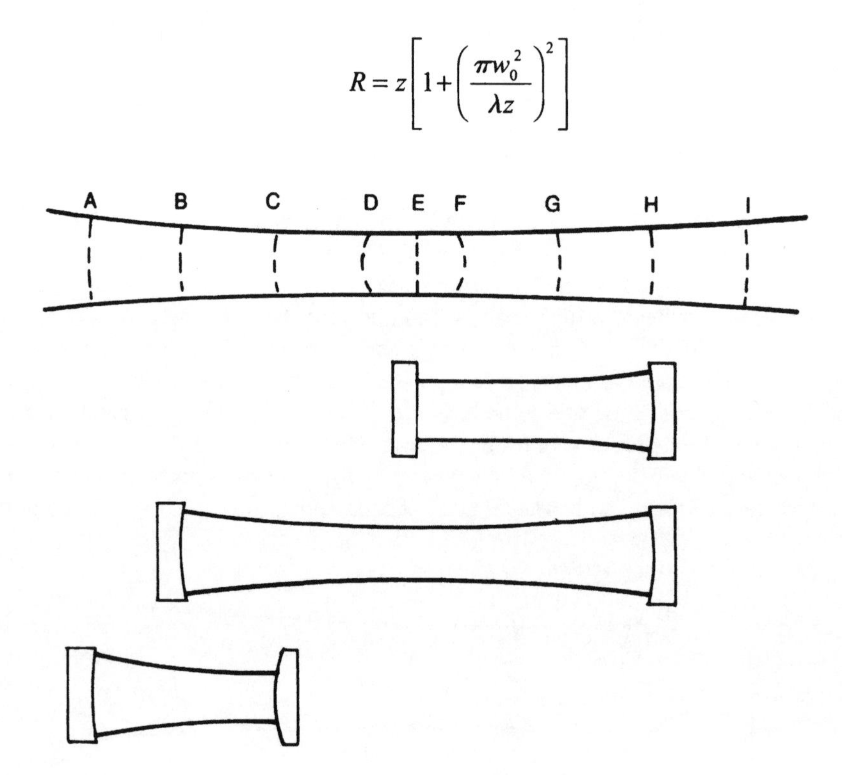

Figure 9.9 A resonator supports a Gaussian beam whose wavefront curvatures fit the mirror curvatures.

Then write down the known parameters, again putting everything in meters:

$$Z = 5 \times 10^{-1} \text{ m}$$

$$\lambda = 5.14 \times 10^{-7} \text{ m}$$

$$w_0 = 2.5 \times 10^{-4} \text{ m}$$

Substitute the values into the equation:

$$R = (5 \times 10^{-1} \text{m}) \left\{ 1 + \left[\frac{\pi (2.5 \times 10^{-4} \text{m})^2}{(5.14 \times 10^{-7} \text{m})(5 \times 10^{-1} \text{m})} \right]^2 \right\}$$

Finally, do the arithmetic:

$$R = (5 \times 10^{-1} \text{ m})[1 + (0.76)^2]$$

$$= (5 \times 10^{-1} \text{ m})(1.58)$$

$$= 0.79 \text{ m, or } 79 \text{ cm}$$

Thus, you would want to use mirrors whose radii of curvature were about 80 cm.

What about the physical size of the mirrors? Of course, they must be larger than the diameter of the beam when it hits the mirrors. Could you calculate that size?

9.4 A STABILITY CRITERION

As we learned in Chapter 8, a stable resonator is one in which rays can be trapped by the curvature of the mirrors; they will bounce back and forth between the mirrors forever. Now that we know about Gaussian beams, a second definition is possible: a stable resonator is one for which a Gaussian beam can be found whose wavefronts fit the curvatures of the mirrors. Obviously, some possible configurations are excluded, such as the one having two convex mirrors. But, in general, it is difficult to determine just from casual inspection whether a resonator is stable. Figure 8.9 shows a resonator with concave and convex mirrors that is unstable, and Figure 9.9 shows a similar-looking concave/convex resonator that is stable. Apparently, the difference depends on the exact curvatures of the mirrors. Moreover, even a resonator that has two concave mirrors is not necessarily stable. In this case as well, stability depends on the exact curvature of the mirrors. How can you tell? Is there some test you can perform to determine whether a particular configuration is a stable resonator?

Fortunately, there is. Otherwise, the only way to know for sure would be to construct the resonator and try to make it lase—a difficult and tedious task. Figure 9.10 shows the parameters you need to calculate the stability of a resonator. The curvatures of the two mirrors are r_1 and r_2, and the spacing between them is ℓ. If the mirror is convex, then its radius is taken to be negative. The condition for stability is

$$0 \leq g_1 g_2 \leq 1$$

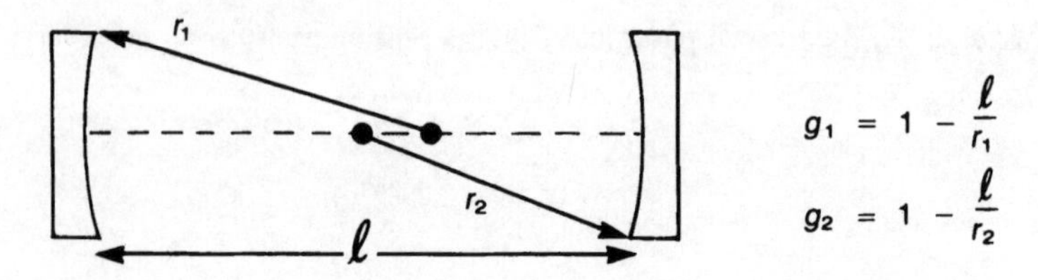

Figure 9.10 The g-parameters for calculating resonator stability.

in which g_1 and g_2 are the so-called g-parameters defined in Figure 9.10. If their product is between zero and one, the resonator is stable. Let us look at some examples.

Figure 9.11 shows a concave/convex configuration. Is it stable? First, calculate the g-parameters:

$$g_1 = 1 - \frac{50\text{ cm}}{-500\text{ cm}} = 1.1$$

$$g_2 = 1 - \frac{50\text{ cm}}{100\text{ cm}} = 0.5$$

Then multiply them together:

$$g_1 g_2 = (1.1)(0.5) = 0.55$$

Because 0.55 is between one and zero, this particular configuration is stable and will support a Gaussian mode or any of the higher-order modes shown in Figure 9.3. Interestingly, stability depends only on the mirror curvature and separation, not on laser gain, laser wavelength, or any other characteristic. The resonator in Figure 9.11 is stable for an He–Ne laser, an Nd:YAG laser, or a carbon dioxide laser.

Figure 9.12 shows a concave/concave mirror configuration. To determine whether it is stable, first calculate the g-parameters:

$$g_1 = g_2 = 1 - 9/3 = -2$$

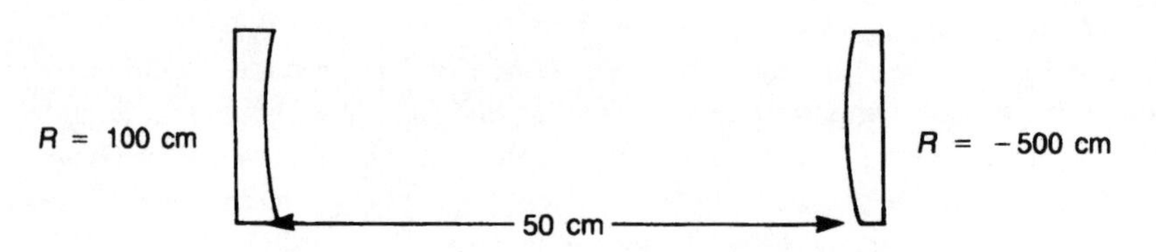

Figure 9.11 An stable concave-convex configuration.

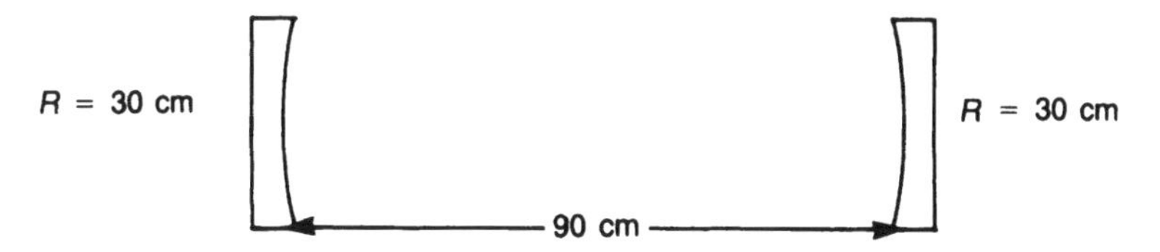

Figure 9.12 An unstable concave-concave configuration.

Then multiply them together:

$$g_1 g_2 = (-2)^2 = 4$$

The product is greater than one, so this configuration is not stable.

It is important to understand that a resonator must not only be stable but also its mirrors must be exactly aligned with each other before it can support laser oscillation. A ray cannot be trapped between the mirrors of any resonator, stable or unstable, if those mirrors are misaligned.

It is also important to understand that the stability of a resonator, as defined here, has nothing to do with how sensitive to misalignment the resonator is. Some resonators can have mirrors tilted by relatively large angles before their output power decreases, and others are completely extinguished by even a small tilt. And the g-parameters do not tell you anything about sensitivity to misalignment; indeed, some of the resonators that are most insensitive to misalignment have the $g_1 g_2$ product exactly equal to zero.

9.5 LONGITUDINAL MODES

The energy stored in a laser resonator has spatial variations not only perpendicular to the laser axis, as shown in Figure 9.1a, but along the axis as well, as shown in Figure 9.1b. But these longitudinal variations are much smaller in scale than the transverse variations. Each longitudinal resonator mode is a standing wave of light, created by the overlap of two traveling waves that are moving in opposite directions. The spatial distribution of energy in one such longitudinal mode is shown in Figure 9.13.

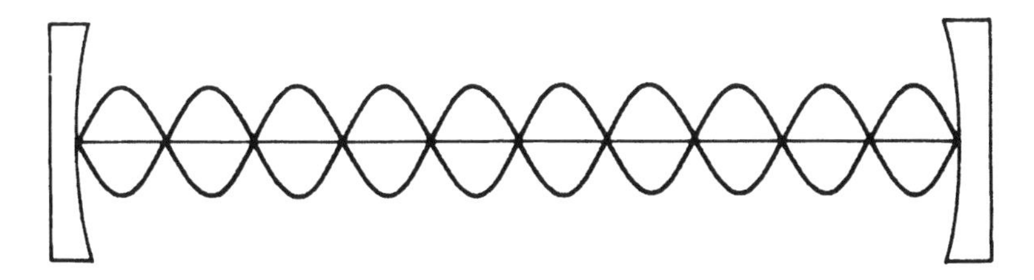

Figure 9.13 Energy distribution in a single longitudinal laser mode.

In Figure 9.13, the wavelength of light is necessarily shown far out of proportion to the size of the mirrors (only five wavelengths between the mirrors are shown). We can calculate how many wavelengths there are between the mirrors of a real laser. Suppose its mirrors are separated by $\ell = 30$ cm and the laser wavelength is 1 μm. Then the number of wavelengths between the mirrors is

$$N = \ell/\lambda = (3 \times 10^{-1}\ \text{m})/(1 \times 10^{-6}\ \text{m}) = 300{,}000$$

Figure 9.13 also shows the so-called boundary condition for a standing light wave in a resonator: there must be a node at both mirrors. (The node of a wave is the point where the wave passes through zero.) Thus, we are led to an important conclusion: Not just any wavelength will work in a resonator. For a wave to work in a given resonator, there must be room for exactly an integral number of half-wavelengths between the mirrors. You can have 600,000 half-wavelengths between the mirrors of the laser in the previous paragraph, or you can have 600,001. But you cannot have anything in between.

If you think about it for a moment, you will see that there is only a very small wavelength difference between a wave with 600,000 half-wavelengths between the mirrors and one with 600,001. That is, you do not have to squeeze each of the 300,000 wavelengths very much to make room between the mirrors for one more half-wavelength. In fact, the difference turns out to be so small that more than one longitudinal mode can oscillate in a laser at the same time. A laser is not perfectly monochromatic, and the amount of imperfection is usually greater than the wavelength difference between longitudinal modes.

Let us calculate the frequency spacing between adjacent longitudinal modes of a laser resonator. In the calculation, we will use some of the concepts about wavelength and frequency introduced in Chapter 2. Begin with the requirement for wavelength: there must be an integral number of half-wavelengths between the mirrors. This can be expressed mathematically by the following equation:

$$n(\lambda/2) = \ell$$

in which n is the integer and ℓ is the mirror spacing. Solve for wavelength:

$$\lambda = \frac{2\ell}{n}$$

From Chapter 2, $f = c/\lambda$, so

$$f_n = n\frac{c}{2\ell}$$

in which c is the speed of light. The frequency of the next mode—the one with $n + 1$ half-wavelengths between the mirrors—is

$$f_{n+1} = (n+1)\frac{c}{2\ell}$$

The difference between these two frequencies is

$$\Delta f \equiv f_{n+1} - f_n = \frac{c}{2\ell}$$

Does this equation look familiar? It is the equation for the resonant frequencies of a Fabry–Perot interferometer, discussed in Chapter 4. Now you know where those frequencies came from: each one is a different standing wave between the mirrors of the interferometer.

It is interesting that the frequency spacing depends only on the spacing between the resonator mirrors and not on the laser wavelength. Hence, the frequency spacing between longitudinal modes of a 30 cm He–Ne resonator is the same as the frequency spacing between longitudinal modes of a 30 cm Nd:YAG resonator.

Let us calculate what that spacing is:

$$\Delta f = \frac{c}{2\ell} = \frac{3\times10^8\,\mathrm{m/s}}{2(3\times10^{-1}\,\mathrm{m})} = 5\times10^8\,\mathrm{Hz}$$

The bandwidth of a typical He–Ne or Nd:YAG laser is many times larger than 500 MHz, so many longitudinal modes can oscillate at the same time in the resonator. Figure 9.14 shows what two simultaneously oscillating longitudinal modes might

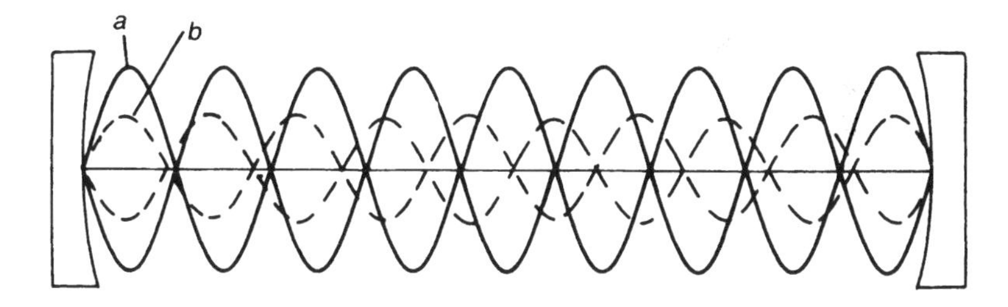

Figure 9.14 Two longitudinal laser modes.

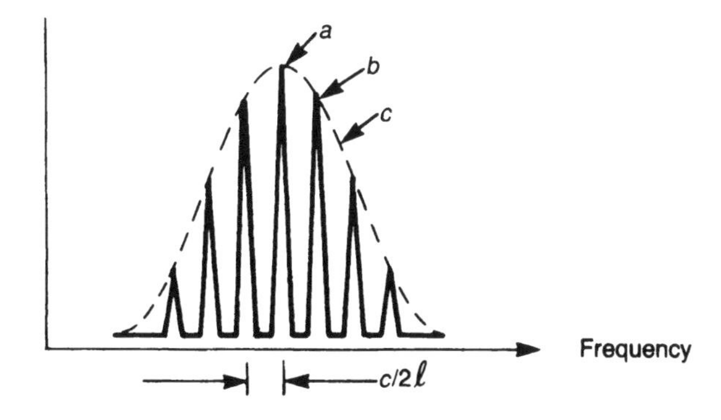

Figure 9.15 The frequency spectrum of longitudinal laser mode.

look like. Figure 9.15 shows what the frequency spectrum of several longitudinal modes looks like. Modes *a* and *b* of this figure correspond to the two modes in Figure 9.14. (Not all the modes in Figure 9.15 are shown in Figure 9.14.)

The shape of the dotted curve in Figure 9.15 is determined by the gain of the active medium, as we will discuss in more detail in the next chapter. There is no mode at *c* in Figure 9.15 because an integral number of wavelengths of light at that frequency would not fit between the mirrors.

QUESTIONS

1. Suppose a Cr:Ruby laser ($\lambda = 694.3$ nm) is used to track an earth satellite. If the beam has a waist diameter of 2 mm at the laser and the satellite is 500 mi straight up, calculate the diameter of the beam when it reaches the satellite. (Warning: do not confuse radius and diameter.)
2. Suppose the beam of the lunar-ranging laser described in this chapter were expanded with a telescope to a 1 m waist. How wide would the beam be at the moon?
3. How wide is the laser beam at the mirrors of the argon-ion laser described at the end of the section on Gaussian beams? Neglect any lensing introduced by the mirror, and calculate the beam radius 25 m from the laser.
4. Design a resonator for an Nd:YAG that has a 0.25 mm waist on the output mirror and 30 cm between mirrors.
5. Suppose two TEM_{00} Nd:YAG lasers are sitting side by side and each has a flat output mirror. One laser has a 0.5 mm waist, and the other has a 1.0 mm waist. Obviously, the 0.5 mm beam will diverge more rapidly than the 1.0 mm beam, so at some distance from the laser both beams will have the same radius. What is that distance?
6. Which of these resonators is stable?

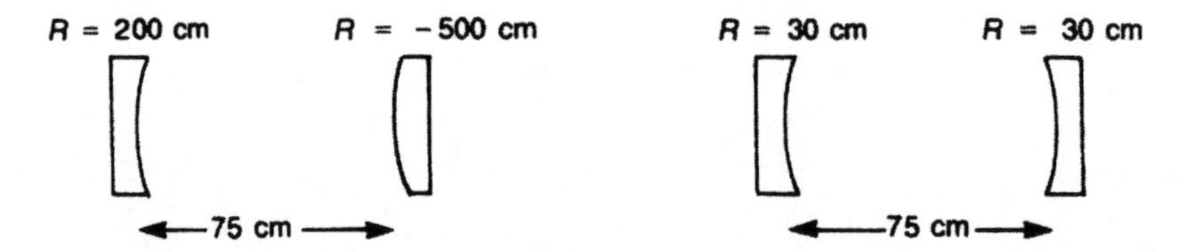

7. What is the frequency spacing between adjacent longitudinal modes of the argon-ion laser described in question 3?
8. Suppose the mirrors of the laser in Figure 9.14 were pulled apart. How would the frequency spectrum in Figure 9.15 be changed? Sketch what the new spectrum would look like, showing any difference in the dotted curve and in the modes themselves. Repeat the sketch for the case when the mirrors are pushed together.

Chapter 10

Reducing Laser Bandwidth

In this chapter, we look at ways to reduce a laser's bandwidth, that is, ways to make it even more monochromatic than it is naturally. A laser is the most nearly monochromatic optical source ever created, but for some applications—in precise spectroscopic studies or for separation of atomic isotopes, for example—a laser's natural bandwidth is just too large. And as we will see in Chapter 19, sometimes a laser with an enormous bandwidth can be narrowed to a particular, desirable wavelength. In all these cases, the techniques described in this chapter are employed to reduce the bandwidth of a laser.

The concept of laser bandwidth was explained in Chapter 5. In this chapter, we examine the different ways of measuring laser bandwidth and discover the mechanisms that cause bandwidth. We look at the devices that are placed inside a laser resonator to reduce its bandwidth and learn how a laser can be forced to oscillate in a single longitudinal mode.

There are several terms in laser technology for laser bandwidth. Bandwidth, linewidth, and spectral width all mean exactly the same thing: the degree of monochromaticity of the laser's output. And, as was explained in Chapter 5, the greater the temporal coherence of a laser, the smaller its bandwidth.

10.1 MEASURING LASER BANDWIDTH

Because the bandwidth of a laser is such an important parameter, it is imperative to have some way of quantifying it. In fact, there are several ways of placing a numerical value on a laser's bandwidth: It can be measured in wavelength, in frequency, in wave numbers, or in coherence length.

A laser bandwidth measured in wavelength is shown in Figure 10.1. In this case, the neodymium laser's peak output is at 1.064 nm, but there is also some light at slightly shorter and slightly longer wavelengths. Note that the bandwidth measurement is made halfway down from the peak of the laser line. This bandwidth is the full-width, half-maximum (FWHM) measurement, and it is the most common measurement of a laser's width.

However, light can also be measured in frequency, so the bandwidth of a laser can be described in frequency, as shown in Figure 10.2. Although Figure 10.2 looks like Figure 10.1, it shows that the laser peaks at 2.8×10^{14} Hz, but there is some

Introduction to Laser Technology, Fourth edition. C. Breck Hitz, J. Ewing, and J. Hecht

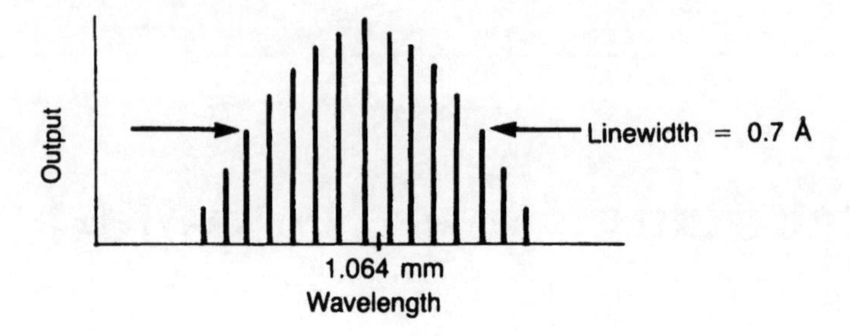

Figure 10.1 Laser output versus wavelength, showing longitudinal-mode structure.

light at slightly higher and slightly lower frequencies. Note that the measured bandwidth is again the FWHM value.

Wave numbers are yet another dimension for measuring laser bandwidth, one that is left over from the early days of spectroscopy and still in general use. The frequency of optical transitions used to be measured in wave numbers. When spectroscopists said a transition occurred at 20,000 wave numbers, they meant that 20,000 of the optical wavelengths would fit into 1 cm, or that the wavelength was 1/20,000 of a centimeter (500 nm). The value is written as 20,000 cm^{-1}, meaning 20,000 wavelengths per centimeter, but it is read as "20,000 wave numbers" or, sometimes, "20,000 inverse centimeters." And since the frequency of light can be measured in wave numbers, the bandwidth can too.

Figure 10.3 is a nomograph that is useful not only in converting among bandwidth measurements but also among line-center measurements. The line-center conversions are made with the left side of the nomograph, and the bandwidth conversions are made with the right side. Down the middle of the nomograph is a column showing laser wavelength. The example shown is for a laser that has a wavelength of 532 nm and a bandwidth of 0.5 cm^{-1}. By following a horizontal line across the columns to the left, you can see that this laser has a line-center frequency of 5.6×10^{14} Hz or 18,800 cm^{-1}, or a photon energy of 2.3 eV. (There are 6.2×10^{18} eV in 1 J.) On the right side of the nomograph, if you draw a straight, angled line connecting the laser wavelength with its bandwidth in the known dimension, you can read its

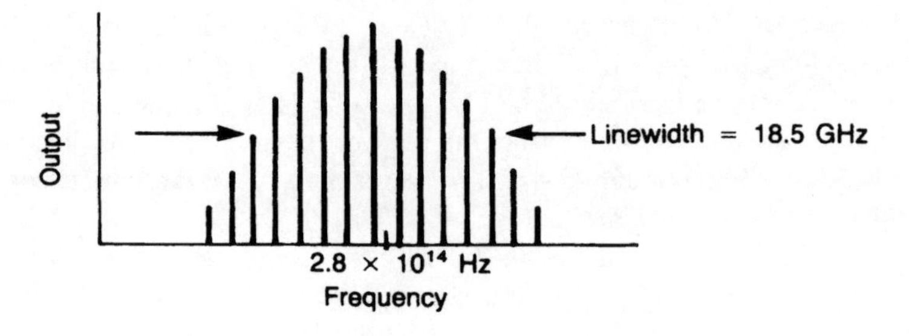

Figure 10.2 Laser output versus frequency.

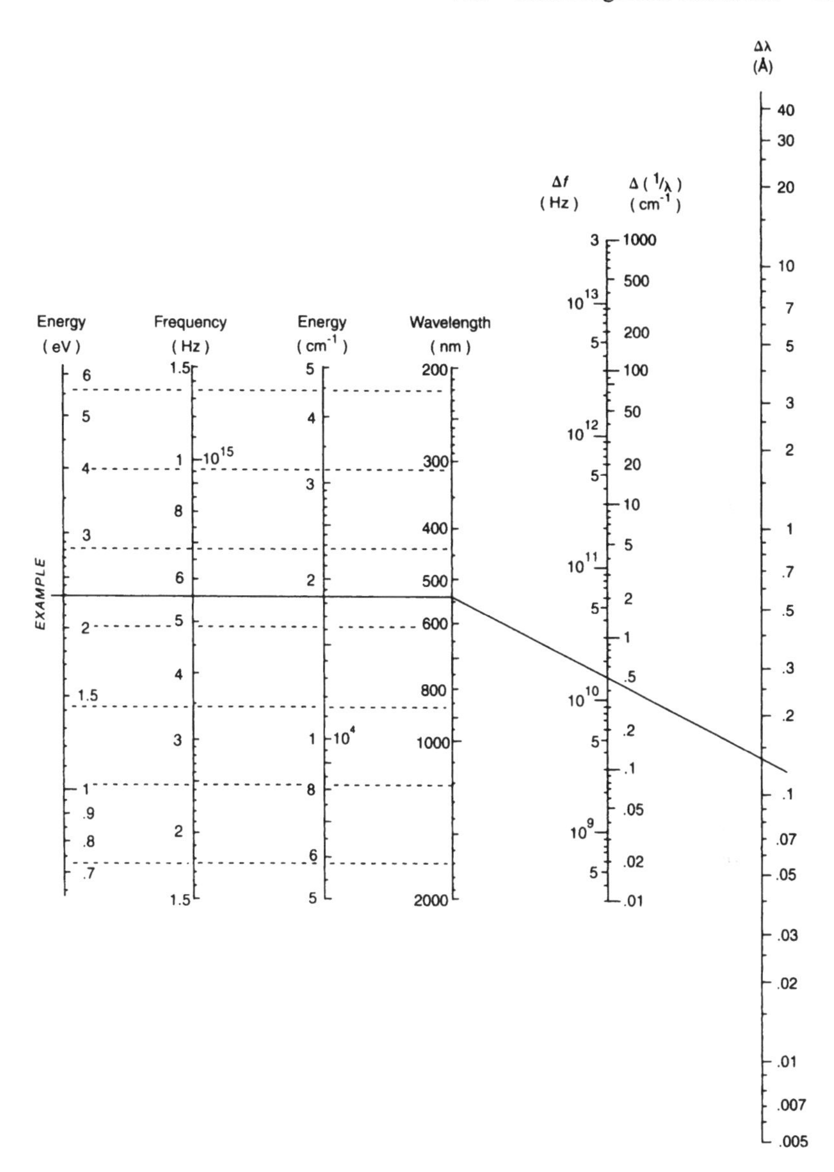

Figure 10.3 An energy nomograph.

bandwidth in the other dimensions. For the laser in the example, the bandwidth of 0.5 cm^{-1} corresponds to 15 GHz or 0.14 Å.

A fourth measurement of a laser's bandwidth is its coherence length. This is the distance over which the laser remains sufficiently coherent to produce interference fringes. It is inversely proportional to the laser bandwidth expressed in frequency or wavelength, and it is equal to the reciprocal of the bandwidth in wave numbers.

10.2 LASER-BROADENING MECHANISMS

Why does a laser have a finite bandwidth? Laser bandwidth derives from the "fuzziness" of the energy levels involved in the stimulated transition. Energy levels of a collection of atoms or molecules are not razor-sharp, like those in Figure 10.4a, but have a definite width to them, as illustrated in Figure 10.4b. Thus, the photons emitted when atoms (or molecules) undergo a transition will not all have exactly the same energy or the same wavelength.

Several mechanisms contribute to the width of energy levels. First, let us look at gas lasers, which have different broadening mechanisms than most solid-state lasers. The atoms (or molecules) in a gas laser are free to bounce around inside the laser tube, whereas the atoms in a solid-state laser are tied down at a particular spot.

Doppler broadening is significant in almost all gas lasers. You experience the acoustic Doppler effect when a car blows its horn as it speeds past you. As the car approaches, the horn sounds high-pitched because the sound source and the sound wave itself are both coming toward you. This means that more waves per second enter your ear than if the car were standing still. When the car moves away from you, on the other hand, the source is moving away from you while the wave is moving in the opposite direction—toward you. So fewer waves per second enter your ear than if the car were standing still. What you hear is a low-pitched horn.

If you were standing in the middle of an intersection with many cars coming at you and moving away from you at different speeds, all blowing their horns, you would hear a broad range of tones, even though all the horns produce the same tone when standing still. The sound from the cars moving away from you would be Doppler shifted down in pitch, while the cars moving toward you would be Doppler shifted up. The faster a car moved, the greater the Doppler shift would be. So you would experience this broad range of tones, despite the fact that all the horns were actually vibrating at exactly the same frequency.

The optical Doppler effect increases the bandwidth of gas lasers. Because the individual atoms are moving about in random directions and at random speeds in the laser tube, their total emission covers a range of frequencies just as the acoustic emission from the cars covered a range of frequencies. The faster the atoms move on the average, that is, the hotter the gas is, the broader the bandwidth. Moreover, there is a relativistic effect in the optical Doppler shift. A relativistic time dilation in the moving reference frame also contributes to the frequency shift.

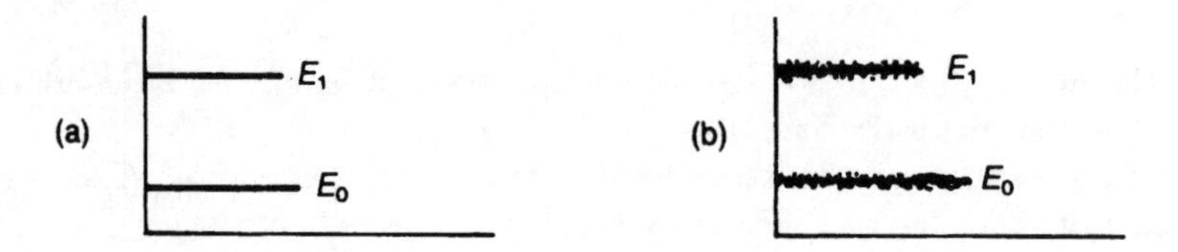

Figure 10.4 Energy levels are not razor-sharp like (a); instead, they are slightly fuzzy like (b).

In a Doppler-broadened laser, the bandwidth of an individual atom (or molecule) is smaller than the laser bandwidth. A single photon might be able to stimulate one atom to emit because that atom happened to be Doppler shifted to the photon's frequency, but it might not be able to stimulate another atom because it had a different Doppler shift than the first. This type of broadening, where the different atoms contribute to the gain at different frequencies within the laser bandwidth, is called inhomogeneous broadening. It is illustrated schematically in Figure 10.5.

Homogeneous broadening is illustrated in Figure 10.6. In a homogeneously broadened laser, each individual atom has a bandwidth equal to the total laser bandwidth. If a particular photon can interact with one of the atoms, it can interact with all of them. In general, it is easier to reduce the bandwidth of a homogeneously broadened laser because all the atoms can still contribute to stimulated emissions at the narrower bandwidth. In an inhomogeneously broadened laser, those atoms that contribute to gain outside the reduced bandwidth cannot be stimulated to emit in the narrowed bandwidth, and, therefore, the total laser power is reduced.

An example of homogeneous broadening in a gas laser is pressure broadening (or collision broadening, as it is sometimes called). One result of the uncertainty principle of physics is that the natural bandwidth of an atom is inversely proportional to the time between collisions. That is, the longer the atom can travel in a straight line without bumping into something (like another atom or the side of the laser tube), the narrower its natural bandwidth. It makes sense that the fewer other atoms there are in the tube, that is, the lower the gas pressure in the tube, the longer will be the average time between collisions. Thus, as the pressure in a laser tube increases, the bandwidth of the laser also increases. This broadening mechanism is homogeneous because it increases the total lasing bandwidth by increasing the bandwidths of the individual atoms.

Doppler broadening and pressure broadening are the most important broadening mechanisms in a gas laser. If the tube contains low-pressure gas, Doppler broadening is predominant; at high gas pressures, pressure broadening becomes more important.

In a solid-state laser, the individual lasing atoms are tied down to the host crystal's lattice points so they cannot move around, bumping into things and being Doppler broadened by their velocity. But there are other broadening mechanisms in

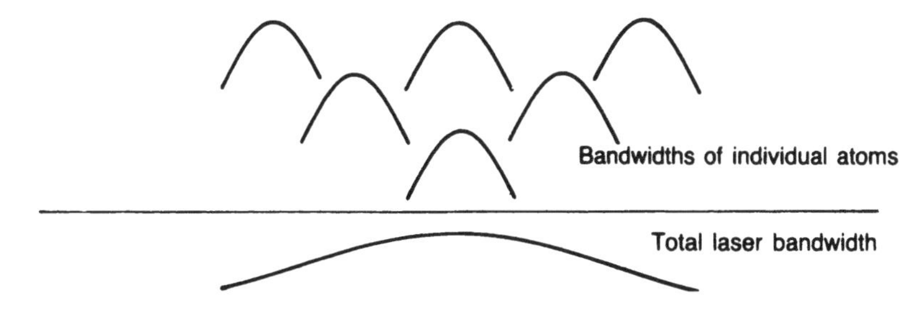

Figure 10.5 In an inhomogeneously broadened laser, individual atoms emit at different frequencies.

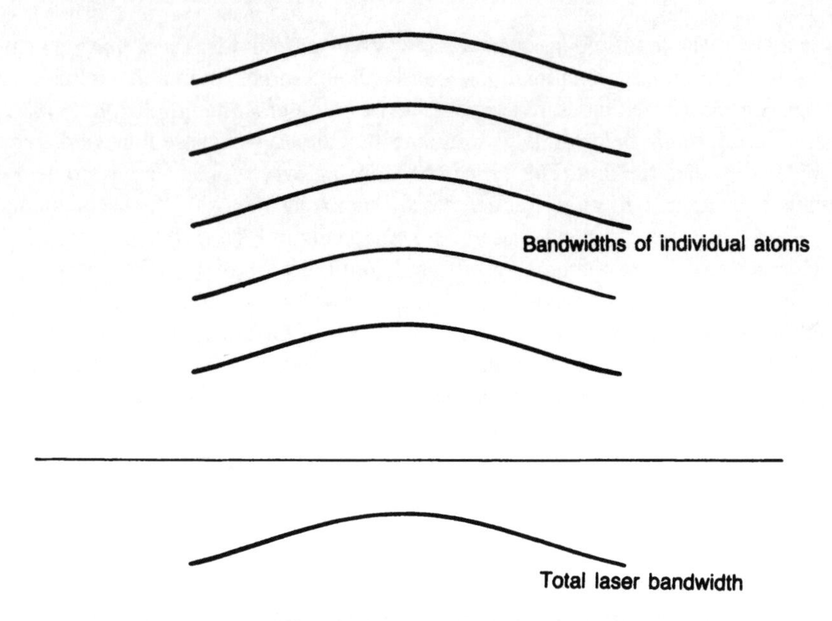

Figure 10.6 In a homogeneously broadened laser, all atoms are the same.

solid-state lasers, the most important of which is thermal broadening. Although the atoms are attached to the crystal lattice, the lattice itself is subject to vibration from thermal energy. This vibration modulates the natural emission frequency of the atoms and thereby broadens it. Thermal broadening is homogeneous because each atom is subject to the same thermal vibration.

When a solid-state laser is operated at very low temperature and the thermal broadening is therefore small, residual broadening results from imperfections of the host crystal. These imperfections are different at various locations in the crystal and give rise to differing electric fields at the active atoms. These fields cause different frequency shifts in the different atoms, so crystal-field broadening is an inhomogeneous broadening mechanism.

10.3 REDUCING LASER BANDWIDTH

The bandwidth of a laser can be narrowed by chilling the active medium to reduce thermal broadening (if it is a solid-state laser) or to reduce Doppler broadening (if it is a gas laser). But chilling is not a very effective way of reducing the bandwidth, and it is often inconvenient. The bandwidth of a gas laser can usually be reduced by reducing the pressure, but, unfortunately, this often reduces the output power because fewer atoms are left to lase. There is another way of reducing a laser's bandwidth, a way that does not cause a disastrous reduction in the laser's output power. Remember that the two conditions necessary for lasing are (1) the existence of a population inversion and (2) a round-trip gain greater than unity. The aforementioned techniques—chilling and pressure reduction—reduce laser bandwidth by reducing the bandwidth of the population inversion. The more effective techniques reduce laser bandwidth by reducing the bandwidth of the laser's round-trip gain. That is, the feedback of the resonator is modified to control the lasing bandwidth.

Suppose a laser has a population inversion that is 4 GHz wide and has mirrors that have reflectivity wider than the population inversion. This situation is diagramed in Figure 10.7a. Because all the light within the bandwidth of the population inversion sees round-trip gain greater than unity, the laser lases over the entire population inversion and the output bandwidth is 4 GHz.

But if the mirrors were replaced by special mirrors with a bandwidth of only 1 GHz, then only part of the light within the bandwidth of the population inversion would see the round-trip gain necessary for lasing. Lasing could occur only in this narrow band, and the bandwidth of the output would be reduced, as diagramed in Figure 10.7b.

That is the fundamental approach to reducing the bandwidth of a laser: you must reduce the bandwidth of the resonator feedback. The approach in Figure 10.7b is not very practical because it is difficult, if not impossible, to make laser mirrors with a 1 GHz bandwidth. Thus, other devices are used to reduce the resonator feedback, but the principle is always exactly what is shown in Figure 10.7b.

An intracavity prism is one common device that reduces the bandwidth of feedback in a resonator. The idea is shown in Figure 10.8. Although the population in-

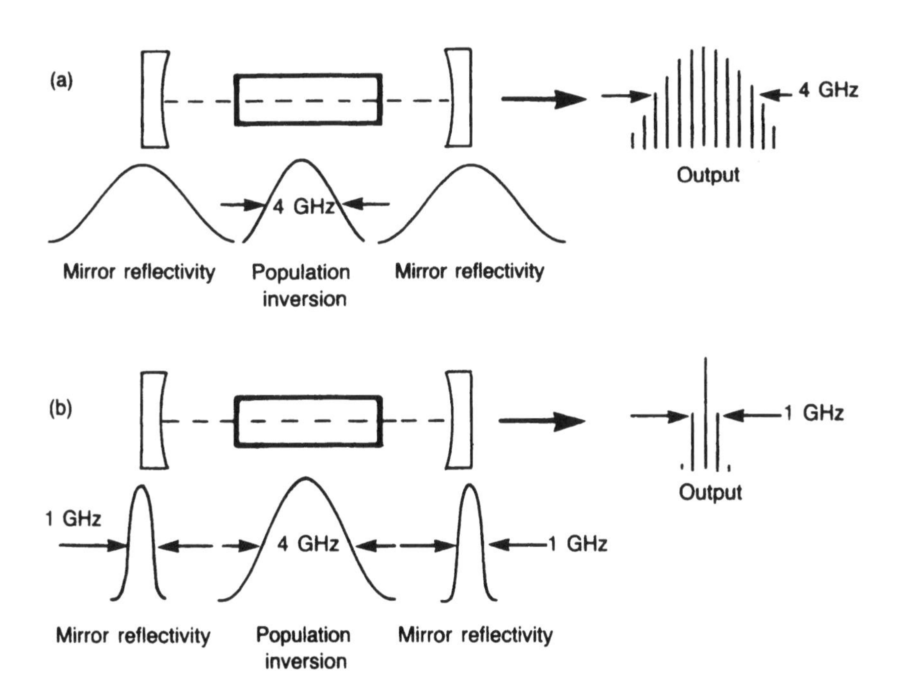

Figure 10.7 (a) If the resonator provides wide-band feedback, then the bandwidth of the output will be as large as the bandwidth of the population inversion. (b) If the bandwidth of the feedback is reduced, the laser bandwidth will be, too.

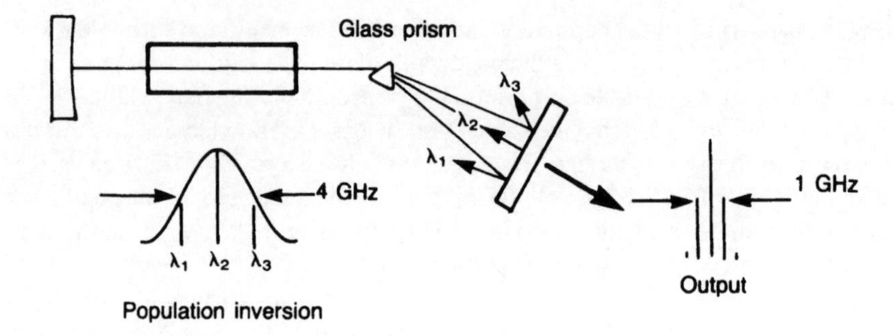

Figure 10.8 An intracavity prism reduces the bandwidth of resonator feedback and, therefore, the laser bandwidth.

version is 4 GHz wide, only light at the center of this bandwidth is bent directly toward the mirror by the prism. Light at the edges of the population-inversion width—λ_1 and λ_3 in Figure 10.8—emerges from the prism at a different angle and cannot be reflected back by the mirror. Hence, this resonator produces round-trip gain only for the narrow band of light at the center of the population inversion, and laser output is restricted to this reduced bandwidth.*

An alternative approach is to replace one of the mirrors with a grating. A grating is an interferometric device that reflects different wavelengths at different angles. When it is aligned correctly at one end of a resonator, it will reflect back to the active medium only light at the center of the population inversion, as shown in Figure 10.9. Thus, the lasing bandwidth is again reduced to the bandwidth of the resonator's round-trip gain.

What would happen if you put the bandwidth-reducing device outside the laser resonator? For example, Figure 10.10 shows a prism in the output beam of a laser and an aperture that passes only a narrow bandwidth. At first glance, you might think this arrangement preferable because the straight-line resonator would be more easily aligned than a bent resonator. What is the catch?

The catch is that you are losing most of the laser light in Figure 10.10. All the light that hits the edge of the aperture is lost, and the only useful output is the small fraction that passes through the aperture. When the bandwidth-limiting device is inside the laser, however, nearly as much laser power can be produced in a narrow bandwidth as in a wide bandwidth.

You can do better in a homogeneously broadened laser than in an inhomogeneously broadened laser because every atom in the population inversion can still contribute to the laser output. (In Figure 10.6, every atom can be stimulated by light at the center frequency.) In an inhomogeneously broadened laser, some of the atoms are unable to contribute to the reduced-bandwidth output and the laser power is reduced. (In Figure 10.5, some atoms cannot be stimulated by light at the center frequency.) But even in an inhomogeneously broadened laser, it is better to do the bandwidth reduction inside the resonator because some of the atoms outside the las-

*Several prisms in a series are necessary to reduce the bandwidths of some lasers.

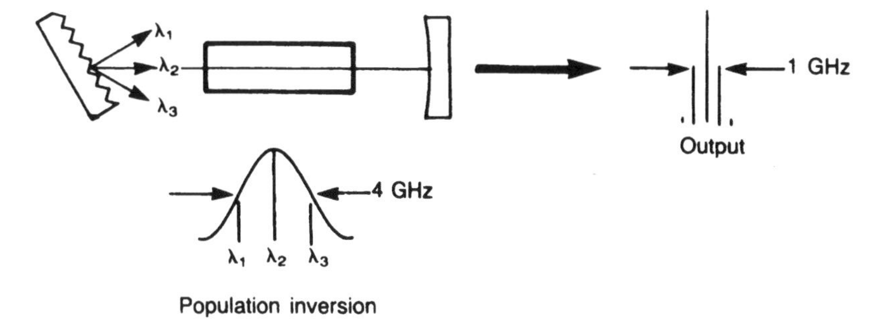

Figure 10.9 An intracavity grating will also restrict a laser's bandwidth.

ing bandwidth may eventually contribute to the laser gain. For example, they could collisionally transfer their energy to atoms that can emit within the lasing bandwidth.

Based on what has been said so far, it might seem that a homogeneously broadened laser could be restricted to a narrow bandwidth with no loss of output power, because as many atoms can contribute to the narrow-band output as to the wide-band output. In reality, it does not work that way for several reasons. For one thing, the insertion of an extra element, such as a prism, into a laser resonator always causes some reduction in output because there is no such thing as a perfect (lossless) optical element. And then there is an effect called *spatial hole burning,* discussed in the next section, that also plays a role in lowering the power from a narrow-band laser.

A birefringent filter is another device that reduces laser bandwidth by narrowing the bandwidth of the resonator's round-trip gain. Recall from question 5, Chapter 3, that a half-wave plate retards one component of polarization 180° with respect to its orthogonal component so that the effective polarization of the light passing through the plate is rotated by 90°. A full-wave plate, then, retards one component by 360°, which produces no change in the effective polarization of the light passing through the plate.

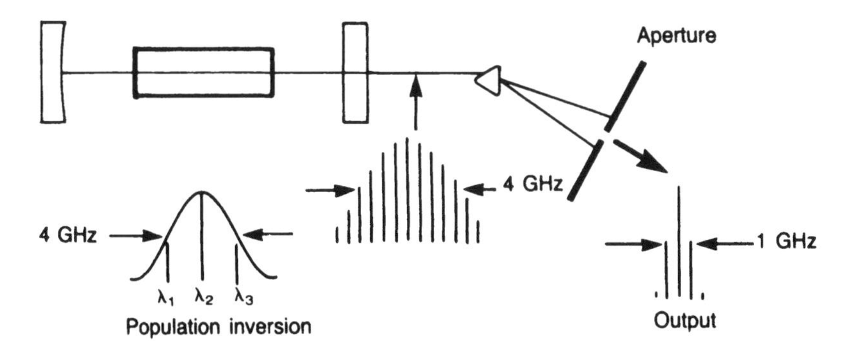

Figure 10.10 One technique to obtain narrow-bandwidth light from a laser.

Suppose a full-wave plate were placed inside a resonator at Brewster's angle, as shown in Figure 10.11. Because the plate is dispersive, that is, it has different refractive indices at different wavelengths, it is exactly a full-wave plate at only one wavelength. Light at slightly offset wavelengths will experience a slightly different retardation, perhaps 359° or 361°. So the light at the offset wavelengths will be slightly elliptically polarized after passing through the plate. Now look at Figure 10.11 again. The plate at Brewster's angle will reflect part of the elliptically polarized light out of the resonator, but light at the central wavelength will be perfectly plane polarized and will not be reflected out of the cavity. So the round-trip gain for light at the offset wavelengths will be lower than that for the central wavelength. If it is sufficiently lower, the bandwidth of the laser will be reduced.

In practice, a birefringent filter is a complex combination of waveplates, polarizers, and other elements. Its purpose is to function as a narrow-band filter, transmitting only a narrow spectrum of light.

10.4 SINGLE-MODE LASERS

Each transverse mode and each longitudinal mode of a laser oscillates at a different frequency, and in an unrestricted laser numerous modes of both types oscillate simultaneously. In Chapter 9, we saw that an intracavity aperture can force a laser to oscillate in a single transverse mode. The ultimate narrow-bandwidth laser oscillates in only a single transverse and longitudinal mode. Normally, the techniques introduced in the previous section are not restrictive enough to force a laser to a single mode.

If an aperture is placed in the resonator of an otherwise unrestricted laser, the laser will oscillate in a comb of frequencies corresponding to the different longitudinal modes of a single transverse mode, as shown in Figure 10.12. Here the laser output at any frequency is determined by the product of the laser gain, the mirror reflectivity, and the resonator mode structure. If you want to know the output from the laser at a particular frequency, you must multiply the gain, reflectivity, and mode structure at that frequency. For some frequencies, the product is zero because the mode structure is zero, so there is no output at those frequencies.

If the bandwidth of the laser is reduced, perhaps with a prism, fewer modes oscillate. This situation is diagramed in Figure 10.13, in which the output is again shown as determined from the product of population inversion and feedback components.

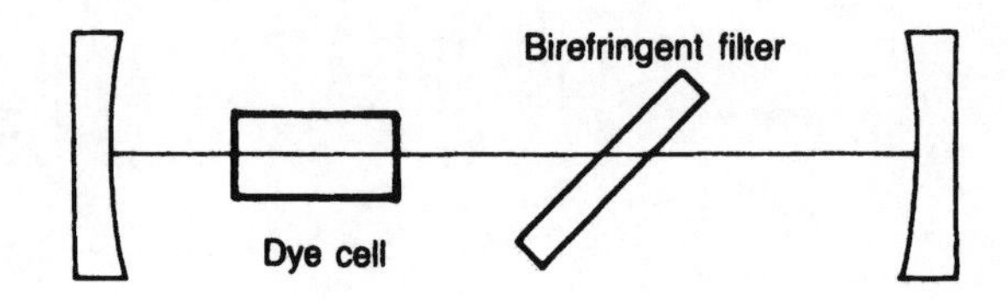

Figure 10.11 A birefringent filter inside a dye-laser resonator.

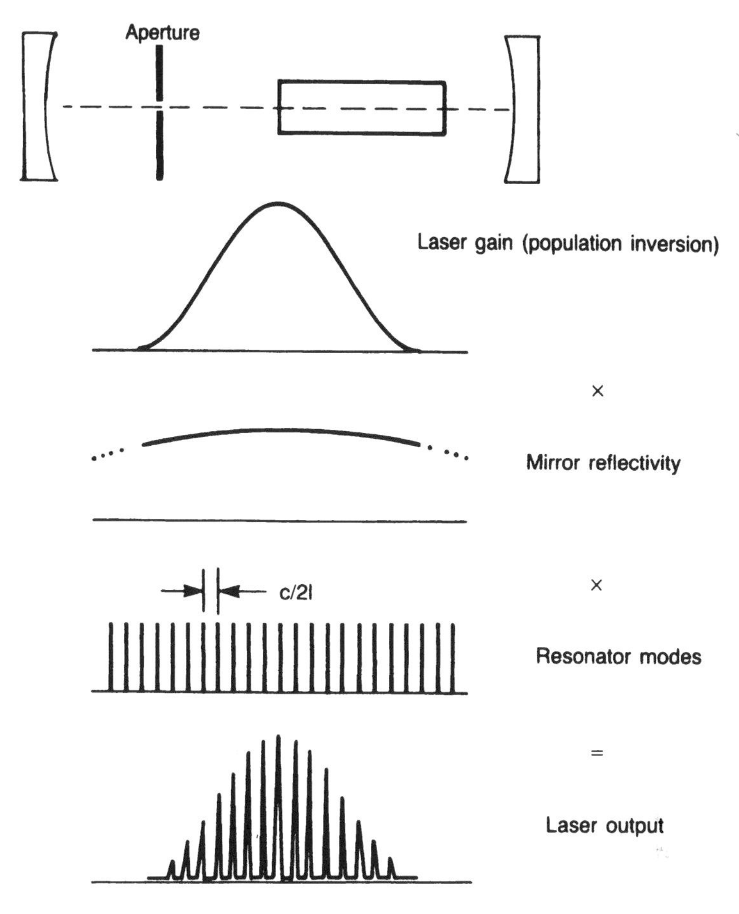

Figure 10.12 A laser restricted to a single transverse mode can oscillate in many longitudinal modes.

To restrict a laser to a single mode, it is usually necessary to place an etalon inside the resonator. An etalon is nothing more than two surfaces that act like a Fabry–Perot interferometer. Remember that the transmission peaks of a Fabry–Perot are separated by $c/2L$, in which L is the distance between the reflecting surfaces. If this distance is small, then there is a relatively large frequency spacing between adjacent transmission peaks. These peaks (which are the longitudinal modes of the etalon) act in concert with the longitudinal modes of the resonator to extinguish all but one of the laser's longitudinal modes, as shown in Figure 10.14.

In practice, an etalon often is a piece of optical-quality glass fabricated with great care to ensure that the surfaces are parallel. The surfaces may be coated to enhance reflectivity or they may be uncoated. (Reflective coatings increase the etalon's finesse, the ratio of transmission separation to width. Coatings would be required to achieve the relatively high etalon finesse shown in Figure 10.14.) Many single-mode lasers use more than one etalon to ensure that the laser is restricted to one mode.

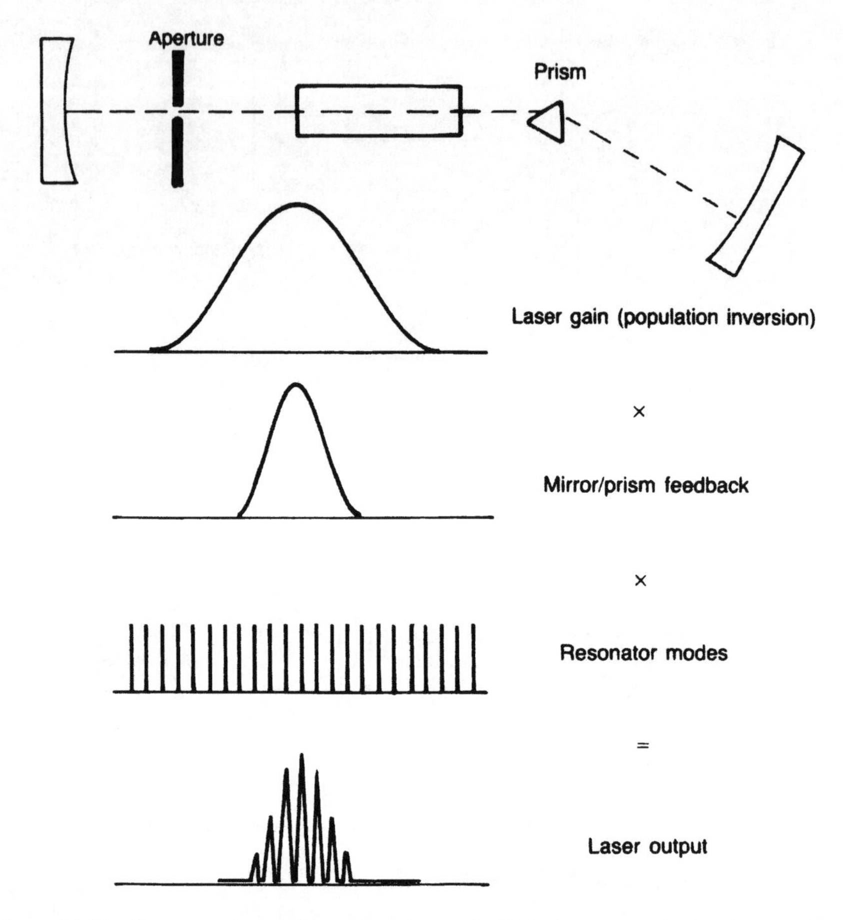

Figure 10.13 Fewer modes oscillate when the bandwidth of resonator feedback is restricted.

A single-mode laser is often called a single-frequency laser. But a single-wavelength laser is something else. The term usually refers to a laser, like an ion laser, that can lase on more than one transition but has been artificially restricted to one transition.

Suppose that you had a perfect, lossless etalon, one whose flawless surfaces scattered absolutely no light and whose magical bulk material absorbed absolutely none. If you placed this device inside a homogeneously broadened laser, you might expect to get as much output from the resultant single mode as you would been getting from all the longitudinal modes combined. After all, every atom that contributed to the output before could still contribute to the single-mode output, right?

Wrong. To understand why, look at the spatial distribution of electrical fields in a single longitudinal mode, as diagramed in Figure 10.15. At the nodes of the standing wave, there is no electrical field, so the atoms located right at the nodes cannot be

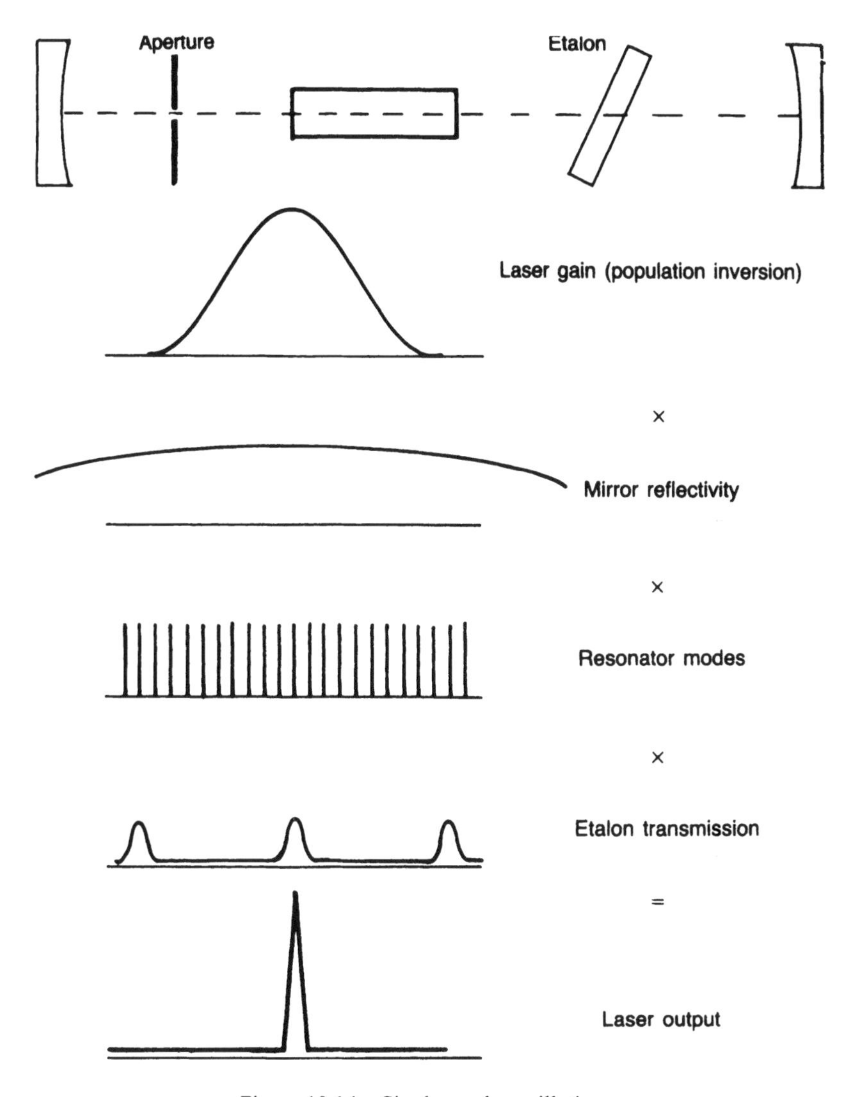

Figure 10.14 Single-mode oscillation.

stimulated to emit their energy. In fact, the single mode will "burn holes" in the population inversion at those locations where the electric field is greatest. And the atoms at the nodes of the single mode cannot contribute to its output, even though they can emit at the correct frequency.

In practice, a single etalon usually will not force a homogeneous laser to oscillate in a single mode. The gain from atoms located at nodes of the preferred mode becomes so great that one or more additional modes will oscillate despite the etalon. You can force single-mode oscillation by adding a second (and maybe a third) etalon to the resonator, but only at the cost of reduced output power. The output

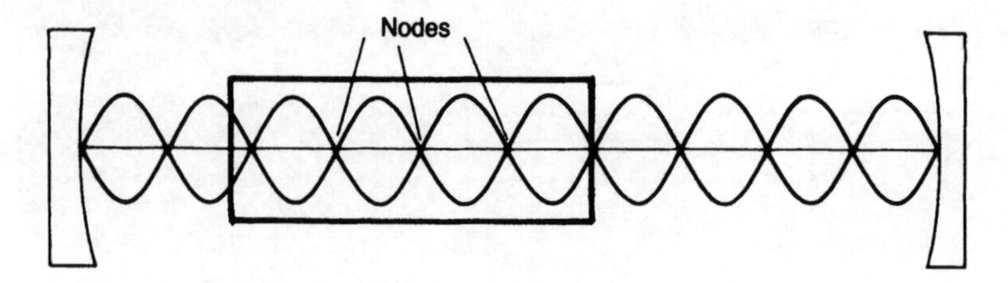

Figure 10.15 Atoms at nodes of standing waves cannot contribute to laser gain.

power is reduced because the atoms located at the nodes of the single mode cannot be tapped for their energy.

QUESTIONS

1. Find the optical frequency of an argon laser whose wavelength is 514.5 nm from the nomograph (Figure 10.3) and from direct calculation. Suppose the laser has a 0.4 cm^{-1} bandwidth. Calculate the bandwidth in frequency and wavelength, and confirm your calculation from the nomograph. An argon laser can also produce a blue line at 488 nm. If it does and its bandwidth is still 0.4 cm^{-1}, what is its bandwidth in frequency and wavelength?
2. Spectral intensity is a measure of laser power within a given bandwidth. For example, a 1 W laser with a 1 nm bandwidth has a spectral intensity of 1 W/nm.
   - What it is the spectral intensity of a 500 mW Nd:YAG laser whose bandwidth is 0.3 Å?
   - Suppose a filter outside the laser reduces the bandwidth to 0.1 Å but reduces the power to 170 mW. What is the spectral intensity?
   - Suppose a prism inside the laser reduces the bandwidth to 0.1 Å but reduces the power to 450 mW. What is the spectral intensity in this case?
   - Which is the better way to obtain narrow-band output from the laser?
3. It is possible to force a laser to oscillate in a single longitudinal mode just by pushing the mirrors close enough together. Why does this work? How close must you push the mirrors of a laser whose unrestricted bandwidth is 1.0 GHz to force it into single-mode oscillation?
4. Consider a dye laser whose output has been tuned to 600 nm. The laser is analyzed with a scanning spherical-mirror interferometer, and its spectral width is determined to be 0.5 Å. Express this spectral width in (a) frequency and (b) wave numbers. If the laser cavity is 1 m long, about how many longitudinal modes can oscillate?
5. Spatial hole burning is a problem with single-frequency lasers because the standing wave "burns holes" in the population inversion. The problem is not the

holes themselves, but the space between the holes where the population inversion cannot contribute to laser gain.

A solution to the problem of spatial hole burning is to design a laser with three or more mirrors—a ring laser—in which all the light travels around the ring in only one direction. Why does this avoid spatial hole burning? (Hint: What causes the standing wave in a two-mirror laser? Is there a standing wave in a ring laser?)

Chapter 11

Q-Switching

Q-switching (Q stands for quality factor) is the first of several techniques for producing pulsed output from a laser that we will examine. Pulsed lasers are useful in many applications in which continuous-wave (cw) lasers will not work because the energy from a pulsed laser is compressed into little concentrated packages. This concentrated energy in a laser pulse is more powerful than the natural-strength energy that comes from a continuous-wave laser.

Q-switching is a technique used almost exclusively with optically pumped, solid-state lasers. As you will see as you peruse this chapter, energy must be stored in the population inversion of a Q-switched laser. The spontaneous lifetime of dye lasers, and of most gas lasers, is too brief to allow significant energy storage in the population inversion, so these lasers simply cannot be Q-switched. Carbon dioxide lasers can be (and occasionally are) Q-switched, but Q-switched CO_2 lasers have not proved particularly useful for anything.

We begin this chapter by looking at the way the output of a pulsed laser is measured. It is more complicated than simply measuring the average power output of a cw laser. Then we explain in detail the concept of Q-switching: storing energy inside the laser and suddenly letting it all out in a giant pulse.* We conclude with a discussion of the four different types of Q-switches that can make a laser store its energy for emission in pulses.

11.1 MEASURING THE OUTPUT OF PULSED LASERS

Measuring the output of a cw laser is fairly simple because the energy flows smoothly and constantly from the laser, as shown in Figure 11.1. But with a pulsed laser, you want to know the answers to questions such as: Are there a lot of little pulses or a few big ones? How tightly is the energy compressed in the pulses?

When you measure the output of a cw laser, you measure the amount of energy that comes out during a given period of time. The energy is measured in a dimension called joules, and time is measured in seconds. The rate at which energy comes from the laser, that is, the number of joules per second, is the power of the laser, measured in watts.

*In the early days of Q-switching (the 1960s), Q-switched pulses were routinely called "giant pulses." The term is rarely used today.

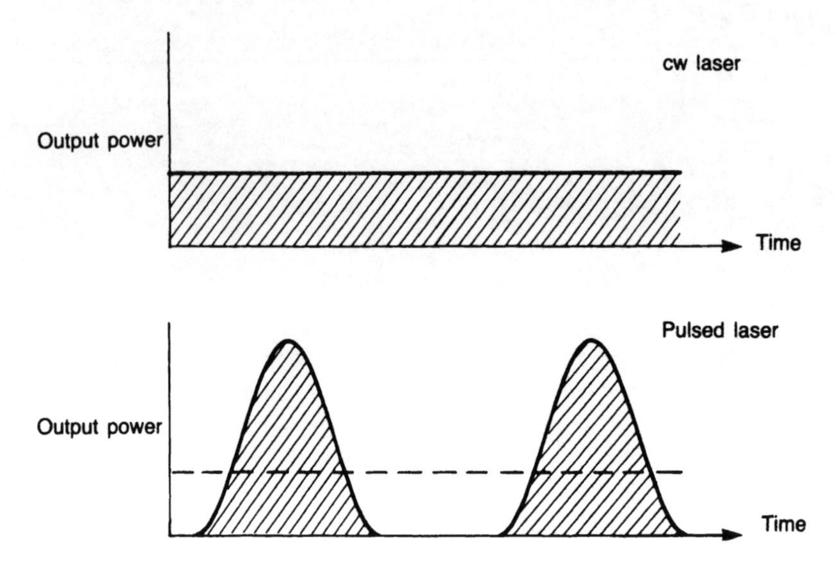

Figure 11.1 Energy (shaded area) is produced in concentrated packages by a pulsed laser. Because the energy is concentrated, its peak power is greater than that of a cw laser.

Thus, to a physicist the words power and energy have different meanings. Energy is measured in joules and is defined as the ability to do work (such as moving or heating something). Power, on the other hand, is the rate of expending energy and is measured in joules per second (watts). For example, a 100 W lightbulb uses 100 J of electrical energy every second it is on. If you leave it on for 5 min, you have used 30,000 J.

Is your electric bill based on the power you have used that month or on the energy? Clearly, it does not make sense to talk about how much power you have used during a month because power is the rate of using something. Therefore, your electricity bill, for so many kilowatt-hours, is a bill for energy. You use a kilowatt-hour of electrical energy when you expend energy at the rate of 1 kW for 1 h. Leaving ten 100 W bulbs on for an hour would do it, or you could use a 4 kW clothes dryer for 15 min. You have used a kilowatt of power (1,000 J/s) for 1 h, or you've used a total of 3,600,000 J. A kilowatt-hour of electricity costs about 10 cents, depending on where you live, so you can see that a joule is not a whole lot of energy.

There are two power measurements for a pulsed laser: peak power and average power. The average power is simply a measurement of the average rate at which energy flows from the laser during an entire cycle. For example, if a laser produces a single half-joule pulse per second, its average power is 0.5 W. The peak power, on the other hand, is a measurement of the rate at which energy comes out during the pulse. If the same laser produces its half-joule output in a microsecond-long pulse, then the peak power is 500,000 W ($0.5 \text{ J}/10^{6} \text{ s} = 500{,}000 \text{ J/s}$).

The pulse repetition frequency (prf) is a measurement of the number of pulses the laser emits per second. The period of a pulsed laser is the amount of time from the beginning of one pulse to the beginning of the next. It is the reciprocal of the prf. The duty cycle of a laser is the fractional amount of time that the laser is producing output, in other words the pulse duration divided by the period.

For example, let us consider a flash-pumped, Q-switched Nd:YAG laser that produces 100 mJ, 20 ns pulses at a prf of 10 Hz.

The average power is equal to the pulse energy divided by the pulse period:

$$P_{\text{average}} = \frac{\text{Energy / pulse}}{\text{Period}} = \frac{10^{-1}\,\text{J}}{10^{-1}\,\text{s}} = 1\,\text{J/s} = 1\,\text{W}$$

On the other hand, the peak power is equal to the pulse energy divided by the pulse duration:

$$P_{\text{peak}} = \frac{\text{Energy / pulse}}{\text{Pulse duration}} = \frac{10^{-1}\,\text{J}}{2\times10^{-8}\,\text{s}} = 5\times10^{6}\,\text{J/s} = 5\,\text{MW}$$

The peak power is *five million times* greater than the average power! Think about what that means. If you take a cw laser with 1 W of output and "repackage" its energy—pack the energy into 10 pulses, each lasting 20 ns—you can create 5,000,000 W of laser power. You have not added any energy at all; you have just repackaged the energy that was already there. And there are a lot of things you can do with 5 MW that you cannot do with 1 W. That is why pulsed lasers are so important.

11.2 Q-SWITCHING

Q-switching is a simple concept. Energy is stored in the population inversion until it reaches a certain level, and then it is released very quickly in a giant pulse. This is analogous to storing water in a flower pot with a hole in the bottom and then releasing the water all at once, as shown in Figure 11.2.

Now the question is: How can energy be plugged up in the population inversion of a laser? In other words, how can you prevent the energy from draining out of the population inversion as fast as it goes in?

To prevent the laser from lasing, you must defeat one of the two requirements for lasing: you must eliminate either the population inversion or the feedback. Obviously, if energy is stored in the population inversion, it does not make sense to talk about eliminating that. But you can eliminate feedback, thereby preventing lasing and thus storing all the extra energy in the population inversion, by blocking one of the laser mirrors.

That is exactly the way a laser is Q-switched. As shown in Figure 11.3, if the normal mirror feedback is present, energy drains out of the population inversion as fast as it is put in. But if feedback is eliminated, energy builds up in the population inversion until feedback is restored, and then all the energy comes out in a single, giant pulse.

Why is it called Q-switching? The Q stands for the quality of the resonator. A high-Q resonator is a high-quality resonator, or one that has low loss. Obviously,

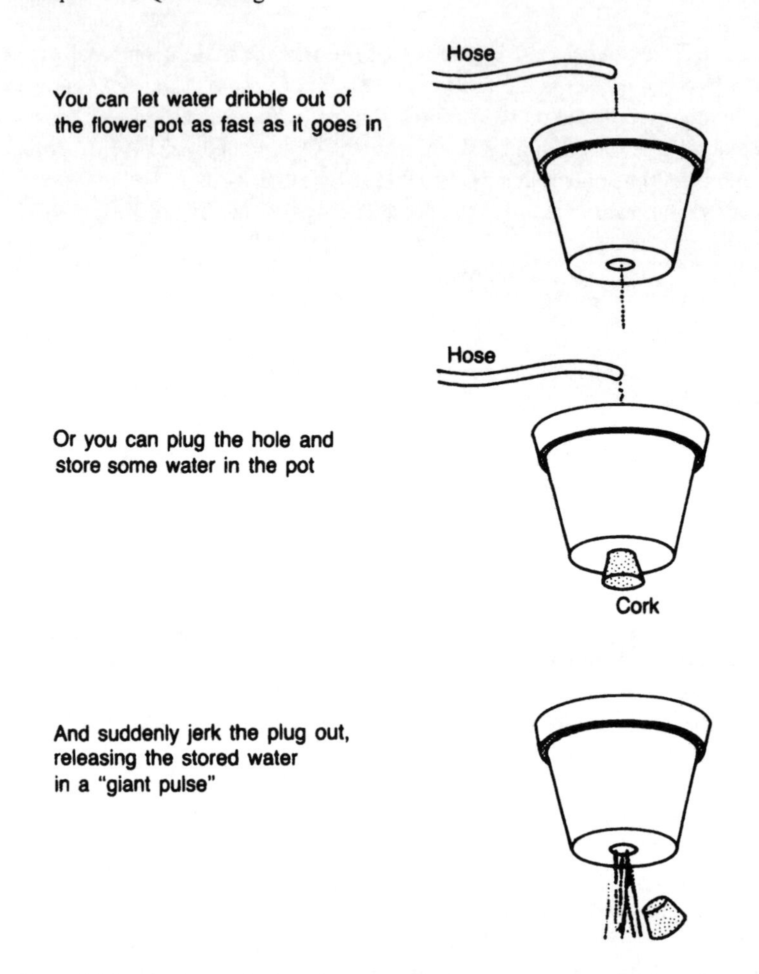

Figure 11.2 Storing water in a flower pot is analogous to Q-switching a laser.

a resonator with a blocked mirror is not very high-Q. But when the mirror is suddenly unblocked, the Q is switched from low to high. Thus, a Q-switched laser is one whose resonator can be switched from low quality to high quality and back again.

Let us take a slightly more analytical look at what goes on inside a Q-switched Nd:YAG laser. You might want to review the discussion of gain saturation and relaxation oscillations in Chapter 8 since a lot of those concepts are similar to the ones here. Figure 11.4 shows, simultaneously, the behavior of four laser parameters. If the flashlamp output lasts a fraction of a millisecond, as shown in the top graph, then the laser gain (or population inversion) quickly becomes greater than the normal intracavity loss shown in the second graph. (Normal means the intracavity loss without the Q-switch.) But the Q-switch prevents the laser from lasing, so the gain increases above the value where it would normally saturate.

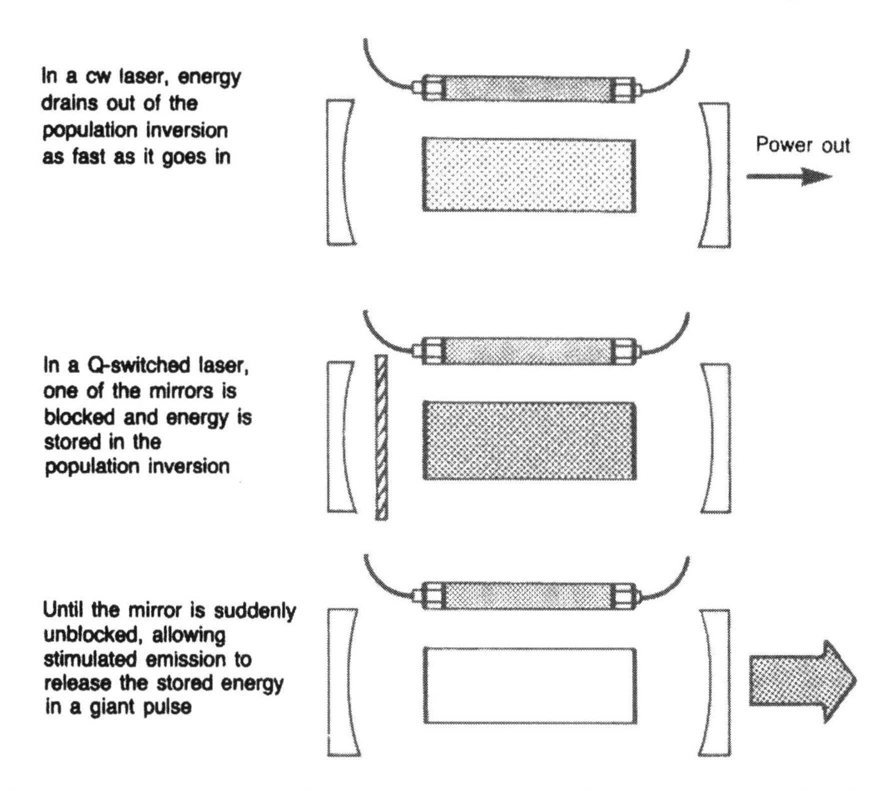

Figure 11.3 Energy is stored in the population inversion of a Q-switched laser. This diagram shows a lamp-pumped solid-state laser.

Eventually, the gain begins to flatten as spontaneous emission takes its toll. The spontaneous lifetime of the excited atoms is several hundred microseconds. After the lamp has been on that long, the first atoms to be pumped up start to emit spontaneously.

If the Q-switch transmission is switched on at this point, the intracavity loss suddenly becomes much smaller than the gain. The laser is far above threshold. Stimulated emission takes place, and circulating power starts building up in the resonator. As the circulating power increases, it saturates the gain, driving it downward. Meanwhile, part of the circulating power is transmitted through the output mirror.

The gain continues to nose-dive as the circulating power increases. When the gain drops below its saturated value, the circulating power must decrease (because the gain is now less than the intracavity loss). But there are still a lot of photons inside the resonator, and these photons bounce back and forth through the population inversion, depleting it further.

Note that the peak of the pulse occurs at the instant when the gain passes through its saturated value. If the gain is greater than the intracavity loss, circulating power (and output power) is increasing; if the gain is less than the loss, circulating power is decreasing.

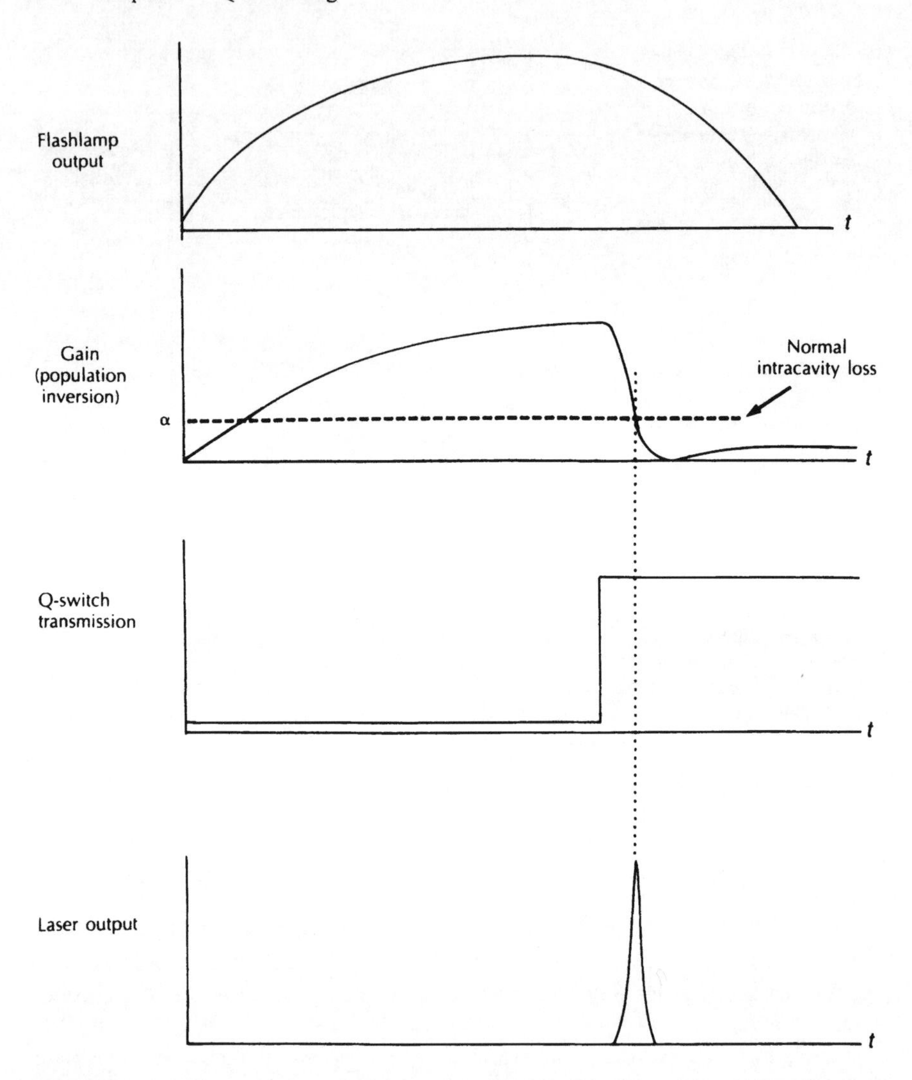

Figure 11.4 A flash-pump Q-switched laser. The total time shown (t) is about one-half millisecond.

After a while, all the photons drain out of the resonator through the output mirror, and the pulse ends. If the lamp has not gone out yet, the laser gain may build up again, but as long as it is less than the loss, there will not be another pulse.

Interestingly, the Q-switch turns the laser on, but it does not turn the laser off. The laser turns off when all the photons have drained out of the resonator. The Q-switch is still "on."

There is a delay between the time the Q-switch opens and the time the laser starts to lase. This is the time it takes the circulating power to build up from noise (spontaneous emission) to a significant value. The length of this delay depends on several things, but an important factor is the gain before the Q-switch is opened. The greater the initial gain, the less time it takes the pulse to build up.

What factors affect the duration of a Q-switched pulse? Gain is one: the faster a pulse can build up from noise, the shorter it will be. A high-gain laser will produce shorter pulses than a low-gain laser.

The other thing that affects pulse duration is how long it takes the photons to drain out of the resonator. The longer it takes them to drain out, the longer the pulse will be. Thus, increasing the spacing between mirrors will keep photons in the resonator longer and increase pulse duration. Likewise, increasing mirror reflectivity keeps photons in the resonator longer, and will usually increase pulse duration. We need to include "usually" in the previous sentence because keeping photons in the resonator can sometimes shorten the pulse. Remember that the more photons you have in the resonator, the more stimulated emission you will have. So keeping photons in the resonator tends to reduce the build-up delay and shorten the pulse. But the draining-out delay usually dominates this effect.

The pulse duration of a Q-switched solid-state laser varies from a few nanoseconds for a high-gain, flash-pumped laser to hundreds of nanoseconds for a low-gain, continuously pumped laser.

11.3 TYPES OF Q-SWITCHES

Placing a beam block in front of the laser mirror, as shown in Figure 11.3, is a straightforward approach to Q-switching a laser, but it is not very practical. The problem is getting the beam block completely out of the beam quickly enough. If the beam is 0.5 mm in diameter and the block must be pulled out in a few nanoseconds, the block must be jerked out with a velocity greater than the speed of sound, which is not very easy to do.

Four types of Q-switches are used in lasers. Mechanical Q-switches actually move something—usually a mirror—to switch the resonator Q. Acousto-optic (A-O) Q-switches diffract part of the light passing through them to reduce feedback from a resonator mirror. The polarization of light passing through an electro-optic (E-O) Q-switch can be rotated so that a polarizer prevents light from returning from a mirror. And a dye Q-switch absorbs light traveling toward the mirror until the intensity of the light becomes so great that it bleaches the dye, allowing subsequent light to pass through the Q-switch and reach the mirror.

11.4 MECHANICAL Q-SWITCHES

A mechanical Q-switch is shown in Figure 11.5. Here, the six-sided mirror spins rapidly and lines up each side with the laser for a very short period. A laser pulse comes out through the other mirror each time one of the six sides of the spinning mirror is aligned. These rotating-mirror Q-switches were fairly common in the early days of lasers, but other types of Q-switches have replaced them for most applications now.

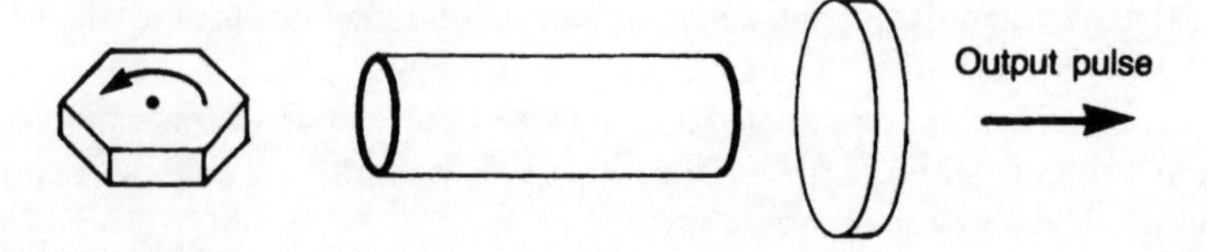

Figure 11.5 The quality of this resonator is switched from low to high when a surface of the spinning mirror is aligned with the other mirror.

Another type of mechanical Q-switch is the frustrated total-internal-reflection (FTIR) Q-switch. This device provides feedback to the laser by total internal reflection from the inner surface of a prism. To reduce resonator Q, a second prism is pushed quickly into optical contact with the reflecting surface of the first prism, frustrating the total internal reflection. Like rotating-mirror Q-switches, these FTIR Q-switches may be interesting from an historical perspective, but they are rarely if ever used today.

Although mechanical Q-switches are conceptually simple, they have several drawbacks that preclude their use in modern lasers. Because they require rapidly moving mechanical parts, the long-term reliability of mechanical Q-switches is poor. Also, it is difficult to synchronize external events with the pulse from a mechanically Q-switched laser. This task might be accomplished, for example, by aligning a small diode laser and detector with a spinning mirror Q-switch so that the detector produces a signal just before the laser pulse is emitted. But this method is awkward and inconvenient and is subject to misalignment.

11.5 ACOUSTO-OPTIC Q-SWITCHES

An A-O Q-switch is a block of transparent material, usually quartz, with an acoustic transducer bonded to one side. This transducer is similar to a loudspeaker because it creates a sound wave in the transparent material, just as a stereo speaker produces a sound wave in your living room (except most A-O Q-switches operate at ultrasonic frequencies).

This sound wave is a periodic disturbance of the material, and any light that happens to be traveling through the material sees this periodic disturbance as a series of slits, just like those in Young's double-slit experiment (see Chapter 4). Thus, the light is diffracted out of the main beam by interference, as shown in Figure 11.6.

The idea, then, is to place an A-O Q-switch inside a laser between the gain medium and the back mirror, as shown in Figure 11.7. If no acoustic signal is applied to the transducer, then the Q-switch transmits all the light without disturbing it and the resonator has a high Q. But when an acoustic signal is applied to the transducer, light is diffracted out of the intracavity beam and the resonator Q is reduced.

The side of the A-O Q-switch opposite the transducer is usually configured to minimize reflection of the acoustic wave, as shown in Figure 11.6. A damping material absorbs most of the sound wave's energy, and what is not absorbed is reflected back off-axis by the oblique surface. If the reflected wave were not minimized, it

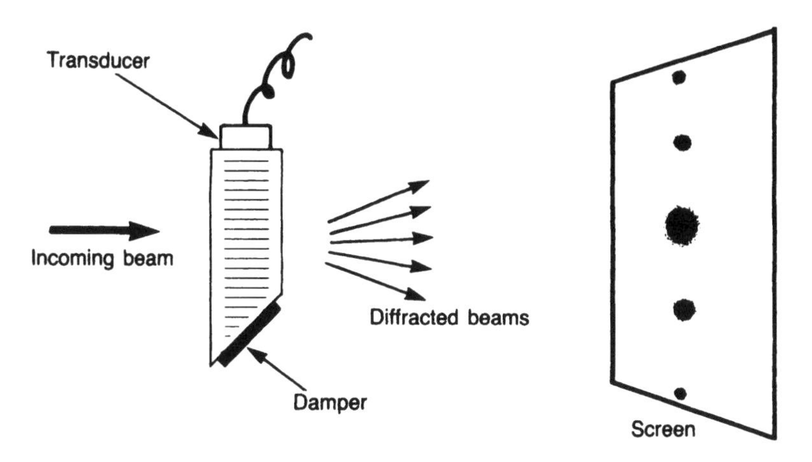

Figure 11.6 The incoming beam of light is diffracted from the periodic disturbance of the sound wave in an A-O Q-switch.

could interfere with the original wave and reduce the diffraction efficiency of the Q-switch. (A-O modelockers, on the other hand, operate at much higher frequencies and depend on acoustic waves traveling in both directions; see Chapter 12.)

The speed of an A-O Q-switch, that is, how quickly it can switch the resonator from low Q to high Q, depends on the sound velocity within the block of transparent material and on the diameter of the laser beam. After all, the switching time is simply the time it takes the sound wave to get out of the way of the beam. Therefore, the smaller the intracavity laser beam is, the faster the speed of the Q-switch.

A-O Q-switches are frequently used in lasers because they are less expensive than E-O Q-switches and their speed is good enough for many applications. A-O Q-switches are easy to synchronize with other events because the pulse is emitted with a constant delay after an acoustic signal is applied to the transducer. The main drawback of A-O Q-switches is their low "hold-off"—their limited ability to keep a high-gain laser from lasing. Only part of the light is diffracted when it passes through the Q-switch, and the remainder is fed back to the laser. If laser gain is great enough, this small feedback can be enough to make the round-trip gain greater than the

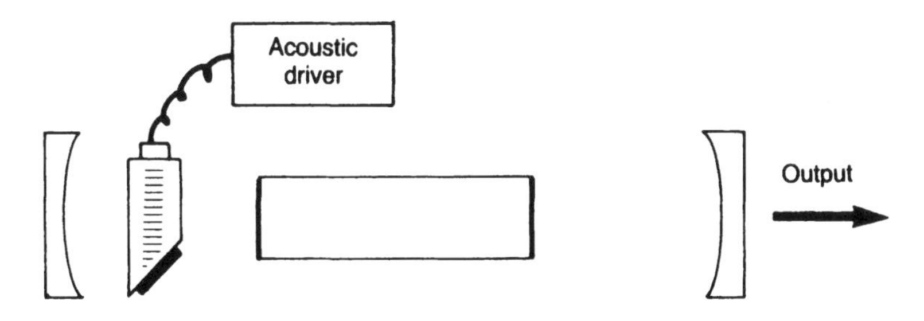

Figure 11.7 An A-O Q-switch placed inside a laser.

round-trip loss, and the laser will lase. Thus, A-O Q-switches can be used only with low-gain lasers.

11.6 ELECTRO-OPTIC Q-SWITCHES

Just as the A-O effect is an interaction between an acoustic field and light, the E-O effect is an interaction between an electric field and light. When an electric field is applied to an electro-optic crystal, the crystal's refractive index changes.

To make an E-O Q-switch, you must have an electro-optic crystal that becomes birefringent when an electric field is applied (or removed). Such a device, called a Pockels cell, can make an optical gate when combined with a polarizer, as shown in Figure 11.8. In Figure 11.8a, no voltage is applied to the crystal, so it exhibits no birefringence and vertically polarized light incident from the left passes unchanged

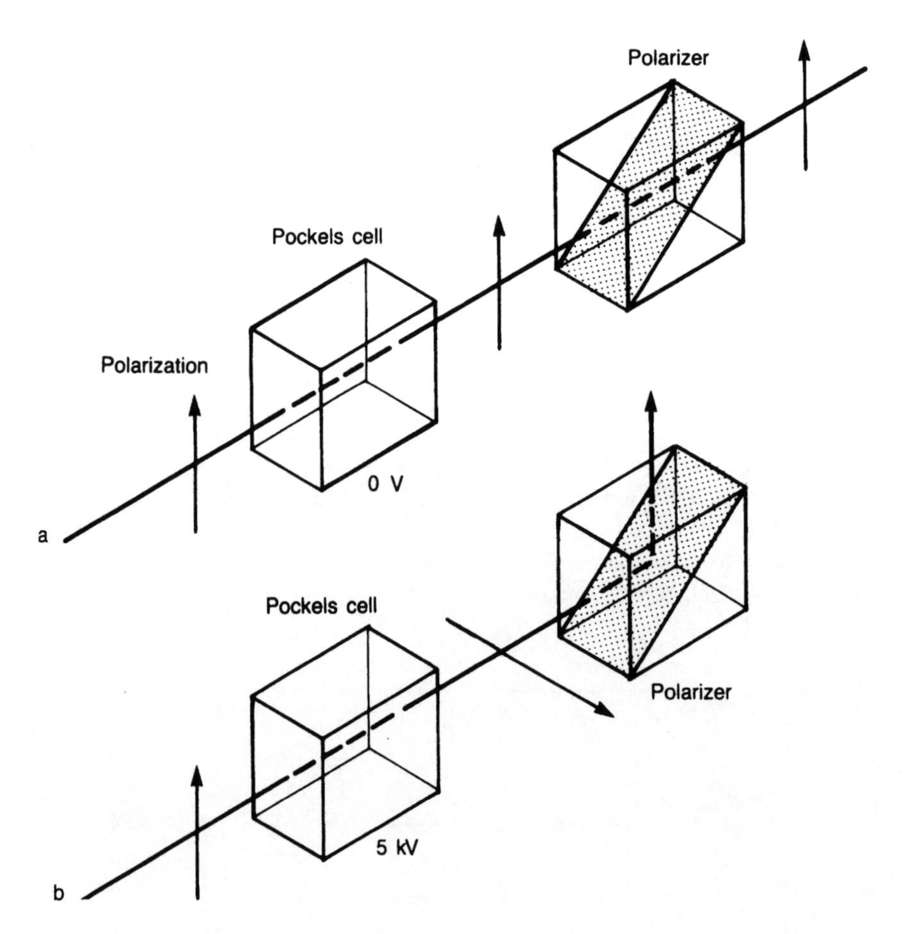

Figure 11.8 When voltage is applied to a Pockels cell, polarization of light passing through it is rotated and the light can be deflected from a subsequent polarizer.

through the crystal and polarizer. In Figure 11.8b, the crystal has become birefringent because a voltage is applied to it, changing its refractive index. If the voltage is chosen exactly right, the Pockels cell behaves as a half-wave plate, rotating the incoming polarization to the horizontal. (These polarization rotations were introduced in Chapter 3.) And the horizontally polarized light is deflected by the polarizer.

Figure 11.8 shows an electrically controllable optical gate. When no voltage is applied, the gate is open and light goes through. When a voltage is applied, the gate is closed and light is deflected.

The optical gate can serve as a Q-switch if it is placed inside a resonator, as shown in the top drawing in Figure 11.9. Light from the laser rod is first polarized by P_1 and then passes through the Pockels cell. If no voltage is applied to the crystal, it exhibits no birefringence, and the light continues through P_2 to the rear mirror. If the half-wave voltage is applied to the crystal, then the polarization is rotated 90° and the light cannot get through P_2.

If you were operating the Q-switched laser shown in the top drawing in Figure 11.9, you would start with the half-wave voltage applied to the Pockels cell while the flashlamp pumped atoms in the laser rod into the upper laser level. When the optimum population inversion had been achieved, you would remove the voltage from the Pockels cell, switching the Q of the resonator from low to high, and producing the Q-switched output pulse.

The chief drawback of electro-optic Q-switches is their high cost. The half-wave voltages for most crystals are in the kilovolt range, and switching speeds are in nanoseconds. That requires sophisticated (i.e., expensive) electronics. One way to

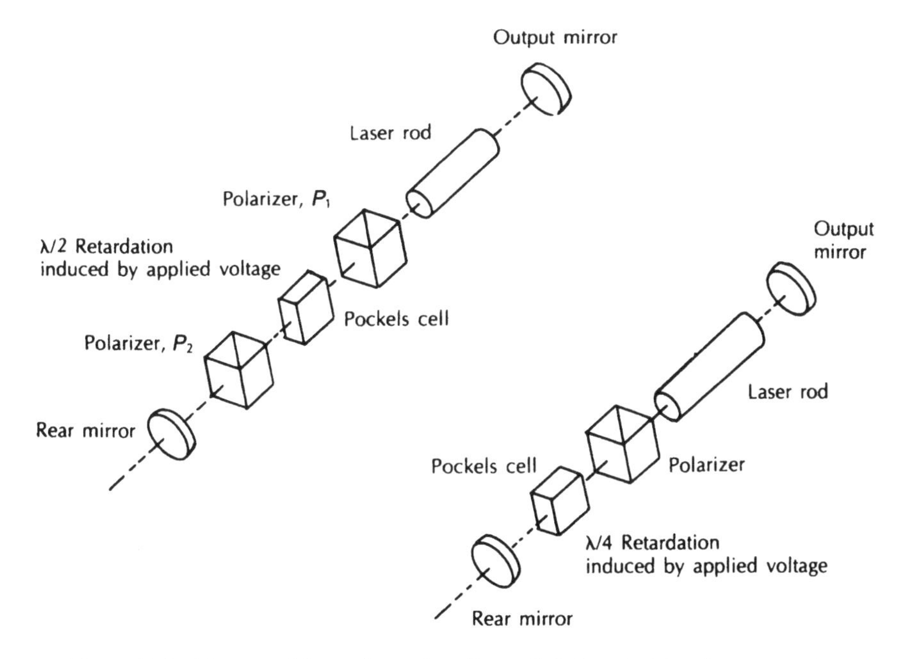

Figure 11.9 Two possible configurations for an E-O Q-switch in a laser resonator.

reduce the cost is shown in the bottom drawing in Figure 11.9. Here, only a quarter-wave voltage need be applied to Q-switch the laser. It is left as an exercise for the student (see question 6) to figure out how this configuration works.

An E-O Q-switch is probably the most effective of the four types of Q-switches. There are no moving parts, not even a sound wave, so it is very fast and reliable. It is also easy to synchronize because the pulse comes out (almost) immediately after the voltage on the Pockels cell is switched. But as mentioned, an electro-optic Q-switch is very expensive. The E-O material from which it is made (usually potassium dihydrogen phosphate or one of its isomorphs) is expensive, as are the power supplies to drive them.

A Kerr cell is similar to a Pockels cell, except a liquid medium instead of a crystal provides the phase retardation. Nitrobenzene is the liquid most commonly used, and the voltage that must be applied is much greater than what is required for a Pockels cell. Optical damage of a Kerr cell tends to be self-healing. However, because they are messy and require very high voltages, these devices are seldom used.

11.7 DYE Q-SWITCHES

Dye Q-switches, also called saturable-absorber Q-switches or passive Q-switches, utilize a dye whose transmission depends on incident light intensity, as shown in Figure 11.10. A cell containing this dye is placed inside the laser and blocks a mirror, as shown in Figure 11.11. But when light emitted from the gain medium becomes intense enough (from both spontaneous emission and stimulated emission), the dye bleaches and light passes through it with little loss.

Dye Q-switches are inexpensive because of their simplicity, but they have several drawbacks, including pulse jitter, dye degradation, and synchronization difficulties. Nonetheless, they find frequent application because they are so simple to use. Dye Q-switches are generally not used with low-gain (cw-pumped) lasers.

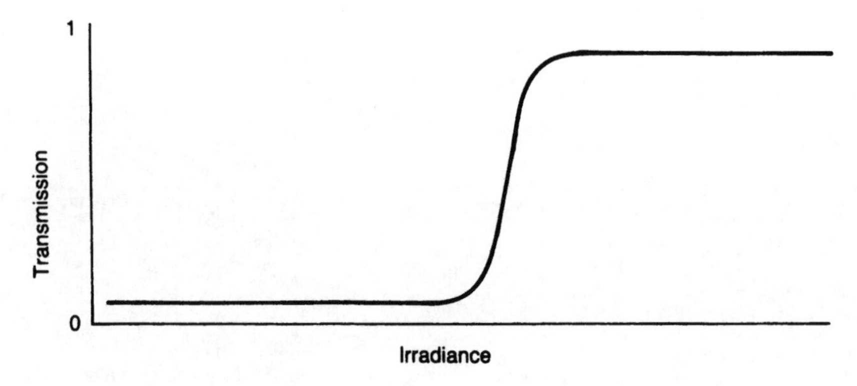

Figure 11.10 The transmission of a saturable absorber increases steeply beyond a certain irradiance.

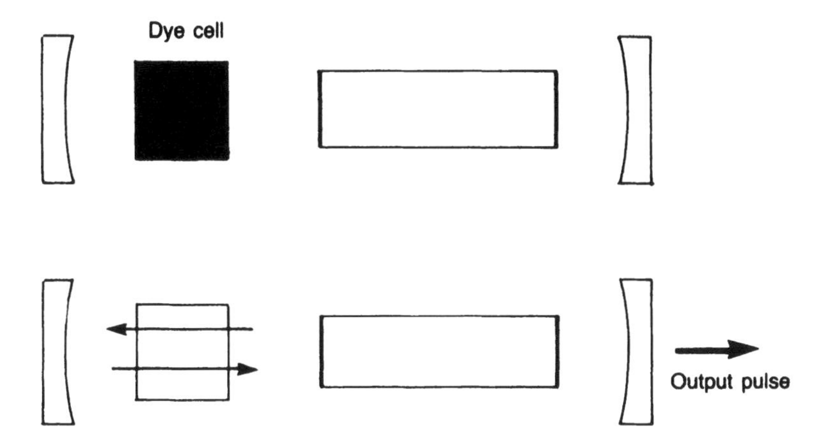

Figure 11.11 Photons emitted by spontaneous and stimulated emission bleach the dye Q-switch so that light can pass through it to the mirror.

QUESTIONS

1. Think about a cw laser that produces a 1 W output. If this laser is Q-switched with no energy loss and produces 10 pulses per second of 100 ns duration, what is the peak power in each pulse?
2. The Q (quality) of a resonator is defined as

$$Q = \frac{\text{energy stored in a resonator}}{\text{energy lost / cycle}}$$

   What is the Q of an Nd:YAG resonator that has 5% round-trip loss?
3. Calculate the switching speed of an A-O modulator if the sound velocity within the modulator is 6×10^3 m/s and the intracavity beam of the laser is 0.5 mm in diameter.
4. Consider a Cr:Ruby laser with 2×10^{19} chromium ions that contribute to the laser action. If each ion emits only one photon per pulse, what is the fractional population inversion required to produce a 1 J pulse? (The fractional population inversion is defined as

$$F = (N_1 - N_0)/(N_1 + N_0)$$

   in which N_1 and N_0 are the populations of the upper and lower laser levels, respectively. (Remember that Cr:Ruby is a three-level system.)
5. Suppose an Nd:YAG laser is Q-switched with a rotating-mirror Q-switch. Calculate the pulse repetition frequency (in Hertz) of the laser if the six-sided mirror rotates at 3200 rpm.

6. When an E-O Q-switch is placed inside a resonator as shown in Figure 11.9, the light passes through the Pockels cell once in each direction and its polarization is rotated 90°. What is the polarization of the light after a single pass through the Pockels cell? (That is, what is the polarization of the light between the Pockels cell and the back mirror?)
7. a) Which takes longer to unload, a moving van full of furniture or a moving van with only several pieces of furniture?
   b) Which takes longer to unload in a Q-switched laser, a population inversion with a lot of gain or a population inversion with only a little gain?
   c) Why the difference?

Chapter 12

Cavity Dumping and Modelocking

Pulsed output can be obtained from many lasers by Q-switching, but this technique will not work with lasers whose upper-state lifetime is too short to store appreciable energy. Another approach—cavity dumping—must be used to obtain pulses from these lasers. And even in lasers that can be Q-switched, there is a lower limit on pulse duration and an upper limit on pulse repetition frequency imposed by the natural time constants of the population inversion. Cavity dumping also can be used to obtain very short- or very high-frequency pulses from these lasers.

But the highest pulse frequencies and the shortest pulses are obtained by modelocking a laser. Any type of laser can be modelocked, even one that has already been Q-switched or cavity dumped. In fact, some lasers are simultaneously Q-switched, cavity dumped, and modelocked.

12.1 CAVITY DUMPING

Cavity dumping is a descriptive name for the process of obtaining a pulsed output from a laser, if you understand that the word *cavity* is used here to mean resonator. One type of cavity-dumped laser is shown in Figure 12.1. Note that both mirrors are maximum reflectors; neither transmits any light. So how does anything get out of the laser? Let's examine what happens when this laser produces a cavity-dumped pulse.

When the lamp flashes, the E-O Q-switch is in its *transmit* mode; that is, light passes through the Pockels cell without any polarization rotation, so the polarizer does not reject the light. As soon as the round-trip gain becomes equal to the round-trip loss, the laser begins lasing and light starts bouncing back and forth between the mirrors. Because neither mirror transmits any light, this circulating power builds up to a high level. When the maximum intracavity circulating power has been obtained, a voltage is applied to the Pockels cell. Then the cell rotates the polarization of light passing through it, and the polarizer ejects all the light in a cavity-dumped pulse, as shown in Figure 12.2.

Thus, the output coupler for a cavity-dumped laser is the cavity dumper itself and not a mirror, as has been the case for all the other lasers we have examined so far. Later we will see additional cases in which the output coupler is not a mirror.

Introduction to Laser Technology, Fourth edition. C. Breck Hitz, J. Ewing, and J. Hecht

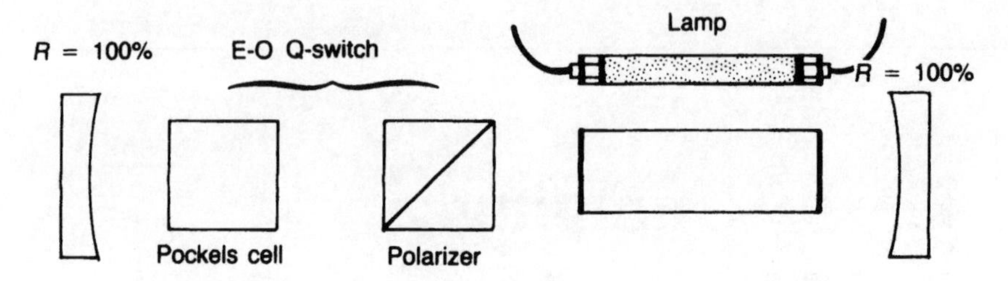

Figure 12.1 A cavity-dumped laser with an E-O cavity dumper.

You may have noticed that a cavity-dumped laser is really just a different type of Q-switched laser, one in which the cavity Q-switches from high to low instead of from low to high. For this reason cavity dumping is sometimes called *pulse-transmission mode Q-switching.* But *cavity dumping* is easier to say.

What is the duration of the cavity-dumped pulse from the laser in Figure 12.2? It depends on how long it takes light to make a round-trip of the resonator, that is, on the resonator length. The longer the resonator, the longer it takes the pulse to get out. The last photon to emerge is the one that passed through the Pockels cell just before the voltage was applied, and it must make one round-trip of the resonator before it comes out. The distance it must travel is $2L$ (L is resonator optical length) and its velocity is c, so the time it takes to do this is $2L/c$. Thus, the duration of the cavity-dumped pulse is the time that elapses between the emergence of the very first photon and the very last one, or $2L/c$.

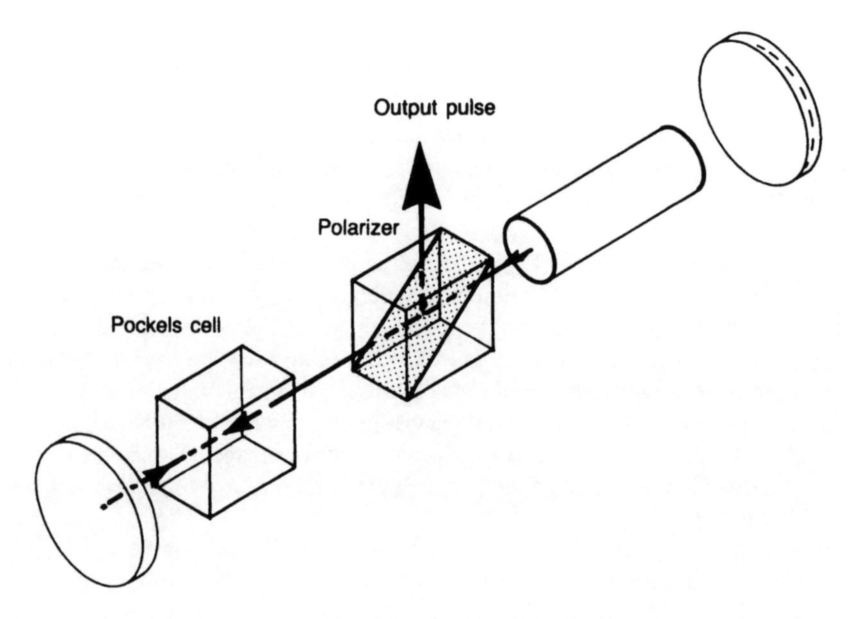

Figure 12.2 When the Pockels cell is biased to rotate the light's polarization, a pulse containing all the intracavity energy is dumped out of the laser resonator.

This pulse duration can turn out to be very short. If you recall that the speed of light is about 1 ft/ns, you can quickly figure that the cavity-dumped pulse from a foot-long laser is about 2 ns long. That is much shorter than the pulse from most Q-switched lasers.

The difference between a cavity-dumped laser and a Q-switched laser is that the energy is stored in the population inversion in a Q-switched laser, whereas it is stored in the optical resonator (cavity) of a cavity-dumped laser. Of course, you can also store energy in a laser's power supply, as is done in flash-pumped solid-state lasers, in some industrial carbon dioxide lasers, and elsewhere. The flower-pot analogy of the previous chapter is revived in Figure 12.3, but this time it shows all three places where energy can be stored: the power supply, the population inversion, and the optical resonator. The electrical input dribbles into the power supply, and if none of the pots is corked, the output dribbles out of the resonator at the same rate. But if any one (or more) of the pots is corked, energy can be stored in that pot and released in a concentrated, high-power pulse. The three parts of a laser corresponding to the three flower pots in Figure 12.3 are shown in Figure 12.4.

A solid-state laser that is flash pumped but not cavity dumped or Q-switched (i.e., one with a cork in only the top pot of Figure 12.3) is known as a *normal-mode* laser. The output pulse consists of several *relaxation oscillation* spikes and is gener-

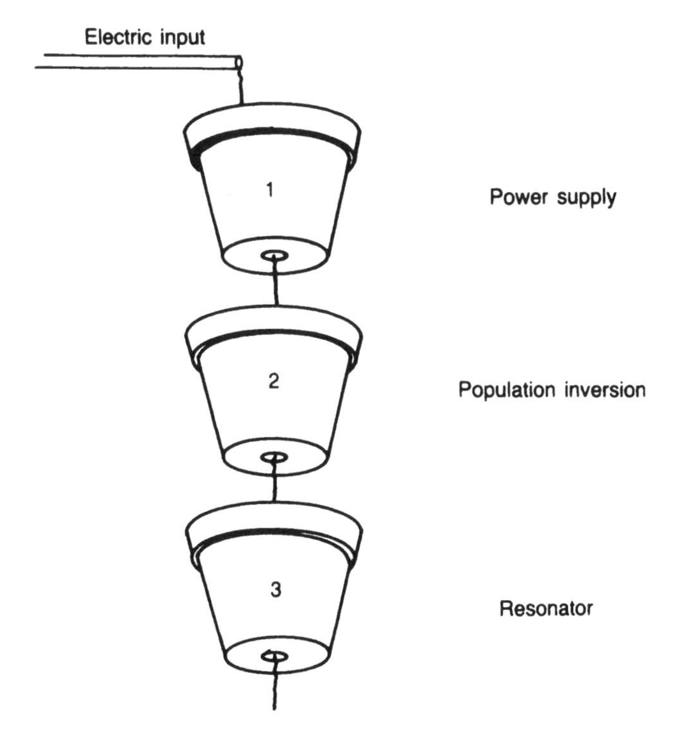

Figure 12.3 Like water flowing form one flower pot to another, energy flows from a laser's power supply to its population inversion to its optical resonator. And energy can be stored in any of these places and released later in a high-power pulse.

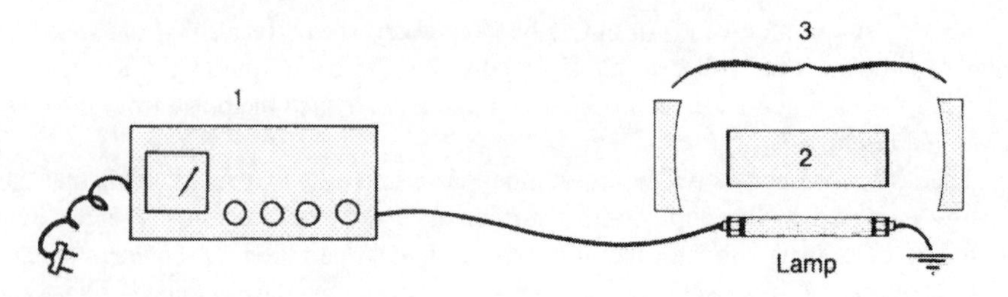

Figure 12.4 The flower pots of Figure 12.3 correspond to a laser's power supply (1), its population inversion (2), and its optical resonator (3).

ally not very repeatable from one pulse to the next. This type of laser is sometimes useful in industrial applications in which only crude energy is needed and in which the refinement of a Q-switch or cavity dumper would be an unnecessary expense.

Things get more interesting when you start putting corks in more than one of the pots in Figure 12.3. For example, a flash-pumped, cavity-dumped laser could be represented by putting corks in both the top and bottom pots. The energy would be compressed first in the power supply, and then the pulse would be concentrated further by storage in the optical resonator. The resulting pulse would be more powerful (i.e., its energy would be more concentrated) than the pulse produced by either flash pumping or cavity dumping alone. If you corked all three pots, you would have a flash-pumped, Q-switched, cavity-dumped laser.

You may be thinking that this is needlessly complicated: Why bother to cork all three pots? After all, can't you achieve the same short pulse length just by corking the third pot (i.e., just by cavity dumping)? The answer is subtle but instructive. Yes, you can achieve the same pulse length just by corking the third pot, but you lose a lot of energy doing it that way.

How long can you store energy in each of the three pots? In the power supply, the energy is stored in capacitors, where it can stay for many seconds (probably many days for that matter). The lifetime of energy stored in the second pot is much shorter, roughly equal to the spontaneous lifetime of the lasing atom. For a solid-state laser, this is usually several hundred microseconds. And the lifetime of energy stored in the third pot is even shorter. It takes a couple of nanoseconds for the photons to make a round-trip of the resonator, and, typically, a few percent are lost on each round-trip. Thus, energy can only be stored in the third pot for a few tens or hundreds of nanoseconds.

So if you are Q-switching a laser, you would like to dump all the available energy into the population inversion in less than a few hundred microseconds. You will lose energy if you try to keep it in the second pot for longer than that, so you pulse the power supply to dump out all its energy in a few hundred microseconds.

Likewise, if you are cavity dumping a laser, you would want to dump all the energy into the resonator in less than a hundred nanoseconds. You will lose energy if you try to keep it in the third pot for longer than that. Thus, you Q-switch the laser and move all the energy from the population inversion to the resonator in less than a hundred nanoseconds.

Of course, when you cork all three pots, you can achieve enormous peak powers, and these powers can easily be great enough to damage the intracavity optical components. Optical damage can occur on the surfaces of mirrors, laser rods, and Q-switches, or it can occur inside transmissive optical elements. Needless to say, avoiding optical damage is a prime consideration when designing high-peak-power lasers.

12.2 PARTIAL CAVITY DUMPING

When all the energy circulating between the mirrors is dumped out of a laser cavity, it is necessary to wait many microseconds or even milliseconds before the energy can be built up again for a second dump. In partial cavity dumping, only a fraction of the energy between the mirrors is dumped out. In terms of our familiar flower-pot analogy, partial cavity dumping corresponds to keeping the bottom pot half-full and pulling its cork out for only the briefest time so that only a small amount of energy is released in the pulse. This arrangement is shown in Figure 12.5.

Of course, energy comes out of the laser on average as fast as it goes in. But because the output energy is compressed into pulses, its peak power is greater. This line of reasoning holds for any pulsed laser.

Ion lasers have upper-state lifetimes too short to allow enough energy storage for Q-switching. (The middle flower pot is too leaky.) If you want to obtain a pulsed output from an argon-ion laser or a krypton-ion laser, you must cavity dump it. Such

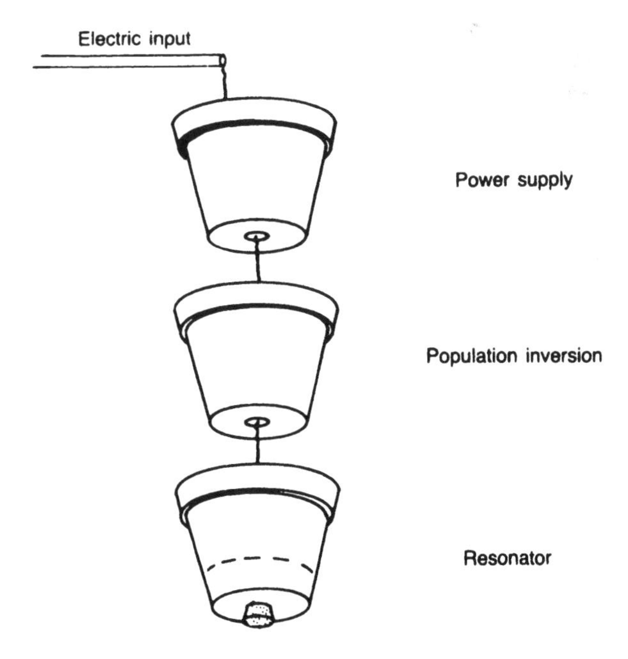

Figure 12.5 In partial cavity dumping, only a fraction of the energy stored in the resonator is let out in each pulse.

a laser is shown in Figure 12.6. Here, an A-O cavity dumper ejects part of the intracavity energy to generate the output pulse train. Typically, the signal to the cavity dumper might be an 80 MHz acoustic signal, chopped at 10 kHz, as shown in Figure 12.7. Then the output would be a 10-kHz train of pulses, diffracted from the A-O cavity dumper.*

12.3 MODELOCKING—TIME DOMAIN

The shortest pulses of light that have ever been generated have come from modelocked lasers. The duration of a Q-switched laser pulse varies from several hundred nanoseconds to shorter than a nanosecond, depending on the laser parameters. Cavity dumping can often produce a shorter pulse than Q-switching, but a modelocked pulse from a dye laser can be shorter than a picosecond, a thousand times shorter than the 1 ns pulse from a cavity-dumped laser.

There are two ways to understand how a modelocked laser works. You can examine what happens in the time domain by thinking about what happens as laser light moves back and forth between the mirrors, or you can examine what happens in the frequency domain by thinking about how the longitudinal modes of the laser interfere with each other. Either way is correct and, in fact, they turn out in the final analysis to be two ways of saying the same thing. But the time-domain picture is easier to understand the first time through. We discuss it first and then explain the frequency domain.

A modelocked laser is shown in Figure 12.8. The optical energy between the mirrors has been compressed to a very short pulse that is shorter than the resonator itself. In a Q-switched or a cavity-dumped laser, the whole resonator is filled with energy, but in a modelocked laser the energy is compacted into a pulse that bounces back and forth between the mirrors. Each time this intracavity pulse bounces off the partially transmitting mirror, an output pulse is transmitted through that mirror.

The energy is compacted into the modelocked pulse in the resonator by the modelocking modulator, which is simply a fast optical gate (e.g., it could be an E-O Q-switch). The gate opens once per round-trip transit time, letting the pulse through. The rest of the time the gate is closed; the only light that can circulate between the mirrors is the light in the modelocked pulse. The modulator is placed as close as possible to the mirror, and it opens only once while the pulse passes through it, reflects off the mirror, and passes through the modulator again. The average power output from the laser is not usually affected by modelocking.

The output of a modelocked laser is a train of very short pulses. The time separation between pulses is the distance traveled by the intracavity pulse between reflections by the output mirror, divided by its velocity. That is, the period between pulses is $2L/c$, in which L is again the optical distance between the mirrors. The frequency

*In the previous chapter, we discussed A-O modulators that produce many diffracted beams. These are known as Raman–Nath modulators. Another kind of A-O modulator, called a Bragg modulator, produces only one diffracted beam. Assuming that you want to obtain the output of a cavity-dumped laser in a single beam, you would use a cavity dumper that operates in the Bragg regime. The difference between the two regimes has to do with the optical and acoustic wavelengths involved and the length of the interaction region.

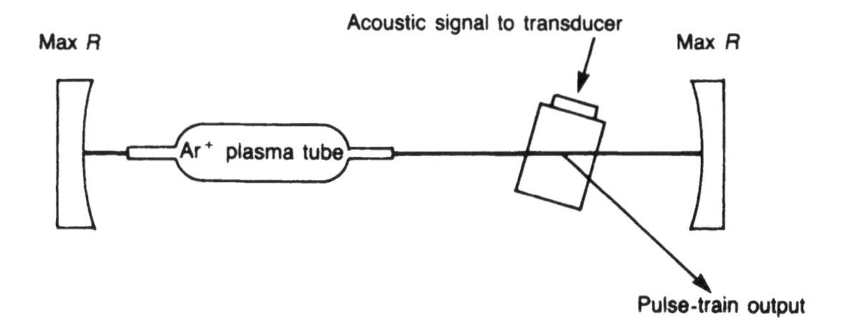

Figure 12.6 An argon-ion laser cavity dumped with an A-O modulator.

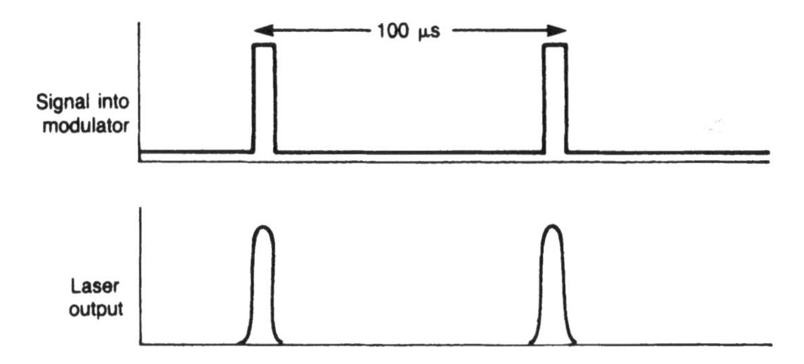

Figure 12.7 Acoustic input signal and optical pulse-train output for partially cavity-dumped laser in Figure 12.6.

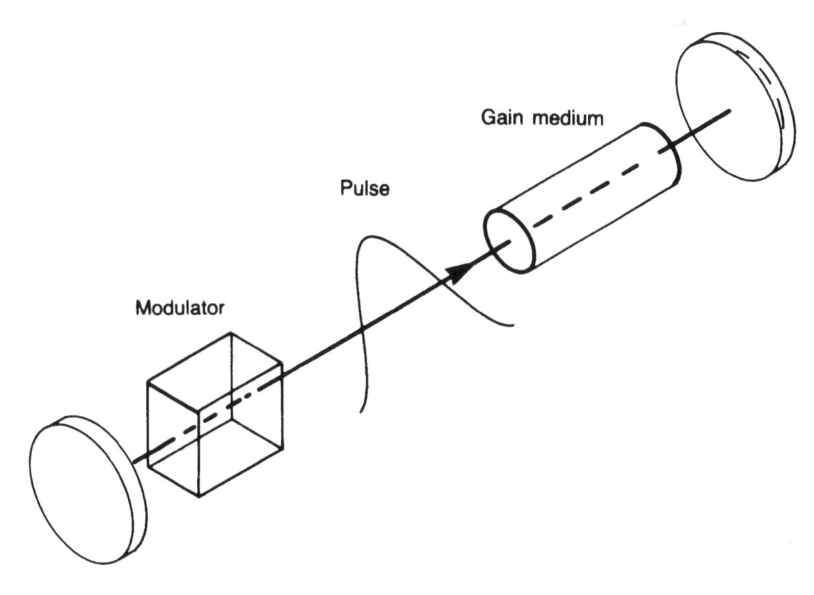

Figure 12.8 A short pulse of light bounces back and forth between the mirrors of a mode-locked laser.

of modelocked pulses is the reciprocal of the period, or $f = c/2L$. Thus, a laser whose mirrors are separated by 30 cm will produce a modelocked pulse train at 500 megahertz MHz—500 million pulses of light per second.

The duration of the modelocked pulses depends on several factors, including the laser's gain bandwidth and the effectiveness of the modelocking modulator (its modulation depth). The greater the laser's bandwidth, the shorter the modelocked pulses can be. Thus, with their enormous bandwidths, some of the tunable solid-state lasers discussed in Chapter 20 are capable of modelocked pulses well into the femtosecond regime. On the other hand, Nd:YAG lasers, with relatively narrow bandwidths, produce modelocked pulses 30–60 ps in duration.

E-O modulators, A-O modulators, and dye cells have all been used to modelock lasers. An A-O modelocker is shown in Figure 12.9. Note that the side of the quartz block opposite the transducer is not configured to minimize reflection of the incident sound wave as it is in an A-O Q-switch. In fact, an acousto-optic modelocker works differently than a Q-switch. The Q-switch is turned off and on by turning the acoustic signal applied to its transducer off and on. A modelocker reflects the sound wave back across the modulator so that a standing wave, like the wave that is formed in a violin string, is produced in the modulator.

If you think about a standing wave for a moment, you realize that it "disappears" twice during each cycle. The violin string is at one instant bowed upward, and an instant later bowed downward. But between those two extremes, the string is flat; it is momentarily a perfectly straight line between its two ends. And this momentarily straight string appears twice per cycle between the extremes of the motion.

Likewise, there are two times per period when the elastic medium of a standing-wave A-O modulator is not perturbed by the acoustic wave. During these times, light is not diffracted from the modulator, and a modelocked pulse of light can pass through the modulator without loss.

So if you apply a 100 MHz signal to a standing-wave A-O modulator, there will be 200 million times per second when the modulator does not diffract light. If you put the modulator near the mirror of a 75-cm-long laser ($c/2L$ = 200 MHz), it will modelock the laser. The intracavity pulse will pass through the modulator, reflect off the mirror, and pass through the modulator again each of the 200 million times

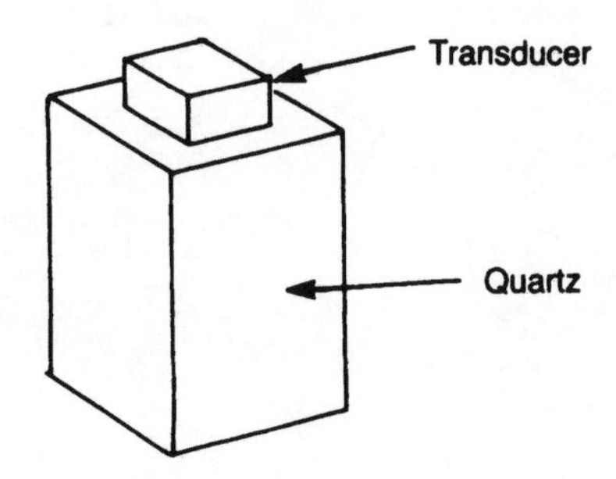

Figure 12.9 An A-O modelocker is a standing-wave device in which sound is reflected back toward the transducer from the far side of the modulator.

per second that the modulator does not diffract light. The output from the laser will be a train of 200 million pulses per second.

E-O modelockers can operate exactly like E-O Q-switches, rotating the polarization of light that is subsequently ejected by a polarizer. Thus, you could drive an E-O modelocker (composed of a Pockels cell and a polarizer) at 100 MHz and substitute it for the A-O modulator in the previous paragraph.

Passive dye cells can also modelock a laser, and the principle is the same as a passively Q-switched laser: the leading edge of the pulse bleaches the dye, so the rest passes through with minimal loss. Another variety of passive modelocker relies on saturated absorption in a semiconductor crystal. Semiconductor saturated-absorption modulators, known in the vernacular as SESAMs, are probably the most common approach to modelocking lasers today.

In all the examples we have discussed, there has been a modulator that does something to degrade light that passes through the modulator at the wrong time. Thus, the only photons that can survive inside the resonator are those that band together in a pulse and sneak through the modulator when it is open. But it is also possible to modelock a laser by modulating the gain rather than the loss.

When you modulate a laser's gain, you turn the gain "on" only when the photons that have banded together in a pulse are in the gain medium. The rest of the time the gain is off. Again, the only photons that can survive in the resonator are those that have banded together in a pulse. Whether you modulate the effective loss or the effective gain, the result is the same: the circulating power is compacted into a pulse bouncing back and forth between the mirrors.

A synchronously pumped (synch-pumped) dye laser is an example of a laser modelocked by gain modulation. The dye laser is optically pumped by modelocked pulses from a second laser, so gain in the dye laser is created only once per round-trip transit time in the dye laser. To ensure proper timing, the length (and hence modelocking frequency) of the pump laser must be exactly the same as that of the dye laser.

12.4 MODELOCKING—FREQUENCY DOMAIN

Modelocking is as curious a name as cavity dumping is descriptive. What modes are locked, and what does it mean to lock modes? To understand why it is called modelocking, you have to understand the frequency-domain viewpoint as well as the time-domain viewpoint.

The longitudinal modes of the resonator are locked together in phase when you modelock a laser. (That is why you will sometimes hear it referred to as phase locking.) Suppose you have a laser with three longitudinal modes oscillating simultaneously, as shown schematically in Figure 12.10. Of course, the waves are moving inside the resonator at about the speed of light. Figure 12.10 only shows what they look like at one instant in time. Most places in the cavity will be like point A in Figure 12.10: the three modes add together to produce a very small total intensity, that is, they interfere with each other destructively.

But at one (or maybe more) place in the cavity, all three modes will be at their maximums and they will add up to a large total. And because all three waves are

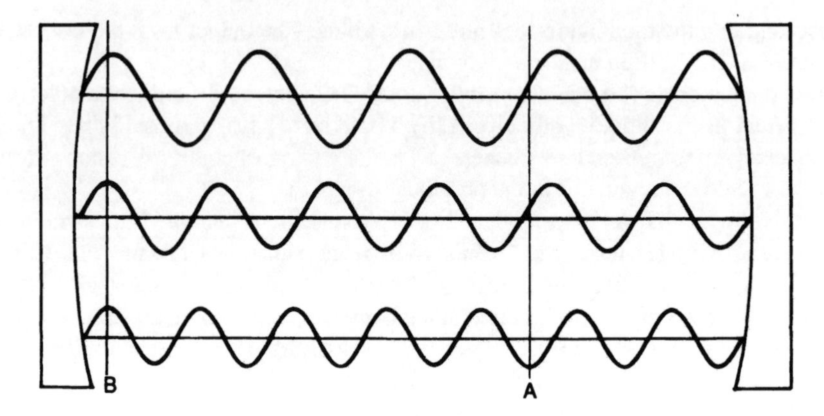

Figure 12.10 Three longitudinal modes are shown spatially displaced for clarity. The three modes add constructively at A.

moving at the speed of light, the spot where they add constructively also moves along at the speed of light. As you have already figured out, constructive interference at this spot is what creates the modelocked pulse.

The problem is that in a free-running laser—one that's not modelocked—the three modes in Figure 12.10 will not stay in phase with each other. Resonator perturbations will cause some modes to stop oscillating, and when they start again, they will have a different phase. So it is necessary to lock them together in phase to produce a modelocked pulse. That is what the modelocking modulator does. It transfers energy among all the modes, and this transferred energy contains phase information that prevents the modes from shifting phase with each other.

It turns out that the time-domain viewpoint and the frequency-domain viewpoint are two equivalent ways of looking at the same thing, but showing that equivalence requires a level of mathematics (Fourier analysis) beyond the scope of this book.

12.5 APPLICATIONS OF MODELOCKED LASERS

Modelocked lasers do not usually have very high peak power. The pulses are very short, but there are so many of them that no one pulse can contain much energy. Thus, unamplified modelocked lasers are not found in applications requiring high peak powers. Instead, modelocked lasers are used when very short pulses are needed. Chapter 20 will go into significant detail about some of the applications of these short pulses.

One application of modelocked lasers is in ranging. A modelocked pulse is reflected from a distant object like a satellite, and the time it takes to return to the transmitter is carefully measured. Since the pulse moves at the speed of light, the distance to the object can be readily calculated. Q-switched lasers are also used in ranging. But because their pulses are several meters long, the precision of a Q-switched laser-ranging system cannot be much better than a meter. The pulse from a

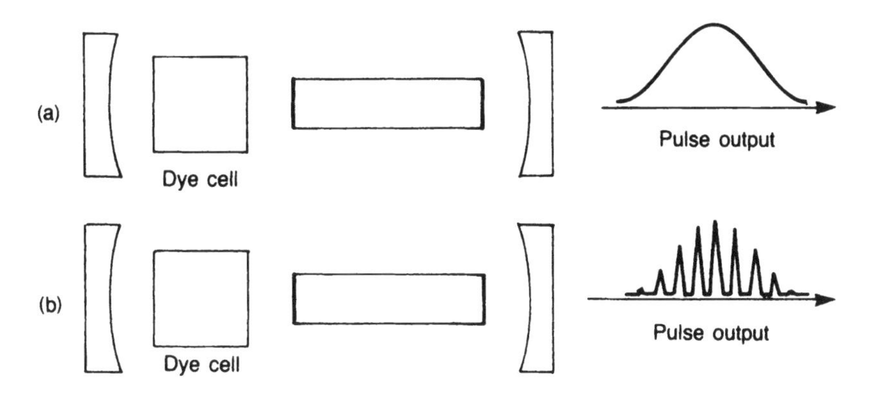

Figure 12.11 A dye cell can Q-switch a laser (a) or it can simultaneously Q-switch and modelock the laser (b).

modelocked laser, on the other hand, is only several centimeters long, so ranging with much greater accuracy is possible.

Modelocked lasers are also used as spectroscopic tools to investigate very fast phenomena. Spectroscopy is a whole science in itself. Basically, it involves studying matter by observing how light interacts with it. If the matter that you are studying changes very quickly, say, in a fraction of a nanosecond, then the probe you are using to study it must also be that fast. The short pulse from a modelocked laser is one of the few probes that can be used to investigate very fast chemical or physical reactions.

12.6 TYPES OF MODELOCKED LASERS

Modelocking can be combined with any of the other techniques discussed in this chapter to produce pulsed lasers, or it can be used by itself to produce an unending train of modelocked pulses. For example, a passively modelocked and Q-switched laser is shown in Figure 12.11. In Figure 12.11a, the dye only Q-switches the laser. In Figure 12.11b, the dye concentration has been changed so that it now modelocks and Q-switches the laser.

QUESTIONS

1. Sometimes a laser is simultaneously Q-switched and cavity dumped. What would be the advantage of such a laser over one that was only cavity dumped or only Q-switched? How would you design such a laser if you could use only a single E-O Q-switch? Sketch the resonator showing the Q-switch and the mirror transmissions, and sketch a plot of the voltage applied to the Q-switch as a function of time.
2. Suppose a laser were simultaneously cavity dumped and modelocked. What would the output of such a laser look like? Sketch a diagram of the resonator showing the intracavity devices you might use.

Chapter 13

Nonlinear Optics

In the previous three chapters, we discussed modifying the spectral and temporal characteristics of a laser. That is, we talked about how to reduce a laser's spectral width and how to change the temporal shape of its output by several pulsing techniques. In this chapter, we examine nonlinear optics, the technology that can change the wavelength of light produced by a laser.

Strictly speaking, you do not need a laser to produce nonlinear optical effects; they are a manifestation of classical physics and no quantum mechanics is required. But as a practical matter, these effects require such high optical intensities that they are difficult to produce without lasers.

Nonlinear optics is a very useful technology because it extends the usefulness of lasers by increasing the variety of wavelengths available from a given laser. Wavelengths both longer and shorter than the original can be produced by nonlinear optics. It is even possible to convert a fixed-wavelength laser to a continuously tunable one.

In this chapter, we begin with a discussion of second-harmonic generation (SHG), which is probably the single most important type of nonlinear effect. Phase matching, which is absolutely necessary for any efficient nonlinear interaction, is explained in the context of second-harmonic generation. Then we take a look at several other nonlinear effects, including higher harmonic generation and mixing, and parametric oscillation.

13.1 WHAT IS NONLINEAR OPTICS?

Nonlinear optics is a completely new effect, unlike anything discussed before in this text. Light of one wavelength is transformed to light of another wavelength—an impressive feat.

This transformation is completely different than, say, a piece of red glass "transforming" white light to red light. The red light was already present in the white light before it hit the piece of red glass. The glass only filters out the other wavelengths; it does not generate a new wavelength. But in nonlinear optics, new wavelengths are generated. A classic example is shown in Figure 13.1, in which the second harmonic—green light at a wavelength of 532 nm— is generated from the 1064 nm beam of infrared light from a Nd:YAG laser.

Introduction to Laser Technology, Fourth edition. C. Breck Hitz, J. Ewing, and J. Hecht

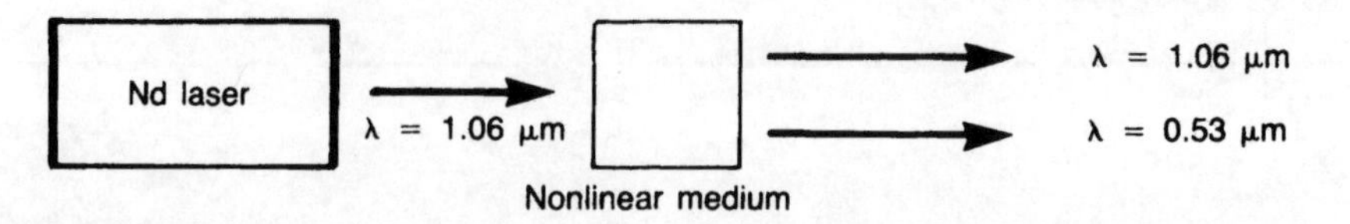

Figure 13.1 In SHG, part of the light passing through the nonlinear medium is converted to light of one-half the original wavelength.

It is important to note that only part of the 1.06 μm light is converted to the second harmonic; part is unchanged. In many cases, it is very important to maximize this conversion efficiency, and in the next section we describe how this can be done.

How is the new wavelength of light created? To gain an intuitive, albeit nonrigorous, understanding of what happens in nonlinear optics, think about the electrons in a nonlinear crystal. (Nonlinear effects can also occur in liquids and gases, but crystals are most common. The explanation we give here would also hold, with minor modifications, for a nonlinear liquid or gas.) These electrons are bound in *potential wells,* which act very much like tiny springs holding the electrons to lattice points in the crystal, as shown in Figure 13.2. If an external force pulls an electron away from its equilibrium position, the spring pulls it back with a force proportional to displacement. The spring's restoring force increases linearly with the electron's displacement from its equilibrium position.

The electric field in a light wave passing through the crystal exerts a force on the electrons that pulls them away from their equilibrium positions. In an ordinary (i.e., linear) optical material, the electrons oscillate about their equilibrium positions at the frequency of this electronic field. A fundamental law of physics says that an oscillating charge will radiate at its frequency of oscillation, so these electrons in the crystal "generate" light at the frequency of the original light wave.

If you think about it for a moment, you will see that this is an intuitive explanation of why light travels more slowly in a crystal, or any dielectric medium, than in a vacuum. Part of the energy in the light wave is converted to motion of the electrons,

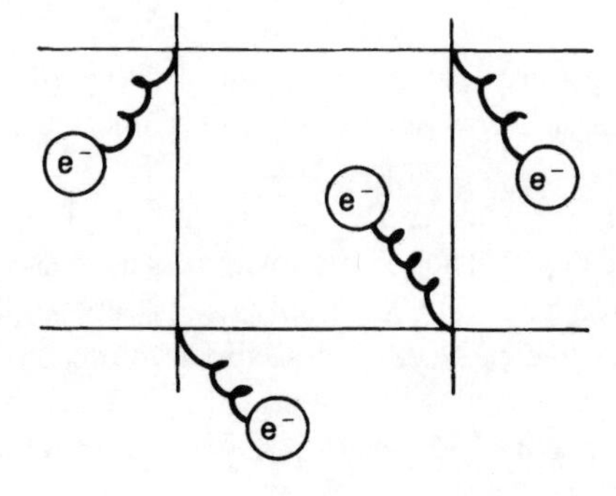

Figure 13.2 Electrons in a nonlinear crystal are bound in potential wells, which act something like springs, holding the electrons to lattice points in the crystal.

and this energy is subsequently converted back to light again. But the overall effect is to retard the energy as it moves through the crystal because it takes a detour into the motion of the electrons.

How is a nonlinear material different from the linear material we have been discussing? You can think of a nonlinear material as one whose electrons are bound by very short springs. If the light passing through the material is intense enough, its electric field can pull the electrons so far that they reach the ends of their springs. The restoring force is no longer proportional to displacement; it becomes nonlinear. The electrons are jerked back roughly rather than pulled back smoothly, and they oscillate at frequencies other than the driving frequency of the light wave. These electrons radiate at the new frequencies, generating the new wavelengths of light.

The exact values of the new wavelengths are determined by conservation of energy. The energy of the new photons generated by the nonlinear interaction must be equal to the energy of the photons used. Figure 13.3 shows the infrared and second-harmonic photons involved in the second-harmonic generation process of Figure 13.1. You can think of the nonlinear process as welding two infrared photons together to produce a single photon of green light. The energy of the two 1.06 μm photons is equal to the energy of the single 532 nm photon.

Another nonlinear process is diagramed in Figure 13.4. In optical mixing, two photons of differing wavelengths are combined into a single photon of shorter wavelength. What is the new wavelength generated? Recall that photon energy is given by $E = hc/\lambda$. Conservation of energy requires that

$$\frac{hc}{\lambda_1} + \frac{hc}{\lambda_2} = \frac{hc}{\lambda_3}$$

Therefore, the new wavelength is

$$\lambda_3 = \frac{\lambda_1 \lambda_2}{\lambda_1 + \lambda_2}$$

So far we have mentioned two of the three requirements for nonlinear optics: intense light and conservation of energy. The third requirement is conservation of mo-

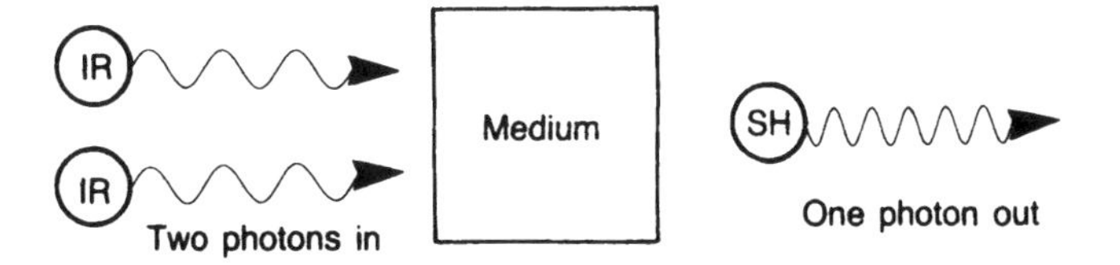

Figure 13.3 You can think of SHG as a welding process: two photons are welded together to produce a single photon with the energy of both original photons. IR = infrareds. SH = second harmonic.

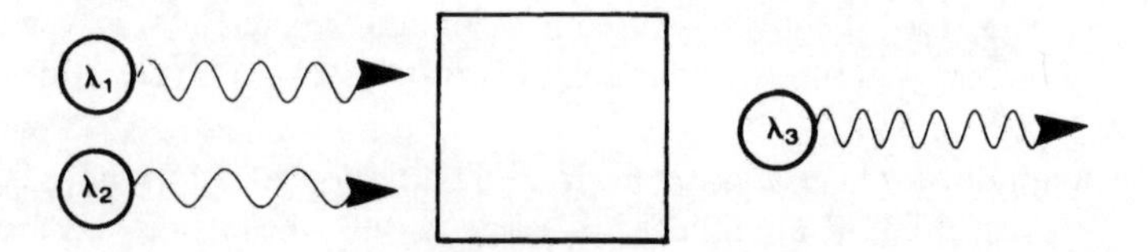

Figure 13.4 Optical mixing is similar to SHG, except that the original photons have different energies.

mentum, and that is fulfilled by phase matching, which is discussed in a later section of this chapter.

13.2 SECOND-HARMONIC GENERATION

Second-harmonic generation (SHG), or frequency doubling, is the most common and probably the most important example of nonlinear optics. It is relatively straightforward compared to other nonlinear interactions, and it can have a relatively high conversion efficiency. (The conversion efficiency can be defined from Figure 13.1 as the ratio of second-harmonic power generated to infrared, or fundamental, power input.)

The conversion efficiency of SHG depends on several factors, as summarized in the proportionality

$$P_{SH} \propto \ell^2 \frac{P_f^2}{A} \left[\frac{\sin^2 \Delta\phi}{(\Delta\phi)^2} \right]$$

in which P_{SH} is the second-harmonic power, ℓ is the length of the nonlinear crystal, P_f is the fundamental power, A is the cross-sectional area of the beam in the nonlinear crystal, and the quantity in brackets is a phase-match factor that can vary between zero and one. (The $\propto$ symbol indicates a proportionality: The quantities on either side of the symbol are proportional to each other, but not necessarily equal.) Obviously, it is important to ensure that this factor be as close to unity as possible, but we will postpone a discussion of how this is done until the next section.

Let us take a look at how the factors in the preceding proportionality affect the harmonic conversion efficiency. For example, Figure 13.5 shows two identical experiments, except that in the second experiment the nonlinear crystal is twice as long. With the 1 cm crystal, the conversion efficiency is $10^{-9}/10^{-3} = 10^{-6}$. What happens with a 2 cm crystal?

Contrary to what you might expect, the 2 cm crystal does not generate twice as much second harmonic; it generates four times as much because the second-harmonic power is proportional to the *square* of the crystal length. In other words, 4 μW of 532 nm light is generated in the second experiment. The rest of the light passes through the hypothetical perfect crystal unchanged. (In a real crystal, part of the light would be absorbed and converted to heat.) Thus, in second-harmonic gen-

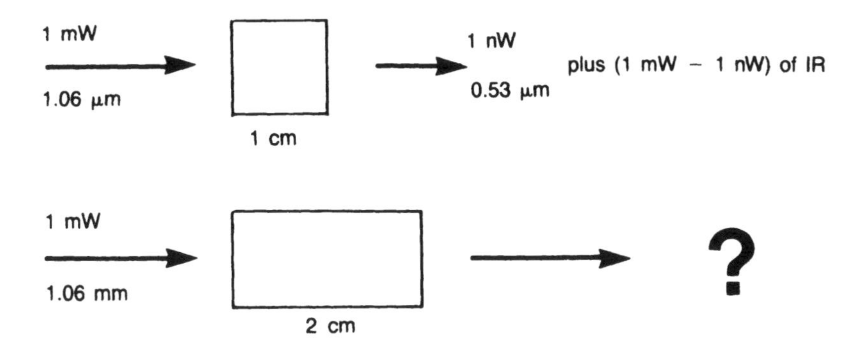

Figure 13.5 If the crystal length in the upper experiment is doubled, how much second harmonic is generated?

eration, there is such a thing as a free lunch—sort of. With twice as much crystal, you get four times as much second-harmonic output. The drawback is that a 2 cm nonlinear crystal can often cost more than four times as much as a 1 cm crystal.

In Figure 13.6 we have gone back to the 1 cm crystal, but now the second experiment has twice as much incoming fundamental power. How much second-harmonic is produced in this case?

Since second-harmonic power is proportional to the square of the fundamental power, four times as much second-harmonic is generated in the second experiment. As was the case when the crystal length was doubled, 4 μW of second-harmonic power is generated in the second experiment in Figure 13.6.

And this fact—that the second-harmonic power generated is proportional to the square of the fundamental power—can be used to advantage with a pulsed laser. The first experiment in Figure 13.7 shows a 20 ns, 2.5 MW (peak power) pulse of 1.06 μm light that is frequency doubled with 10% efficiency in a nonlinear crystal. In the second experiment, the same amount of fundamental energy (2.5 MW × 20 ns = 50 mJ) is compacted into a 10 ns pulse, creating a pulse with twice as much peak

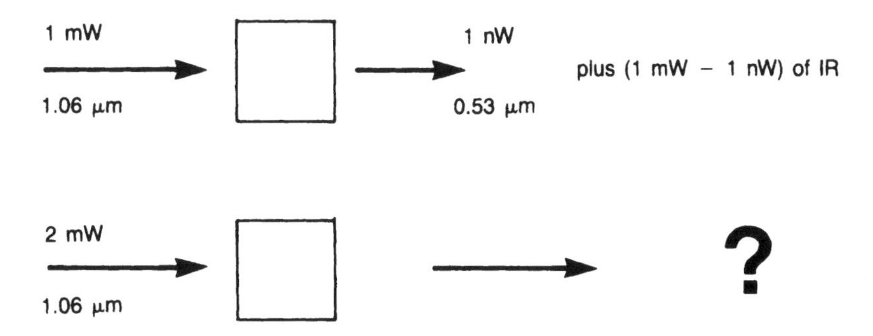

Figure 13.6 If the fundamental power in the upper experiment is doubled, how much second harmonic is generated?

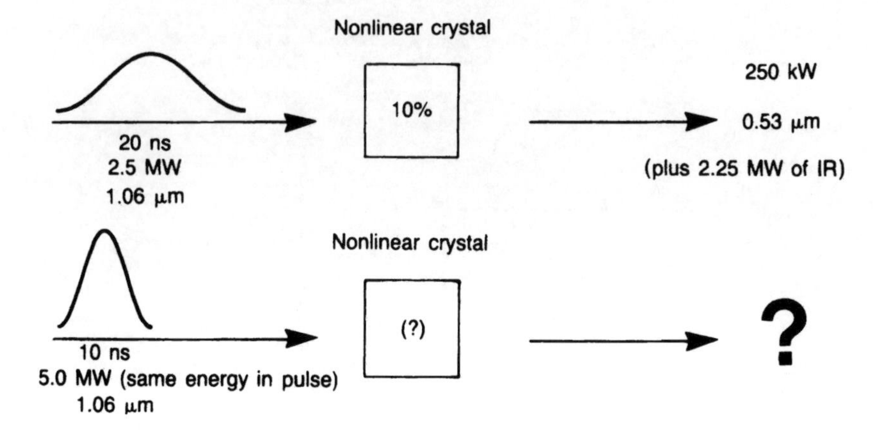

Figure 13.7 If the pulse energy in the upper experiment is compacted into a pulse half as long, how much second harmonic is generated.

power as in the first experiment. How much second-harmonic power is produced in this case?

As before, we have doubled the fundamental power, so the second-harmonic power is quadrupled to 1 MW in Figure 13.7. Another way to look at this is to solve the proportionality above for conversion efficiency:

$$\frac{P_{SH}}{P_f} \propto \ell^2 \frac{P_f}{A}\left[\frac{\sin^2 \Delta\phi}{(\Delta\phi)^2}\right]$$

Now you can see that conversion efficiency is proportional to the fundamental power. Because the fundamental power was doubled in the second experiment, the conversion efficiency must be 20%. And if 20% of the 5 MW input is frequency doubled, then the second-harmonic power is 20% of 5 MW, or 1 MW.

There is still one more thing you can do to boost the conversion efficiency of second-harmonic generation, and it is shown in Figure 13.8. Suppose the beam has a 1-mm radius in the first experiment, and in the second experiment a lens shrinks the beam to a 0.5-mm radius. Now what is the second-harmonic power?

Again, it is 4 μW. Why? Because the second-harmonic power is inversely proportional to beam area, which is proportional to the square of the beam radius ($A = \pi r^2$). So by reducing the beam radius by a factor of two, you reduce the beam area by a factor of four and increase the second-harmonic power by a factor of four.

These are precisely the kinds of things you do when you want to increase the conversion efficiency of SHG: You get a longer crystal, you increase the fundamental power, and you focus the beam into the crystal. Unfortunately, there are limitations on all these tricks. If you focus too tightly or increase the fundamental power too much, you may damage the expensive nonlinear crystal.

And there is another limitation on focusing. Remember from Chapter 9 that a tightly focused Gaussian beam has greater divergence than one that is less tightly

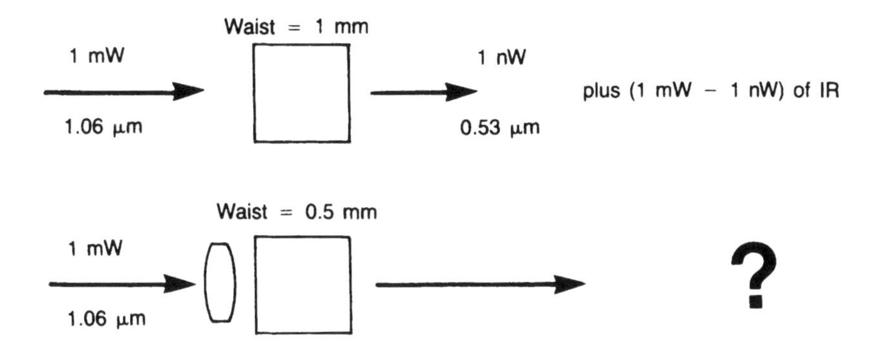

Figure 13.8 If the beam in the upper experiment is focused to one-half its original diameter, how much second harmonic is generated?

focused. It turns out that a diverging beam is less efficient at SHG than a collimated one. (It has to do with phase matching.) So even if you do not damage the crystal, you still cannot focus as tightly as you might like to into a nonlinear crystal.

13.3 BIREFRINGENT PHASE MATCHING

None of the things discussed in the previous section makes any difference if the last term of the proportionality (the phase-match term) is equal to zero. Phase matching is vital to any nonlinear interaction, but we discuss it here in the context of SHG. A similar concept holds for any other nonlinear interaction.

First, let us understand the problem, then we'll look for a solution. Figure 13.9 shows the problem. If the second-harmonic power generated at point B is out of phase with the second-harmonic power generated at point A, they will interfere destructively and result in a total of zero second-harmonic from the two points. What is worse, if the crystal is not phase matched, the second harmonic generated at nearly every point in the crystal will be canceled by a second harmonic from another point. Practically no second harmonic will be produced, no matter how tightly you focus or how long the crystal is.

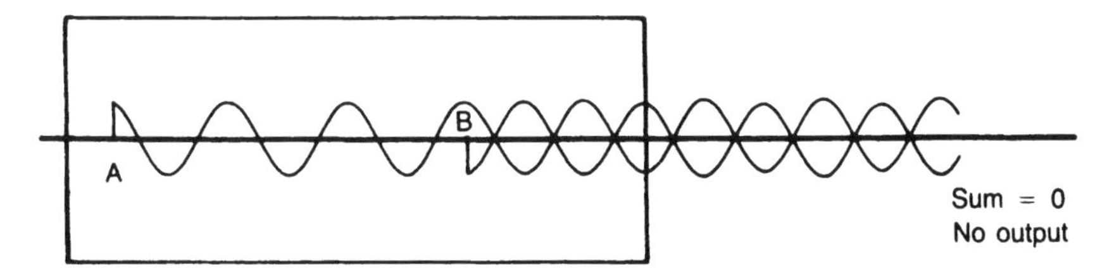

Figure 13.9 If a nonlinear crystal is not phase matched, harmonic light generated at one point will interfere destructively with that generated at another point.

How does the second-harmonic get out of phase with itself in the first place? That is, the second-harmonic from point B is generated from the same fundamental wave that generated the second-harmonic at point A. Obviously, the fundamental wave stays in phase with itself as it propagates through the crystal. Why isn't all the second-harmonic automatically generated in phase with itself?

Dispersion is the answer to that question. Remember that the refractive index of the nonlinear crystal is slightly different for the two wavelengths. Therefore, although the second-harmonic wavelength is exactly half as long as the fundamental wavelength in a vacuum, that is not true inside the crystal. The frequency of the second-harmonic is still exactly twice that of the fundamental, but the wavelength relationship between the two is altered by dispersion. If you are confused at this point, it might be helpful to go back and review the text relating to Figures 3.5 and 3.6.

Figure 13.9 has been redrawn to include the fundamental wave in Figure 13.10. Again, note that the second-harmonic from point A is exactly out of phase with that from point B; also note that the effect of dispersion from Figure 3.6 has been shown in Figure 13.10. A final, important thing to notice about Figure 13.10 is that it shows that the second harmonic is being generated in a polarization orthogonal to the fundamental.

We said we would first try to understand the problem, then look for a solution. Dispersion is the problem. It causes the phase between the fundamental and the second harmonic to shift slightly as the two travel along together inside the nonlinear crystal. Eventually, the phase shift becomes large enough so that new second harmonic light is generated exactly 180° out of phase with the original second harmon-

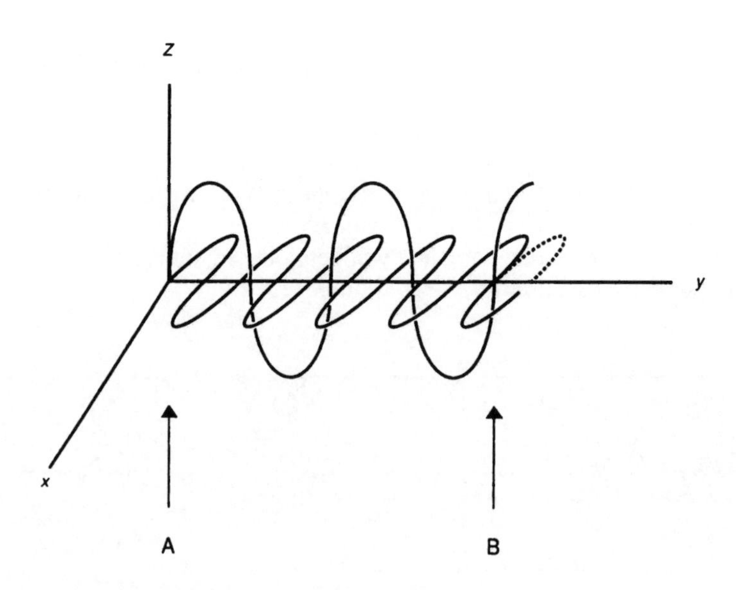

Figure 13.10 The (horizontally polarized) second-harmonic wave generated at point A is exactly out of phase with the wave generated at point B (shown dotted). The fundamental wave that creates the second-harmonic wave is vertically polarized.

ic. (In Figure 13.10, it only takes several wavelengths for this to happen; in reality, dispersion is a small effect and the full 180° phase mismatch typically requires many wavelengths.)

The solution to this problem hinges on the fact that the second harmonic is generated in a polarization orthogonal to the fundamental. Remember that in a birefringent crystal, the two orthogonal polarizations experience different refractive indices. (Review the discussion of birefringence in Chapter 3 if necessary.) The dispersion problem is solved by choosing a nonlinear crystal whose birefringence exactly compensates for its dispersion.

That was an important sentence. Let us put it another way. There are two things that can cause the waves in Figure 13.10 to experience different refractive indices. First, they are different wavelengths, so dispersion will cause them to experience different refractive indices. Second, they are orthogonally polarized, so birefringence will cause them to experience different refractive indices. The trick is to make these two things exactly balance each other: make the index difference due to dispersion be exactly opposite to the index difference due to birefringence. The result is that they both experience the same refractive index.

If dispersion is the problem, birefringence is the solution. But it is not quite that easy because nature does not supply us with crystals that have all the requirements of nonlinear materials, and that just happen to have the right birefringence as well. In practice, what you have to do is find a crystal with the right nonlinear properties, and then fine-tune its birefringence with one of two techniques.

The first technique, temperature tuning, takes advantage of the temperature dependence of some crystals' refractive indices. The nonlinear crystal is placed in an oven (or a cryostat) and heated (cooled) to a temperature at which its birefringence exactly compensates for dispersion.

The second technique, angle tuning, can be used with crystals whose indices are not temperature dependent. The amount of birefringence depends on the angle of propagation through the crystal, so the crystal can be rotated with respect to the incoming beam until the proper birefringence is obtained.

If you understood the baseball/Gouda cheese model of birefringence in Chapter 3, you can use that model to understand the details of how phase matching works in nonlinear crystals. Figure 13.11a shows the baseball and Gouda cheese for light of the fundamental wavelength, and Figure 13.11b shows them for the second harmonic. Remember what these shells are: they are the shapes of the ordinary (the baseball) and extraordinary (the Gouda cheese) Huygens' wavelets expanding from a point source at the center. Before going, on you should convince yourself that you understand why the fundamental wavelets are larger than the second-harmonic wavelets (hint: dispersion). Figure 13.11c shows the second-harmonic wavelets superimposed on the fundamental wavelets.

Here is the crucial point to recognize in Figure 13.11(c). Along the direction defined by the arrow, ordinary light at the fundamental wavelength moves with the same velocity as extraordinary light at the second-harmonic wavelength. And that is precisely the requirement for phase matching. That is, for light traveling through the crystal in the direction defined by the arrow in Figure 13.11(c), the effect of birefringence exactly counteracts the effect of dispersion.

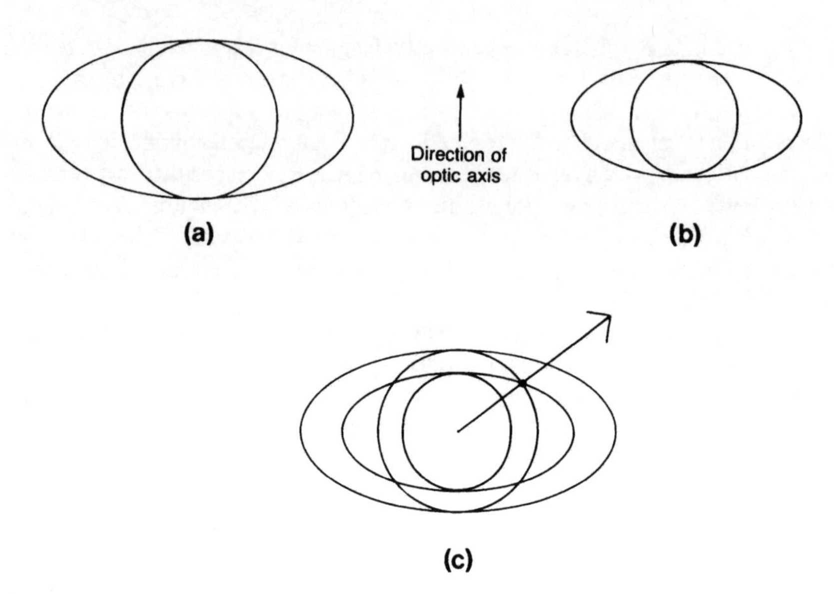

Figure 13.11 The baseball/Gouda cheese wavefronts for ordinary and extraordinary light of fundamental wavelength (a) and harmonic wavelength (b); the two sets of wavefronts superimposed.

That is how angle tuning works in SHG. You draw pictures like Figure 13.11 to figure out what the correct angle is, and then you shine the laser beam into the crystal at that angle. Presto, it is phase matched.

A drawback of angle tuning is beam walkoff. In Chapter 3, you saw that when light propagates at an oblique angle to the optic axis, the extraordinary light walks off from the ordinary. In an angle-tuned nonlinear crystal, the second harmonic walks off from the fundamental, limiting the useful length of the crystal. This effect also reduces the bandwidth over which the interaction can be phase matched and imposes some severe requirements on beam divergence as well. Because angle tuning has such limitations, it is sometimes called "critical" phase matching.

All these problems can be avoided with 90° or "noncritical" phase matching. This special situation is illustrated in Figure 13.12. Phase matching occurs when the light is propagating at a 90° angle to the optic axis, so there is no walkoff. This type of phase matching can be achieved with some crystals, although it is usually necessary to adjust the crystal's temperature to do so.

The foregoing discussion of phase matching has applied only to one type of birefringent phase matching; there is another type that is more difficult to explain intuitively. What has been explained here is so-called Type I phase matching, in which the fundamental light is in the ordinary polarization of the nonlinear crystal and the second-harmonic light is generated in the extraordinary polarization. In Type II phase matching, the fundamental is evenly divided between the ordinary and extraordinary polarizations and the second harmonic is generated in the extraordinary polarization. Unfortunately, there is no nonmathematical explanation of Type II phase matching corresponding to the discussion here of Type I. But it turns out that in

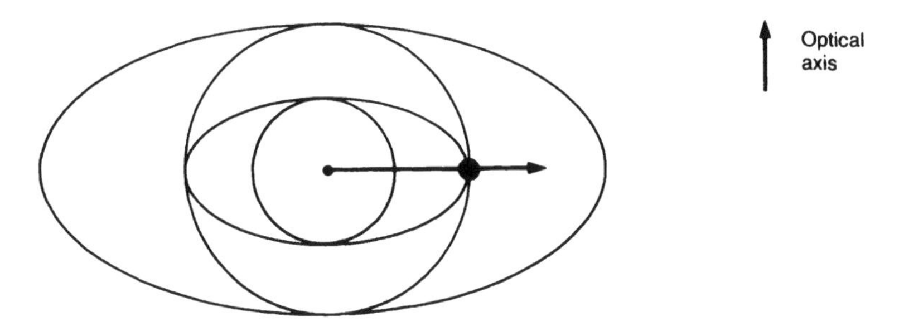

Figure 13.12 Hugens' wavelets for 90° phase matching.

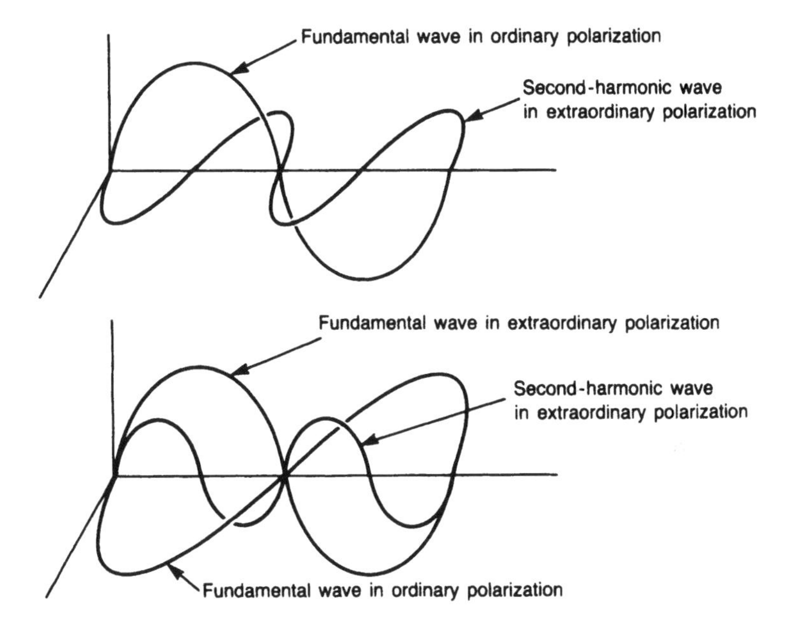

Figure 13.13 For Type 1 phase matching (top), fundamental light is in ordinary polarization and second-harmonic light is in extraordinary polarization. For Type II phase matching (bottom), fundamental light is polarized midway between ordinary and extraordinary directions, and second-harmonic light is polarized as extraordinary.

some cases such as SHG with very high-power, solid-state lasers, Type II phase matching is more efficient than Type I. The orientation of the electric fields for Type I and Type II phase matching are shown in Figure 13.13.

13.4 QUASI-PHASEMATCHING

A completely different approach to avoiding the phase-mismatch problem in second-harmonic generation has evolved in recent years. *Quasi-phasematching,*

(QPM), requires a special nonlinear crystal whose susceptibility reverses periodically. What is "susceptibility"? It describes the crystal's nonlinear response, in this case the second-harmonic fields, created by the fundamental fields. The sign of a nonlinear crystal's susceptibility determines the phase in which the second harmonic is generated. There are two possibilities, and they are illustrated in Figure 13.14. We have arbitrarily named one case "Zig," and the other "Zag."

Two domains of a periodically poled nonlinear crystal are illustrated in Figure 13.15. In Domain 1, the second harmonic is generated in the Zig phase, and in Domain 2 it is generated in the Zag phase. Start with an incoming fundamental wave entering the crystal from the left. At point (a), the second harmonic is generated with the Zig phase.

As the two waves propagate through the crystal, normal dispersion slows the second harmonic with respect to the fundamental. By the time the waves get to (b), the second harmonic lags about a quarter-wave behind. So the second harmonic generated at (b) (which is not shown in the figure) is partially out of phase with the sec-

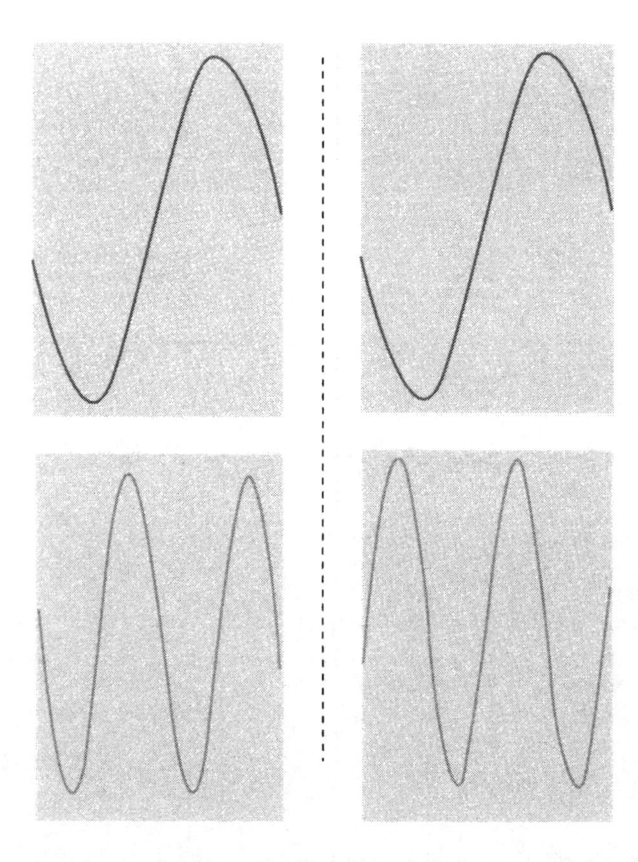

Figure 13.14 The same fundamental wave (top) can generate second-harmonic in the "Zig" phase (bottom left) or n the "Zag" phase (bottom right). Which of these two possibilities happens depends on the sign of the crystal's susceptibility.

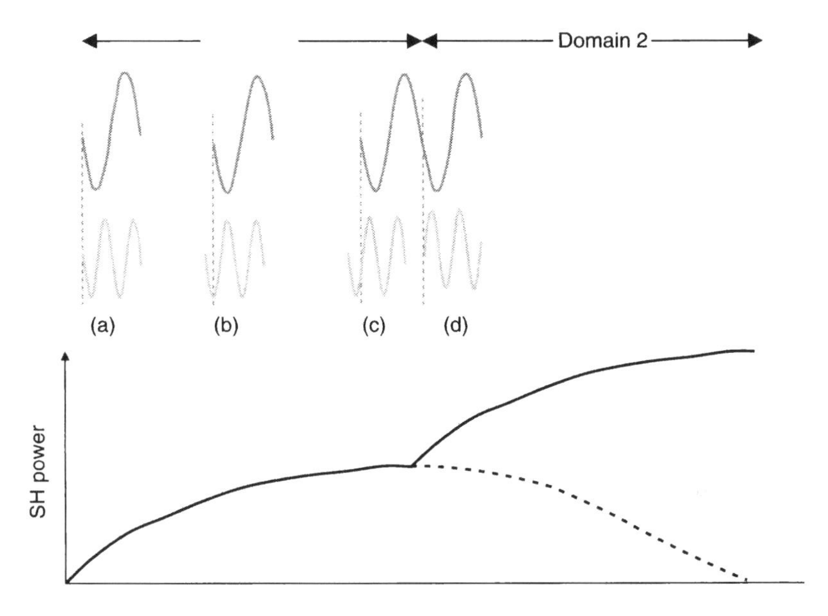

Figure 13.15 At the top of this figure are two domains of a poled nonlinear crystal, and the waves show the relationship between the fundamental and the generated second harmonic at various distances inside the crystal. (A vertical broken line is added for clarity.) The plot at the bottom of the figure shows the accumulated second-harmonic power as the fundamental wave travels across the crystal.

ond harmonic that was generated at (a). The rate at which second-harmonic light is generated, that is, the slope of the second-harmonic power curve at the bottom of the figure, decreases.

By the time the waves got to (c), the second harmonic has fallen a half wave behind, and second harmonic generated at (c) is exactly out of phase with that generated at (a). If nothing changed, the second-harmonic power would diminish as the waves continued farther, as indicated by the broken line in the second-harmonic power plot.

Of course, something does change. The crystal has been poled so the domain change occurs at exactly the right place. As the waves cross into Domain 2, the second harmonic is generated in the Zag phase, so the second harmonic generated at (d), which is shown in the figure, is exactly in phase with that generated at (a).

And now the story repeats itself until the waves reach the right-hand edge of Domain 2, where the poled crystal changes back to the Zig phase. And then it's the same story over and over again—Zig, Zag, Zig, Zag, Zig—until the second harmonic builds up to an acceptable level.

Now let's consider what happens across the entire length of the nonlinear crystal. Figure 13.16 shows three cases. If the crystal in not phasematched at all, the second harmonic repetitively self-destructs, as we explained with the broken line in Figure 13.15. If birefringent phasematching is used, the second harmonic builds up smoothly, scaling as the square of the distance traveled in the crystal.

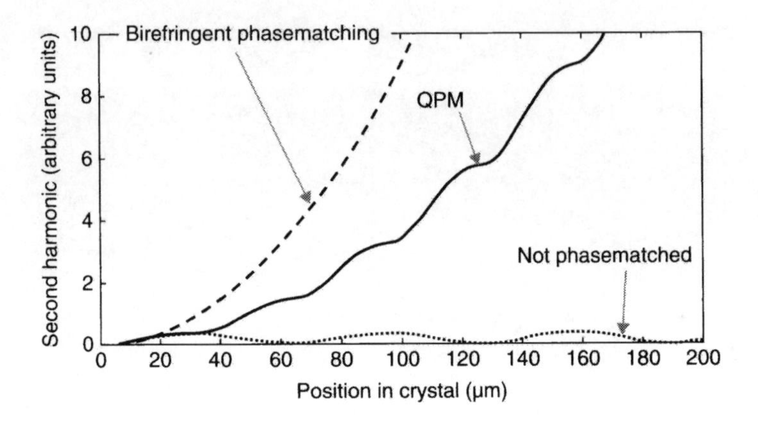

Figure 13.16 The three plots show the second-harmonic power generated with no phasematching, with quasi-phasematching, and with birefringent phasematching.

For quasi-phasematching, the second harmonic builds up haltingly, seemingly reinspired at each domain boundary. But it builds up more slowly than the second harmonic generated with birefringent phasematching. Thus, all other things being equal, birefringent phasematching is clearly superior to quasi-phasematching.

But all other things are often far from equal. For starters, birefringent phasematching requires a birefringent crystal. And not just any birefringent crystal, but one whose birefringence exactly compensates for dispersion between the fundamental and the second harmonic. That is a difficult requirement that does not exist for quasi-phasematching. Quasi-phasematching will work in any material, whether or not it is birefringent.

Birefringent phasematching often requires angle tuning, and as we explained in the previous section, that causes beam walkoff. Again, no such problem exists with quasi-phasematching.

But probably most significantly, quasi-phasematching allows the use of whichever nonlinear coefficient in a given crystal is best for second-harmonic generation. With birefringent phasematching, the beam must be aligned for the correct birefringence, regardless of the nonlinear coefficient. In the case of lithium niobate, one of the more common nonlinear crystals, the nonlinear coefficient available for quasi-phasematching is seven times greater than the one necessary for birefringent phasematching. Since the second-harmonic power scales as the square of the nonlinear coefficient, quasi-phasematching starts out with an advantage of 49 over birefringent phasematching. The net result is, in many cases, that quasi-phasematching will produce more second harmonic than birefringent phasematching.

13.5 INTRACAVITY HARMONIC GENERATION

Normally, a nonlinear crystal is placed in the output beam of a laser as illustrated in Figures 13.4 through 13.7. For intracavity doubling, the crystal is placed between the mirrors, inside the resonator, as shown in Figure 13.17.

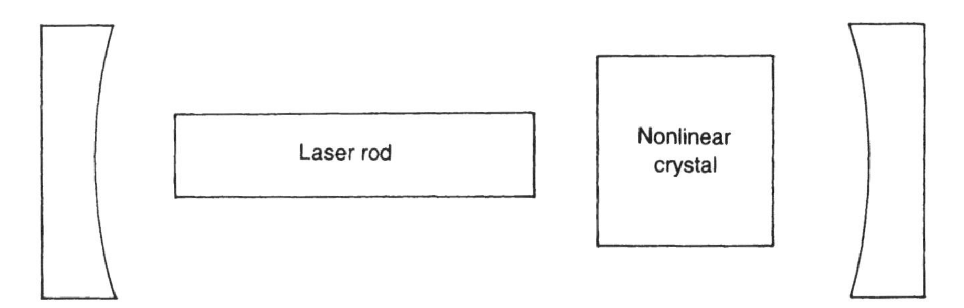

Figure 13.17 Intracavity doubling with the nonlinear crystal inside the resonator.

The efficiency of frequency doubling can be enhanced by placing the nonlinear crystal inside the resonator. As you saw earlier in this chapter, the amount of second-harmonic power generated is proportional to the square of the fundamental power. Pulsed lasers with kilowatts, megawatts, or more peak power can be doubled with reasonable (5% to 50%) efficiency with an external crystal. But continuous-wave lasers with milliwatts or even watts of output are a different story. If all other things were equal, a 1 W cw laser would produce one one-millionth the second-harmonic power of a pulsed laser with 1 KW of peak power.

But the circulating power inside that 1 W laser can be nearly 100 W, or even more if the output mirror is replaced with a maximum reflector. If the nonlinear crystal is placed inside the resonator, the effective conversion efficiency can be much higher.

There are problems with putting a nonlinear crystal inside a laser, though. If the crystal introduces even a small loss—due to imperfect surfaces, for example—it can drastically decrease the circulating power. One percent additional loss can cut the circulating power of some lasers by half, and the advantage of placing the crystal inside the resonator is immediately lost.

A special output mirror is required for internally doubled lasers. The mirror must have maximum reflectivity at the fundamental to keep the circulating power inside the resonator. At the same time, it must transmit all the second harmonic that falls on it.

Another difficulty becomes obvious if you study Figure 13.17. The circulating power is moving in both directions through the nonlinear crystal, so a second-harmonic beam is generated in both directions. One beam passes through the output mirror, but the other beam usually goes to waste. There are ways around this problem, but they entail complex optics and are not straightforward to implement.

13.6 HIGHER HARMONICS

Third-harmonic light can be generated with an arrangement quite similar to the SHG, as shown in Fig 13.18a. But phase-matching requirements make it impossible to generate the third harmonic in a single step in a crystal, so a two-step process is common. As shown in Figure 13.18b, the second harmonic is generated in the first crystal and is then mixed with the fundamental in the second crystal to produce the third harmonic.

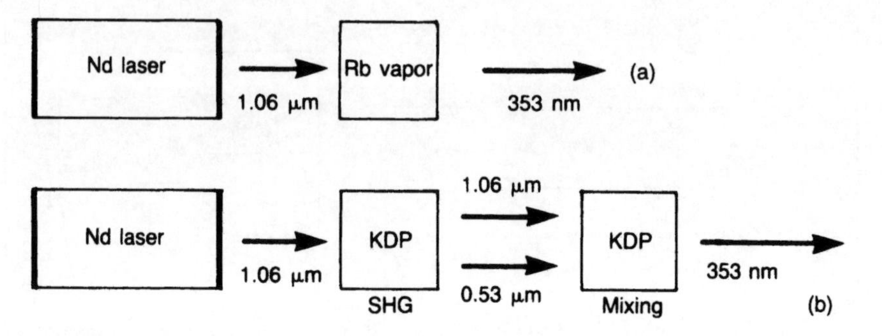

Figure 13.18 (a) Single-step, third-harmonic generation; (b) generation of third-harmonic light by SHG, and mixing.

Fourth, fifth, and higher harmonics can also be generated, but the efficiency of these processes is generally quite low. Even higher harmonics of lasers have been generated experimentally, but the purpose of these experiments was more to investigate the properties of the nonlinear media than to generate useful amounts of short-wavelength light.

13.7 OPTICAL PARAMETRIC OSCILLATION

So far, all the nonlinear interactions we have discussed involve combining the energy of one or more photons into a single, more-energetic (shorter-wavelength) photon. But the process can also work the other way: The energy in one photon can be divided between two new photons. That is what happens in an optical parametric oscillator (OPO), as shown in Figure 13.19.

Conservation of energy must hold between the photons involved; that is,

$$\frac{hc}{\lambda_1} = \frac{hc}{\lambda_2} + \frac{hc}{\lambda_3}$$

An OPO is an oscillator. Unlike the other examples of nonlinear optics we have discussed, an OPO must have mirrors like a laser to form an optical resonator. (But what happens in an OPO is only nonlinear optics; there is no stimulated emission.) Figure 13.20 shows a singly resonant OPO in which only one wavelength, called the

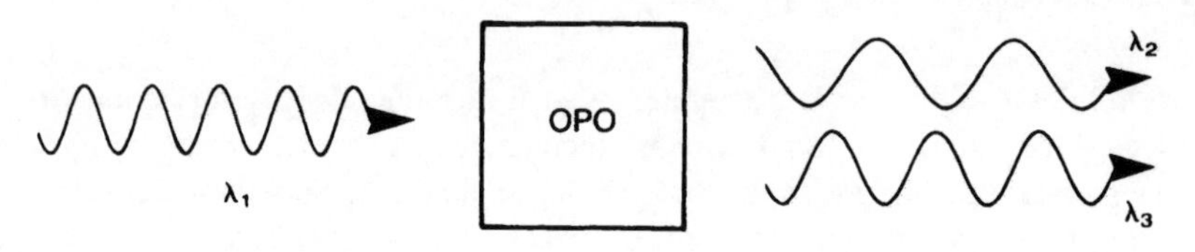

Figure 13.19 An OPO generates two wavelengths from a single-input wavelength.

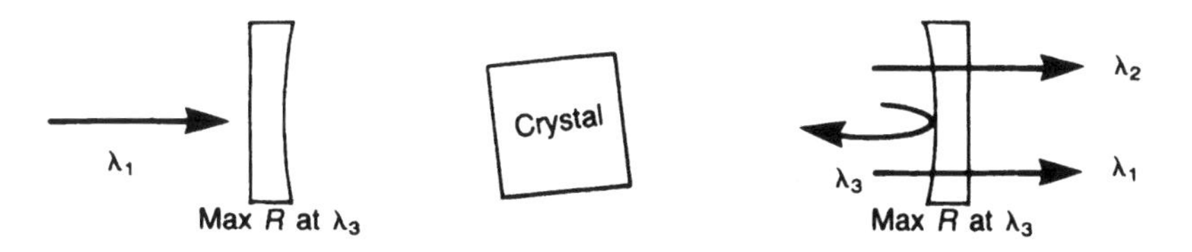

Figure 13.20 In a singly resonant OPO, only one wavelength is reflected from the mirrow.

idler, is reflected by the mirrors. In a doubly resonant OPO, both the pump (λ_1) and the idler (λ_3) are reflected, and only the signal (λ_2) is transmitted.

What determines the output wavelength (λ_2) of the OPO in Figure 13.14? The preceding equation can be solved to yield

$$\lambda_2 = \frac{\lambda_1 \lambda_3}{\lambda_3 - \lambda_1}$$

But λ_2 is not uniquely determined by this equation because λ_3 can have any value. Does this mean that an OPO generates light of many wavelengths?

The answer is yes, but only one wavelength at a time because only one wavelength is phase matched at a time. It is this capability to generate many wavelengths that makes an OPO important: You can tune an OPO to generate the wavelength you want. If you want light whose wavelength is 1.48 μm, for example, you could generate it with an OPO that was pumped by an Nd:YAG laser.

The phase-matched wavelengths of an OPO are changed by adjusting the nonlinear crystal's temperature or angle. Figure 13.21 shows the tuning curve of an angle-tuned, Nd:YAG-pumped $LiNbO_3$ OPO. If the propagation angle (with respect to the crystal's optic axis) is 46°, then the signal and idler wavelengths will be about 1.6 and 3.1 μm, respectively.

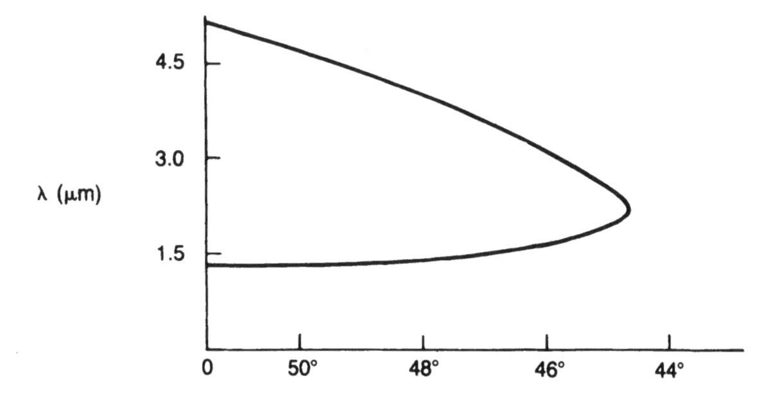

Figure 13.21 Turning curve for an $LiNbO_3$ OPO.

13.8 RAMAN LASERS

In the nonlinear effects discussed so far, energy is swapped back and forth between photons. Two or more photons are "welded" together to produce a single photon that has all the energy of the original photons or, as we saw in the previous section, the energy of a single photon is divided between two photons. But there are other nonlinear effects by which photons exchange energy with the medium through which they are travelling. Raman scattering is one of those effects.

In Chapter 6, we talked about energy levels and how a photon can be absorbed to boost an atom or molecule to a higher energy level. In those examples, the photon's energy was completely absorbed into the atom or molecule, and the atom or molecule was left in an excited state. A short time later, the energy might emerge in the form of another photon.

In Raman scattering, a photon is scattered from a molecule but leaves part of its energy behind. The molecule is left in an excited state, from which it may eventually decay with a radiative or nonradiative transition.* The two processes are shown schematically in Figure 13.22

Raman spectroscopy has been an important tool for investigating vibrational levels of molecules since the early 1900s. An intense light source—originally a high-power mercury lamp but more recently a laser—illuminates a sample, and the frequencies of the scattered light reveal the energies of the sample's vibrational levels.

But we are interested in *stimulated* Raman scattering, which occurs when the incoming photons' intensity exceeds a certain threshold. At that point, the scattered

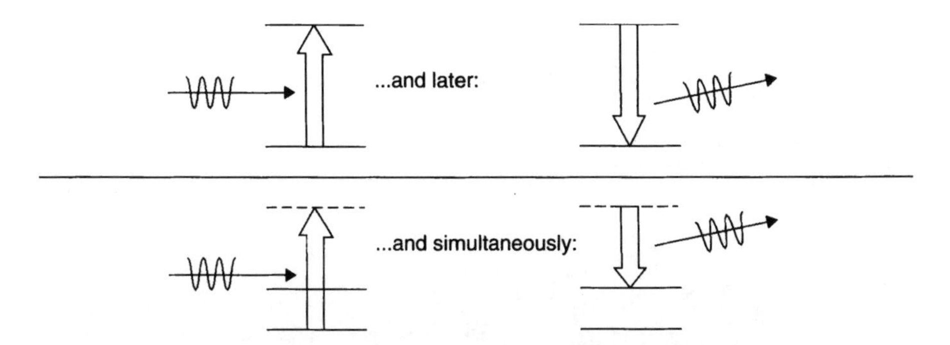

Figure 13.22 In normal optical absorption and emission (top), a photon is absorbed in a molecule and leaves the molecule in an excited state. Later, the molecule may relax back to its ground state by emitting a photon. In Raman scattering (bottom), a photon is scattered off a molecule and is shifted to a shorter wavelength, leaving part of its energy behind in the molecule. Later, the excited molecule may relax back to its ground state by emitting a photon. Note the broken line in the bottom diagrams; there is no a real energy level corresponding to the energy of the incoming photon.

*The process described here is the Stokes component of Raman scattering, In anti-Stokes Raman scattering, an incoming photon is scattered from an excited molecule, and leaves with its own energy plus the energy of the formerly excited molecule.

beam becomes directional, monochromatic, and coherent. If you put mirrors around the medium to feed the scattered light back on itself, you can create a Raman laser. And unlike almost all the other lasers discussed in this book, a Raman laser operates without a population inversion.

Raman lasers are useful for converting the outputs of other lasers to new wavelengths. The amount of wavelength shift depends on the Raman medium, and by selecting the appropriate medium designers can achieve a wide variety of wavelengths. A commercially important example is the pump laser used in erbium-doped fiber amplifiers.* The amplifier needs pump radiation at about 1480 nm, which can be obtained by pumping a Raman laser with the 1.06 μm output of a Nd:YAG laser.

QUESTIONS

1. What is the second-harmonic wavelength of a chromium–ruby laser? What is the third-harmonic wavelength of an Nd:YAG laser?
2. The text explained that an internally doubled laser generates two second-harmonic beams, one in each direction. In Figure 13.17, the beam generated traveling away from the output mirror is wasted. (It is absorbed in the laser rod.) How would you design a resonator to combine the two harmonic beams into a single output beam?
3. Suppose the second crystal in Figure 13.5 were 3 cm long instead of 2 cm. How much second-harmonic light would be generated?
4. The modelocked laser sketched below produces 1 W of light at its fundamental wavelength, and 10 mW of second-harmonic light is normally generated by the nonlinear crystal. But if the modelocking modulator is driven harder, it will produce shorter pulses and 15 mW of second-harmonic light will be produced. Calculate the fractional change in pulse duration caused by driving the modulator harder.

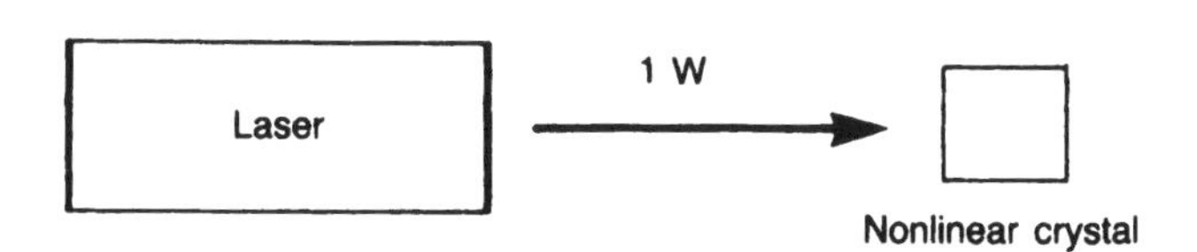

5. Can a 750 nm output (signal) be obtained from an Nd:YAG-pumped OPO? Why? Suppose the Nd:YAG laser is frequency doubled to produce 530 nm of light. Can this light be used to generate a 750 nm signal from the OPO? What is the idler wavelength in this case?
6. An OPO is degenerate when its signal and idler are the same wavelength. What is this wavelength for a normal Nd:YAG-pumped OPO? For an OPO pumped by a frequency-doubled Nd:YAG?

*Erbium-doped fiber amplifiers, or EDFAs, amplify the optical signals in fiberoptic transmission lines. EDFAs are discussed in Section 15.1 and in the introduction to Chapter 20.

Chapter 14

Semiconductor Lasers

Semiconductor lasers, or diode lasers as they are often called, are by far the most ubiquitous of all lasers. They are present everywhere from consumer products such as CD and DVD players and laser pointers to the innards of the global telecommunication network. A decade into the twenty-first century, diode-laser sales are in the range of a billion units per year. Inexpensive red and infrared chips used in consumer products account for most of the units sold, but dollar sales are dominated by high-performance diode lasers designed to transmit high-speed telecommunication signals, and high-power diodes used for materials working or pumping other lasers.

The same basic principles underlie all semiconductor diode lasers, although the details differ widely. A very simple semiconductor laser is shown in Figure 14.1, and you can readily see why diode lasers are so much smaller, lighter, and more rugged than other lasers. What is shown in Figure 14.1 is the entire package; the laser itself is just the small crystalline block with the wire bonded to its top. The laser is formed by the junction of two dissimilar types of semiconductor, and the light emerges from the edge of the block, coming directly from the junction.

14.1 SEMICONDUCTOR PHYSICS

To understand how the semiconductor laser in Figure 14.1 works, you need to learn a little bit about what semiconductors are, and what conductors and insulators are. An electric current is composed of moving electrons, and any material that has "loose" electrons can be a conductor. When a voltage is applied across a copper wire, the "loose" electrons in the wire flow from one end to the other. In an insulator, on the other hand, all the electrons are held tightly in place. When a voltage is applied to an insulator, no current flows because there are no "loose" electrons to move from one place to another. (Extremely high voltages can pull electrons out of atoms in an insulator, like lightning does in air, but normally only low voltages are applied to semiconductors.)

In Chapter 6, we discussed the cloud of electrons that surrounds the nucleus of an atom. These electrons are arranged in *shells,* and it is usually the electrons in the outermost shell that are most likely to shake loose and become carriers of an electrical current. In a metal such as copper, the force holding the outermost electrons to the nucleus is weak, so copper is a good electrical conductor. In a material such as silicon, however, the electrons in the outermost shell are more tightly bound to the

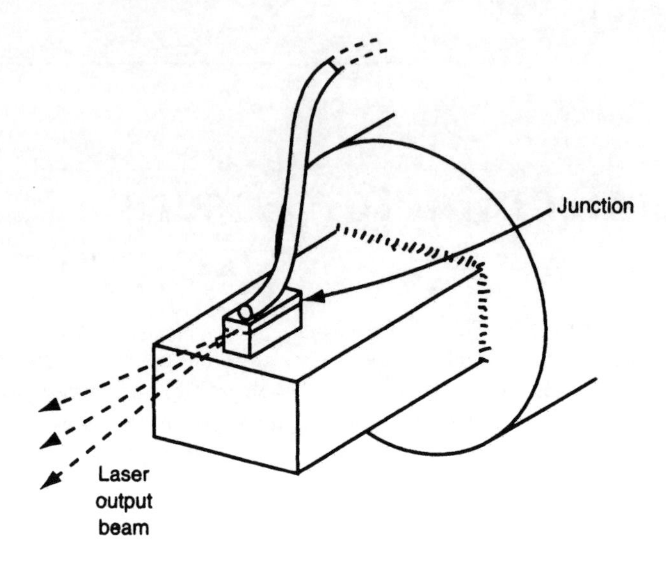

Figure 14.1 The semiconductor laser itself is the tiny, rectangular crystal on the mounting post.

atoms in the crystal, and cannot move as easily through the material as those in copper. Thus, silicon is a semiconductor.

Silicon has four electrons in its outermost shell, and they all form bonds with adjacent atoms in the silicon crystal, as illustrated in Figure 14.2a. An arsenic atom is more or less the same size as a silicon atom, but it has five electrons in its outermost shell. If a silicon crystal is doped with arsenic, that is, a small amount of arsenic is added as an impurity to the silicon crystal as it is grown, then arsenic atoms will replace silicon atoms at some locations in the crystal. This is shown schematically in Figure 14.2b.

Four of the arsenic atom's outermost-shell electrons form bonds with adjacent atoms. But the fifth electron is not tightly attached to the nucleus and can, under the right conditions, become available to conduct a current. The arsenic thereby enhances the conductivity of the silicon semiconductor. The material is called n-type semiconductor, because the current carriers are negatively charged electrons.

But there is a variation on this theme. Instead of doping the silicon with arsenic, which has five outermost-shell electrons, you could dope it with boron, an atom that has only three outermost-shell electrons. Then there would be a *hole* where the fourth electron is supposed to be. And it turns out that such holes can conduct electricity just the way electrons can. This is called p-type semiconductor, because the current carriers are positively charged holes. Of course, you understand that when a hole moves from point A to point B, what really happens is that an electron moves from point B to point A. However, the physics is easier if you think of holes being the things that move, rather than electrons.

Thus, there are two fundamental types of semiconductors. In an n-doped semiconductor, there are extra electrons that are the charge carriers, and in a p-doped semiconductor, there are extra holes that are the charge carriers. Put the two types of

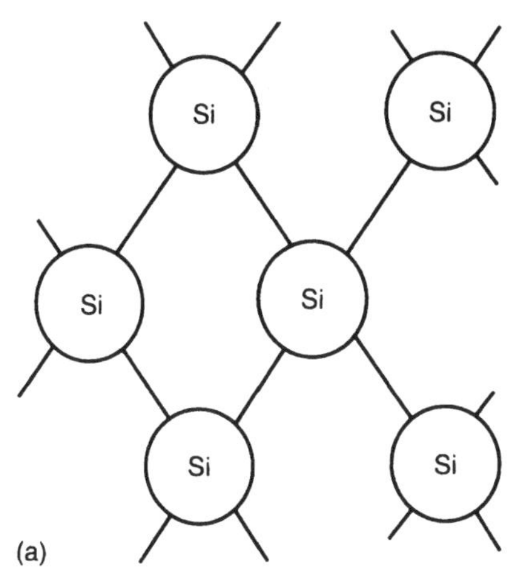

Figure 14.2a In a normal silicon crystal, all four outermost shell electrons are used, forming bonds with neighbors. There are no "loose" electrons, and normal silicon is not an electrical conductor.

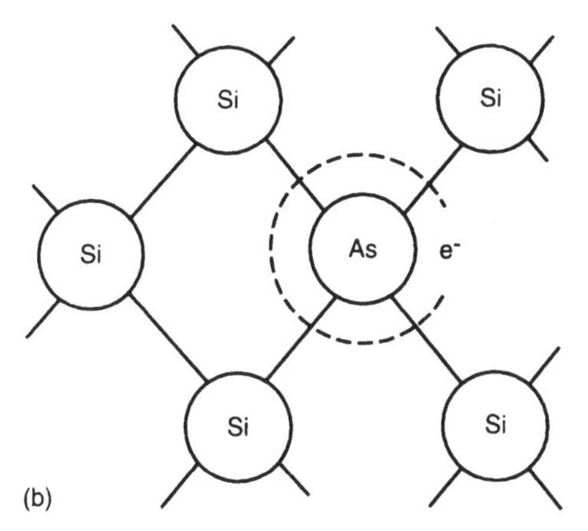

Figure 14.2b When a silicon crystal is doped with arsenic (which has live electrons in the outermost shell), the extra electron is loose and turns the crystal into a semiconductor.

semiconductor together and you get a diode, an electronic device with two terminals, one attached to the n-type and one to the p-type semiconductor. Under the right conditions, a diode can generate laser light, becoming a diode laser.

In a diode laser, laser action occurs at the junction* of an n- and a p-doped material. Let us consider what happens when you bring an n-doped material into contact with a p-doped material, as illustrated in Figure 14.3. The p-doped material has loosely bound positive charges (holes), and the n-doped material has loosely bound

*It is very common to have an *intrinsic*, or undoped, layer of material between the n- and p-type semiconductors at the junction. In this case, the electron-hole recombination occurs in the intrinsic material.

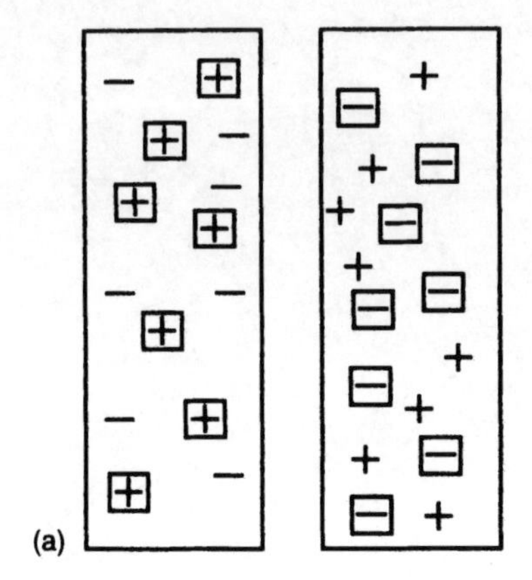

Figure 14.3a Both p-type (right) and n-type (left) semiconductors are electrically neutral, but p-type semiconductors have mobile positive carriers (holes), and n-type semiconductors have mobile negative carriers (electrons). In this figure, mobile carriers are represented as charges without boxes around them.

negative charges (electrons). Hence, initially, some of the p-doped material's holes move across the junction and combine with some of the n-doped material's electrons, and vice versa.

It is very important to understand what happens when an electron and a hole combine. Remember that a hole is a place where there is supposed to be an electron, so when an electron and hole combine, the electron literally falls into the hole, fill-

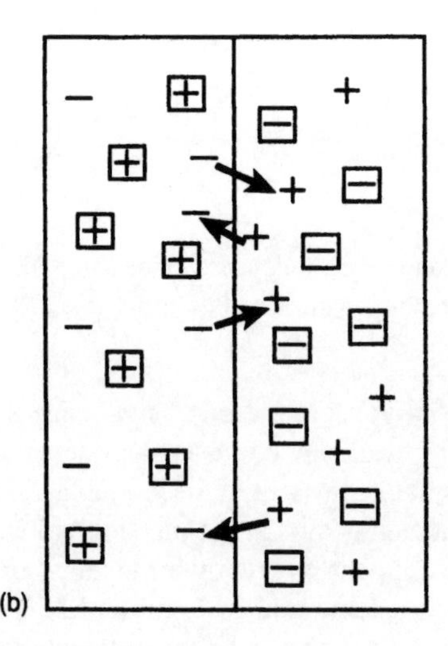

Figure 14.3b When a junction is formed between an n- and a p-type semiconductor, some of the holes and electrons move across the junction to combine with opposite charges in the other material.

ing it. And since the electron loses energy as it falls, at least part of that energy can be liberated as a photon. That is where the photons come from in a diode laser; they are created when an electron loses energy by combining with a hole.

When an electron and a hole combine at a diode junction, the energy is not always released as a photon. Sometimes it is released in the form of heat. In particular, in the silicon crystal we discussed earlier, the energy takes the form of heat, because the total momentum of the electron and hole changes when they combine. Momentum must be conserved during the process, and that is not possible if the energy is emitted as a photon. The bottom line is that silicon diodes do not make good lasers, and that turns out to be a serious problem. As electronic gadgets—cell phones, music players, and so forth—become smaller and smaller, the repulsive force among electrons starts to set limits on how small gadgets can get. Because photons do not interact with each other, engineers and designers would like to use photons to do many of the things electrons do today. But silicon is the universal building material of miniature electronics, so it would be enormously helpful if lasers and other optical components could be made from silicon. To date, the best compromise seems to be a "hybrid" silicon laser, comprised of a lasing semiconductor bonded to a silicon chip, so the silicon is the laser resonator for photons generated in the other semiconductor.

There are semiconductors, such as gallium arsenide, in which the total momentum of an electron and a hole do not change when they combine. So in GaAs, the energy released by the electron falling into a hole can be emitted as light. Semiconductor lasers are made of GaAs and related materials, not silicon.

Let us go back to the junction we have created between an n- and a p-doped material. We saw that initially a number of charge carriers moved across the junction to combine with opposite charges in the other material. But soon, as shown in Figure 14.4a, there is an excess of positive charge on the n-doped material side, and an excess of negative charge on the p-doped material side. This excess charge stops more carriers from moving across the junction.

Thus, when you form a junction between an n- and a p-type semiconductor, you get an initial shower of photons as mobile charges move across the junction and combine with opposite charges, but the show is quickly over. To prolong it, you must inject additional mobile charges, as shown in Figure 14.4b (i.e., you pass an electric current through the junction). This creates a steady-state condition in which the junction continuously emits photons as a current flows through the diode.

In fact, the device illustrated in Figure 14.4b is a light-emitting diode (LED). To turn this LED into a diode laser, you have to recall from Chapter 6 the two things required for laser action: a population inversion and feedback. To create a population inversion, you must have more population at an upper level than a lower level. This usually requires a very high current density, so lots of electrons are falling into holes at the junction. And to provide feedback, you can place a reflector at the ends of the junction. Here nature is kind, because semiconductors such as silicon and GaAs have high refractive indexes, so they reflect much of the light generated in the junction back into the crystal, which in many cases gives adequate feedback for a diode laser. If more feedback is needed, a reflective layer can be coated onto the edge of the laser, or other types of reflectors can be added, as described below.

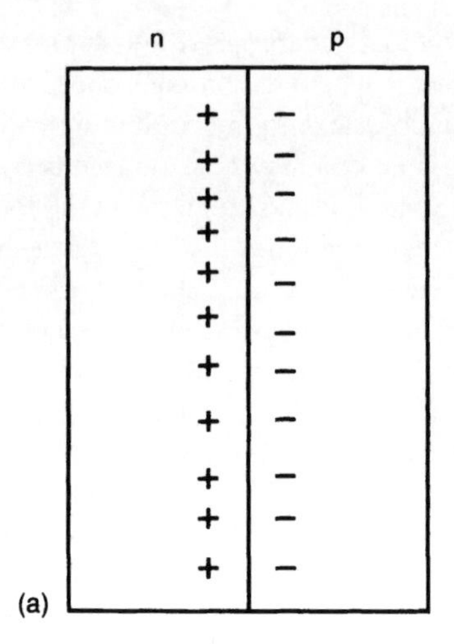

Figure 14.4a After an initial cascade of charge carriers across the junction, a potential barrier builds up and prevents further charge migration.

Interestingly, diode lasers are generally more efficient than LEDs, because an LED depends on spontaneous emission, and, therefore, a given atom must wait for the spontaneous lifetime before it can emit its photon. While it is waiting, the energy can be lost to competing, nonradiative decay mechanisms. In a diode laser, on the other hand, the photons are created by stimulated emission, so the energy does not have to wait as long in the upper laser level.

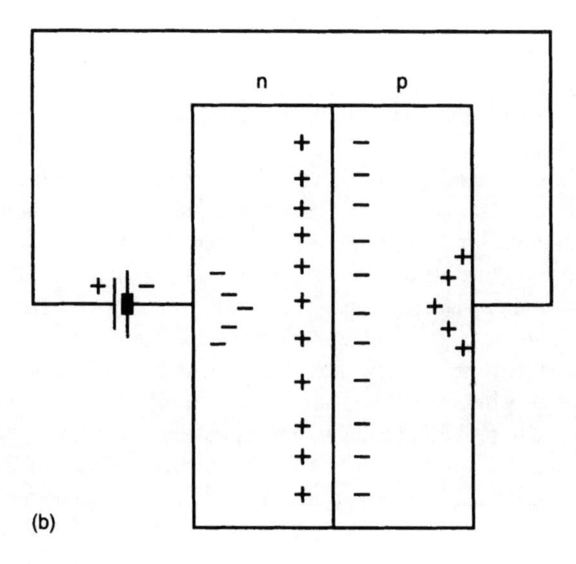

Figure 14.4b By injecting carriers into the junction as shown here, a steady current across the junction can be maintained.

14.2 MODERN DIODE LASERS

The first diode lasers, developed in the early 1960s, looked something like the device illustrated in Figure 14.1. But semiconductor lasers have evolved enormously since then. The laser in Figure 14.1 requires very high current flow to maintain a population inversion, and the heat generated by the steady-state current would quickly destroy the device.

To reduce the current and heat while maintaining a population inversion, modern diode lasers pack the stimulated emission into a small region. Hence, the current density is still great enough to maintain a population inversion, but the total current does not overheat the laser. There are two approaches to increasing the density of stimulated emission: increasing the density of charge carriers and increasing the density of intracavity optical power.

Both techniques invoke the sophisticated semiconductor fabrication techniques that have evolved over the past 30 years. These techniques allow the complex structures illustrated in Figures 14.5 through 14.9 to be grown, literally one molecule at a time, from the basic raw materials. Today, methods such as molecular-beam epitaxy and metal–organic chemical vapor deposition allow the creation of semiconductor structures that are only several atoms thick.

One way to increase the density of charge carriers is to use a stripe electrode, as illustrated in Figure 14.5. Instead of injecting the current over a wide area of the diode's surface, current is injected only along a narrow stripe, resulting in a much higher concentration of charge carriers inside the diode.

The laser in Figure 14.5 confines the current to a small region with its stripe electrode. This is called current confinement in the plane of the junction. Look closely and you can see several layers of different composition within the diode laser, not just a single pair of layers as in Figure 14.1. Such designs are called double heterostructure* lasers because they include layers with different composition above and below the active layer (also called the junction layer) where electrons and holes recombine and emit light.

The double heterostructure does two very important things that concentrate both current carriers and light in the active layer.

First, the compositions of the layers are chosen so current carriers have slightly more energy in the layers above and below the active layer than in the active layer itself. This tends to hold the carriers longer in the low-energy trap, making them more likely to recombine and emit light.

Second, the layers above and below the active layer also have a higher refractive index than the active layer, so they serve to guide light in the plane of the active layer. That means that photons are totally reflected off the interface between the materials and, therefore, are confined to the active layer, with thickness d which serves as a waveguide for light. (Total internal reflection is discussed in Chapter 3.)

*Heterostructure means a structure containing two or more semiconductor layers with different composition. They are sometimes called heterojunctions because there is a junction between the two layers, but we avoid this because the word *junction* also is used to denote the layer in which electrons and holes recombine and emit light.

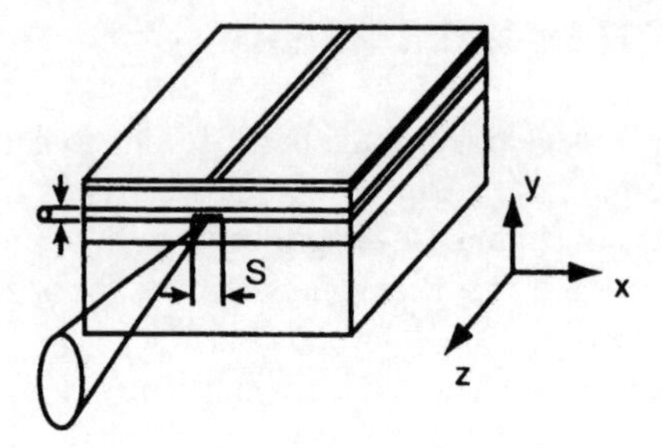

Figure 14.5 This laser has a stripe electrode on the top to restrict the current flow to a narrow region, and a double heterostructure (see text) to confine the photons.

A more sophisticated method to further increase the density of charge carriers involves the quantum-mechanical nature of carriers in very thin layers. If the dimension *d* in Figure 14.5 is very tiny (a few tens of nanometers), quantum mechanical effects become important and the charge carriers are trapped in the thin region. Such super-thin structures are called quantum wells and are used widely in modern semiconductor lasers. Multiple quantum wells can be stacked within the active layer, increasing output power.

Double heterostructures and quantum wells confine photons vertically, within the active region, but the photons are still free to spread out sideways. Index-guiding structures, like the one shown in Figure 14.6, stop this spreading. Here, the laser has been fabricated with a low-index material on both sides of the active regions, as well as above and below the active region. Now the photons cannot spread out in any direction. Note that Figure 14.7 also illustrates the stripe electrode, a few micrometers thick; a narrow quantum well; and a double heterostructure.

14.3 DIODE LASER BANDWIDTH

We have talked about confining the generated photons in three-dimensional space, but confining them in wavelength or frequency can also be beneficial. Chapter 10 discussed some of the techniques for reducing a laser's bandwidth, and the same principles apply to diode lasers. In particular, the best way to reduce a laser's bandwidth is usually to reduce the bandwidth of the laser's feedback.

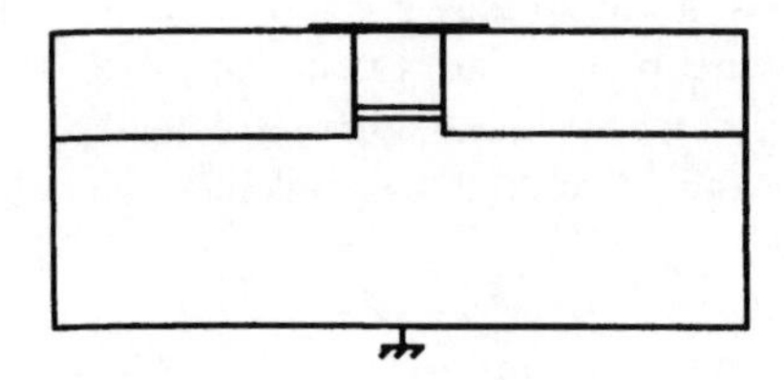

Figure 14.6 The lower refractive index of InP in the blocking regions prevents the laser photons from spreading outside the micrometer-wide active region.

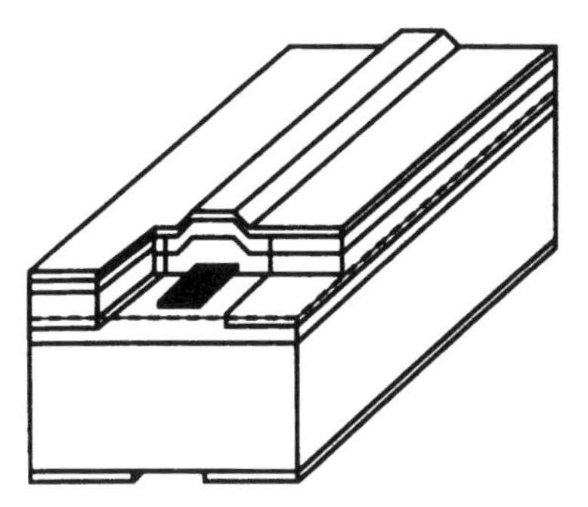

Figure 14.7 Frequency-selective feedback is provided by the optical gratings along the length of this distributed-feedback laser.

As you saw in Chapters 8 and 9, a simple resonant cavity allows a laser to emit light at a number of different modes, each with its own wavelength. Many semiconductors operate like this, emitting light across a range of a few nanometers.

Suitable optics can limit emission to a much narrower range. In Chapter 10 we discussed using gratings for this purpose, and small, built-in gratings are often used with semiconductor lasers. Figure 14.7 shows a semiconductor laser with a grating fabricated into the structure. Such lasers are sometimes called distributed-feedback (DFB) lasers because the feedback, or reflectivity, is distributed over the length of the grating, rather than occurring all at once at a mirror. The wavelength that is fed back is determined by the period of the grating. Usually, a DFB laser has a grating fabricated into the entire length of the laser. A variation referred to as a distributed Bragg reflector (DBR) has a distinct grating fabricated into the substrate on each side of the active area.

An alternative approach is to the *external cavity* diode laser, in which one or both mirrors are separate from the semiconductor, and the end faces of the diode laser are coated so they do not reflect light back into the semiconductor. Then the laser diode is placed in a separate resonator like those described in Chapter 10. Such lasers are capable of very narrow bandwidth and good frequency stability, and also can be made tunable in wavelength.

14.4 WAVELENGTH OF DIODE LASERS

The wavelength of a diode laser depends on the amount of energy released in the form of a photon when an electron and a hole combine. In the argot of semiconductor physics, this is the semiconductor's bandgap energy. For semiconductors composed of two elements (binary), the bandgap is fixed at a given value. The bandgap energy of GaAs, for example, corresponds to a photon wavelength of 904 nm. Other binary semiconductors have different bandgap energies.

Virtually all practical diode lasers are made from compounds like gallium arsenide, called III–V semiconductors because they contain elements from Groups III and V of the periodic table. Mixing three (ternary) or four (quaternary) of these elements together gives a compound with a bandgap energy that depends on the relative concentration of elements. The ternary semiconductor gallium aluminum arsenide (GaAlAs) can have a bandgap energy corresponding to photons from 620 to 900 nm. The exact wavelength of a GaAlAs laser depends on the relative amounts of gallium, aluminum, and arsenic in the crystal, with more aluminum giving shorter wavelengths. Although GaAlAs compounds can be made with bandgaps corresponding to wavelengths of 620 to 900 nm in theory, practical lasers are limited to a smaller band, from about 750 to about 880 nm.

Diode lasers can also be made from other semiconductor compounds. The most important of these are listed in Table 14-1. Indium–gallium–nitride (InGaN) diode lasers have larger bandgaps and emit at blue and violet wavelengths at the short end of the visible spectrum; they are used in BluRay disc players. Visible-red diode lasers are made from AlGalnP. Diode lasers used in telecommunications systems are made from InGaAsP, which emits from about 1.1 to 1.65 μm. Replacing phosphorous with antimony yields InGaAsSb compounds emitting longer infrared wavelengths.

You can create ternary and quaternary semiconductors with a wider range of bandgap energies, corresponding to photons from the mid infrared (several microns) to the ultraviolet. However, several factors limit diode lasers to operating in a more limited range. Some compounds tend to release recombination energy as heat rather than photons, like silicon. Some compounds are hard to grow without crystalline defects that lead to failure when they conduct the high currents needed to drive a diode laser.

Moreover, some semiconductor compositions are difficult to fabricate on available substrates. As you saw earlier, a diode laser is a complex assemblage of many layers of semiconductors. Fabricating it requires matching the spacing of its crystalline lattice with that of a substrate that can be grown in large volumes, typically a binary semiconductor such as GaAs, InP, or GaN. Buffer layers can be deposited on top of the substrate to accommodate small mismatches, but some compositions cannot be grown with the crystal quality needed for laser operation.

Developers are working on these problems, particularly on extending the operation of InGaN lasers to green wavelengths, but, at this writing, that technology remains in the laboratory.

TABLE 14-1 Major diode laser compounds and wavelengths

Material	Wavelength range	Ranges and applications
GaInN/GaN	375–480 nm	UV, violet, and blue; BluRay
AlGaInP/GaAs	620–680 nm	Red; laser pointers, DVD
GaAlAs/GaAs	750–880 nm	Near-IR; CD-ROM, laser printers
GaAs/GaAs (pure)	904 nm	Near-IR pulsed
InGaAs/GaAs	905–1050 nm	Near-IR laser pumping; direct diode applications
InGaAsP/InP	1100–1650 nm	Near-IR, Fiber-optic systems
InGaAsSb/GaSb	1.85–4 μm	Mid-infrared

14.5 DIODE ARRAYS AND STACKS

Narrow-stripe diode lasers are limited to modest milliwatt-class power by the small volume of the active layer. Power can be increased somewhat by using broad single stripes, but making high-power diode lasers requires combining the output of many laser stripes. Although the resulting diode-laser devices have poor beam quality, they generate high powers with high efficiency, and are widely used for pumping other lasers and for applications such as heat treating, which do not require a concentrated high-quality beam.

Diode lasers can be assembled in a number of ways. The first step, illustrated in Figure 14.8, is a one-dimensional or linear array of several laser stripes emitting from the edge of a monolithic chip, delivering several times more power than a single laser stripe. Such a monolithic, one-dimensional array is often called a *bar*.

The next step is to group several bars together. They may be placed side by side as well as stacked atop each other, as shown in Figure 14.9. Output power of such edge-emitting stacks can exceed a kilowatt, continuous wave. Stacks operating at such power levels include provisions for active cooling, such as microchannels fabricated in the structure.

Commercially, these high-power lasers may be designed for quasi-CW (or QCW) power output as well as for CW operation. In Chapter 11, we talked about peak power and average power, and the QCW power of a diode laser is analogous to the peak power. Such lasers are sometimes run at relatively low duty cycles—10% or less—to avoid overheating. So a quasi-CW laser rated at 1000 W with a 10% duty cycle would have an average power of 100 W.

14.6 VERTICAL CAVITY, SURFACE-EMITTING LASERS

One of the most fundamental problems with the diode lasers we have discussed so far is the highly divergent, elliptical beam inherent in their emission from the edge of a chip. In Chapter 5, we explained that the divergence of a laser beam is inversely proportional to the beam size at the source—the smaller the source, the larger the divergence. But the laser illustrated in Figure 14.5, for example, is an extremely small light source, perhaps 1 μm wide. And it is much smaller in the vertical direction than in the horizontal. Thus, the beam that emerges from the device in Figure 14.5 diverges much

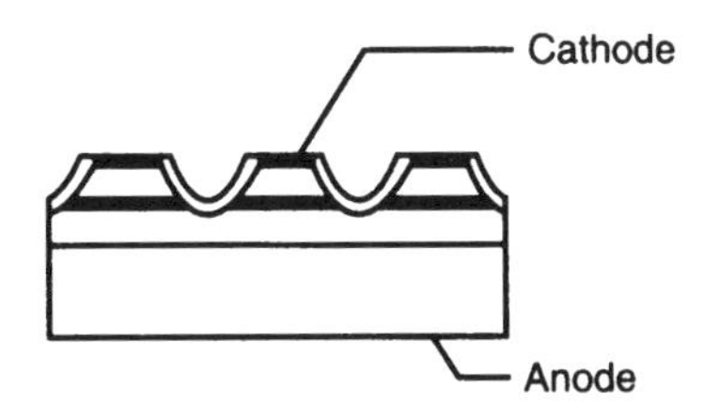

Figure 14.8 A one-dimensional diode-laser array. The original layered structure was grown on the substrate, and then the individual lasers were etched out chemically.

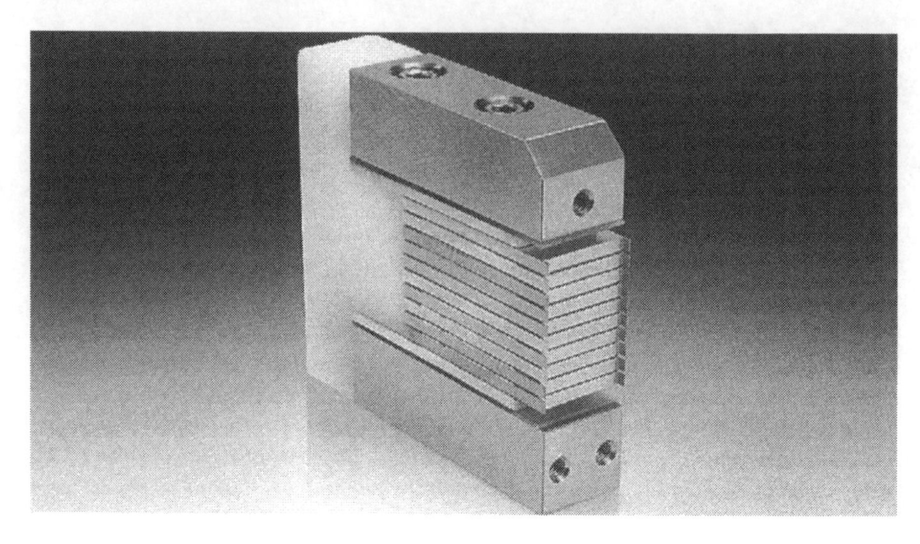

Figure 14.9 These stacks of one-dimensional arrays are capable of very high power output. (Photo courtesy of Coherent Inc.)

more rapidly in the vertical direction than in the horizontal. The vertical divergence is typically tens of degrees, and the horizontal divergence is several degrees.

The highly divergent, elliptical beam can be corrected, to an extent, with a cylindrical lens, but the inherent problem of a small, elliptical source can never be completely rectified. And the diode lasers we have discussed so far have other fundamental limitations. Although they are extremely small, their resonators are still hundreds of micrometers in length—long enough to support multiple longitudinal modes. Unless the laser's bandwidth is artificially reduced (e.g., by fabricating a DFB structure), mode hopping among these modes produces an instability in both the amplitude and frequency of the laser's output.

Because the output beam emerges from the edge of the cleaved crystal, and the crystal is not cleaved until the end of the manufacturing process, it is not possible to test the devices optically during manufacture. This limitation tends to drive up the price of manufacturing.

The vertical cavity, surface-emitting laser (VCSEL, pronounced like *vixel*) avoids these shortcomings. In conventional diode lasers, as we have already discussed, the cavity, or resonator, is in the horizontal plane. In a vertical-cavity laser, the cavity is along the vertical direction. The two approaches are illustrated conceptually in Figure 14.10. In the vertical-cavity laser, the mirrors are located above and below the population inversion, instead of on either side. The horizontal-cavity laser is an edge emitter, whereas the vertical-cavity laser is a surface emitter.

You can readily see the advantage of such a design. For one thing, it immediately eliminates the problem of a divergent, elliptical beam caused by a small, irregular emitting surface of an edge emitter. The emitting area of a surface emitter is round and many times larger.

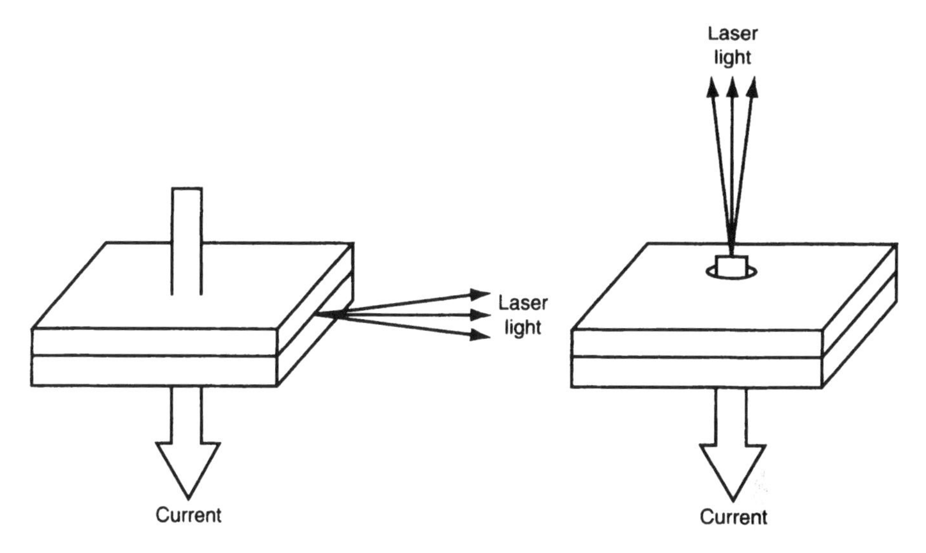

Figure 14.10 In a VCSEL (right) the resonator is vertical, and the light emerges from the surface of the laser rather than from the edge of a conventional diode laser (left).

Figure 14.10 makes it clear that the resonator of a VCSEL is shorter than a conventional laser's. In fact, it is so short that the spacing between longitudinal modes is too great for more than one mode to oscillate. (Longitudinal modes are spaced by $c/2\ell$, as discussed in Chapter 9.) Thus, the mode-hopping instability of conventional lasers is eliminated.

Moreover, the manufacturing difficulties are reduced because it is possible to test all the VCSELs on a wafer before dicing the wafer into chips. Very high densities, tens of millions of diodes per wafer, can be achieved, further driving down the cost of individual diodes. Figure 14.11 shows the detailed structure of a VCSEL. The mirrors are thin layers of semiconductors, fabricated to the proper thickness for constructive interference of the light reflected from different interfaces.

VCSELs inherently are lower-power lasers than edge-emitters because the resonant cavity in a VCSEL includes a much smaller volume where recombination takes place than does an edge-emitter, which typically has a cavity hundreds of micrometers long. However, VCSELs also have a very low threshold for laser operation, making them attractive for low-power laser applications because they need little current.

14.7 OPTICALLY PUMPED SEMICONDUCTOR LASERS

A variation of the VCSEL structure is the optically pumped semiconductor laser illustrated in Figure 14.12. Here, the P-type mirror stack that serves as the output coupler in Figure 14.11 has been replaced by a standard laser mirror that is some distance from the rest of the diode structure. This configuration, with the output mirror

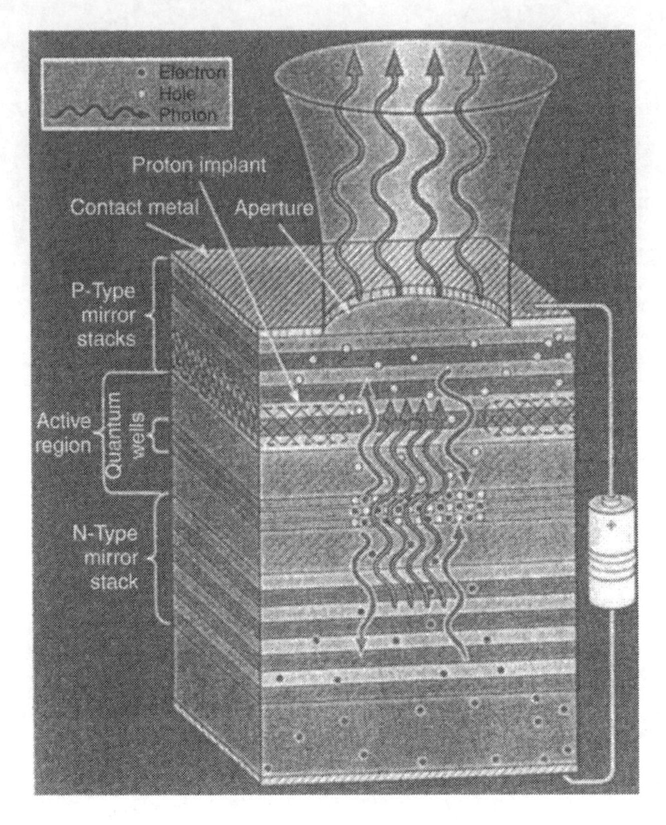

Figure 14.11 A structural view of a VCSEL. The mirrors for the vertical resonator are formed by layers of semiconductors above and below the active region.

physically separate, is often called a VECSEL, where the acronym stands for Vertical *external* cavity, surface-emitting laser. (The VECSEL name applies whether or not the laser is optically pumped. And you are right; the cavity in Figure 14.12 is horizontal, not vertical. Such is the ambiguity of the laser vernacular.)

The quantum wells in Figure 14.12 are extremely thin layers in which electrons and holes are trapped to increase their concentration and enhance their interaction. But the important thing to notice about Figure 14.12 is that the laser is pumped by the photons from another laser, not the electrical current that creates the population inversion of all the other lasers in this chapter.

There are several advantages to optical pumping, in comparison to the usual technique of electrical pumping. The intracavity beam in an electrically pumped laser can often be larger, diluting the intracavity power density and alleviating heating issues. Forcing an electrical current through a conventional diode is also a significant source of heat, one that is eliminated in an optically pumped laser. Avoiding the need to guide current through the laser eases design requirements, allowing better laser performance.

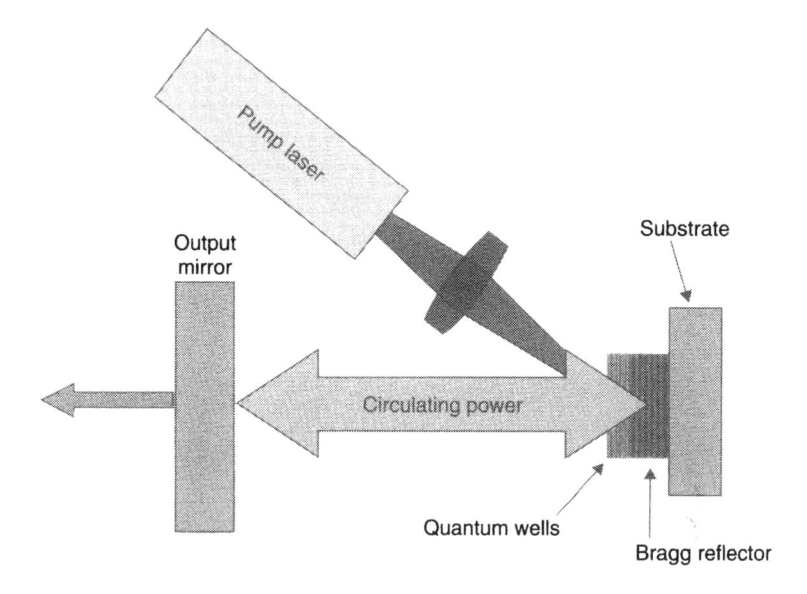

Figure 14.12 An optically pumped semiconductor laser is pumped by another laser rather than an electrical current.

Optically pumped semiconductor lasers (also called OPSLs) in some ways resemble the optically pumped, thin-disk lasers that will be discussed in the next chapter. Those thin disks are excellent in many applications, but there are also advantages for the optically pumped semiconductor lasers. Their absorption of pump light is more efficient for one thing, and they can produce a broader range of output wavelengths than thin-disk lasers.

14.8 QUANTUM CASCADE LASERS

The final variation on the semiconductor laser theme is not a diode, but is powered by electrical current flowing through it. Called the quantum cascade laser, it requires a bit more explanation.

A quantum cascade laser produces light by making electrons flow through a series of quantum wells in which they emit light. The quantum wells are sandwiched between barrier layers engineered to trap electrons in the quantum wells. When a voltage is applied across the quantum wells, electrons fall into the first one, where they occupy an energy level well above the ground state, an excited state in which they can be stimulated to emit a photon. Then the electron tunnels through the barrier layer and drops into the next quantum well, where it also drops into an excited state in which it can be stimulated to emit a photon before it tunnels into the next quantum well where the process repeats. This is the cascade in its name, with a single electron emitting many photons.

The transitions the electrons make in the quantum well are within the conduction band, not from the conduction band into the valence band, so they are lower in energy than in a conventional diode. Thus quantum cascade lasers emit at longer wavelengths than conventional diode lasers, from about 3 to 60 μm or in the terahertz band at even longer wavelengths. They operate best at wavelengths from about 4 to 14 μm, a range of mid-infrared wavelengths where few other lasers are available.

Chapter 15

Solid-State Lasers

The term *solid-state* has a special meaning in the laser world. The semiconductor lasers described in the previous chapter are considered solid-state devices in the world of electronics, but not in laser jargon. The term *solid-state laser* usually is reserved for the optically pumped lasers discussed in this chapter and also covered in Chapter 20, and (often) for the fiber lasers covered in Chapter 16.

The crucial difference is in how population inversions are produced. As you learned in Chapter 14, semiconductor lasers are tiny lasers in which flow of an electrical current through the laser directly creates the population inversion. In this chapter, we address larger lasers in which light from an external source excites the active medium to create a population inversion. That light source can be anything: another laser, a lamp, or (in a few rare instances) even the sun. If the light source is a laser, it might be a semiconductor laser, or an array of semiconductor lasers like the ones depicted in Figure 14.9, or it might be a gas laser or even another solid-state laser. Strictly speaking, these lasers should be called optically pumped solid-state lasers because their energy comes from the external light, but in common use they are simply called solid-state lasers. Semiconductor diode lasers, which could be considered electrically pumped solid-state lasers, are called semiconductor or diode lasers. Illogical as it may be, we adopt that usage throughout the remainder of the text.

Interestingly, optical pumping is not limited to solid-state lasers. Occasionally, gas lasers are optically pumped, and optical pumping is very common in liquid dye lasers. Certain semiconductor lasers are pumped optically rather than electrically. But the discussion in this chapter is limited to solid-state lasers in which the solid is neither a semiconductor nor a fiber. The next chapter covers solid-state lasers in which the active medium is an optical fiber. Chapter 20 discusses optically pumped solid-state lasers that have a very broad tuning range. These optically pumped lasers have developed their own niche of applications and are characterized by unique laser properties.

15.1 SOLID-STATE LASER MATERIALS

The active medium of a solid-state laser consists of a passive host material, typically a crystal, and the active ion, and it is these components that give the laser its name. An Nd:YAG laser, a very common solid-state laser, for example, consists of a

Introduction to Laser Technology, Fourth edition. C. Breck Hitz, J. Ewing, and J. Hecht

crystal of YAG with a small amount of Nd added as an impurity. YAG is short for yttrium aluminum garnet and has the chemical formula $Y_3Al_5O_{12}$. The crystals are usually grown with the so-called Czochralski technique. The Nd^{3+} ions, in the form of Nd_2O_3, are mixed into the host material, YAG. Then the two materials are melted together at quite high temperature, and a single crystal is drawn out of the temperature-controlled melt. The population inversion is created in the Nd ion (Nd^{3+}), and this ion generates the photon of laser light. Typically, the Nd ion concentration is about 0.1–1% as dense as the metal ions of the host crystal or glass.

Table 15.1 lists the most important optically pumped solid-state laser systems. It lists the lasing ion and the solid-state host in which the lasing ion is a constituent, along with wavelength for lasing, pumping techniques used, and major applications. In each case, the active laser species is a metal ion, usually a triply charged positive ion, that is a minority constituent in a crystalline or glass host. The solid-state host often is a simple rod cut from a synthetically grown crystal (or glass), but it also may be a slab, thin disk, or optical fiber. Table 15.1 also gives the main wavelengths for lasing, but in many cases the ion/host combination also can produce other wavelengths. In addition, Table 15.1 gives the pump sources for the various ion/host systems. Diode pumping offers higher wall-plug efficiency, but lamp pumping is more cost-effective for most (nonmilitary) high-power or high-energy applications.

The first solid-state laser, indeed the first laser ever, was the ruby laser. A flashlamp produced a population inversion in a cylindrical rod cut from a large single crystal of Al_2O_3 containing small amounts of chromium in the form of Cr_2O_3. After the ruby laser was demonstrated, other ion/host combinations were investigated. The ones that have been successful in commercial applications are listed in Table 15.1.

The names used for solid-state lasers can be confusing to those encountering solid-state lasers for the first time. Sometimes the laser is named for its host material. For example, an Nd:YAG laser is sometimes referred to simply as a YAG laser. This is ambiguous, of course, because YAG also can be the host for other lasing ions, as shown in Table 15.1. By contrast, the ruby laser, which uses synthetic crystals, is named for the gem found in nature. However, the same host (Al_2O_3) doped with titanium becomes a Ti:sapphire laser.

The Nd:YAG laser is the most prevalent of today's solid-state lasers, though other hosts and other lasing ions such as Yb^{3+} are finding greater use, especially with diode pumping. These lasers can be found in machine shops welding heavy metals, in the surgical suite performing delicate surgery, in the research laboratory making precise spectroscopic measurements, and on a satellite that orbited Mars measuring the detailed topography of that distant planet. Nd:YAG lasers are engineered to have the distinct characteristics needed for each of these diverse applications. The welding laser, for example, is likely to be a large, lamp-pumped laser, whereas the surgical system is likely to be a smaller, diode-pumped system that is frequency doubled (Chapter 13) to produce green light. The spectroscopic laser could also be a diode-pumped laser, perhaps utilizing some of the techniques discussed in Chapter 10 to reduce its bandwidth. And the Mars-orbiting laser was Q-switched (Chapter 11) to produce short pulses for distance measurement.

The Nd ion can be added as an impurity to a glass matrix, producing the ion/host combination of an Nd:glass laser. Although the same triply ionized Nd ion does the

Table 15.1 Major solid-state lasers and applications

Laser ion	Host material	Main laser λ	Pumping technique	Applications
Nd^{3+}	YAG YLF YVO_4 Glass	1.064 μm 1.055 and 1.047 1.064 1.055	Diode arrays near 808 nm or lamps	Workhorse laser for industrial, scientific, medical, and military uses. Glass can be cast in large sizes ideal for large pulse energy such as laser fusion.
Yb^{3+}	YAG Glass fiber	1.03 μm 1.02 to 1.06	Diode arrays near 940 nm (YAG) or 970 nm for Glass fiber	High-power industrial uses. Ultrafast industrial and medical for fiber
Er^{3+}	Glass Glass fiber	1.54 μm	Lamps or diodes for bulk glass lasers; diodes for fiber laser	Eye-safe rangefinders (military). Fiber-based systems for telecom
Er^{3+}	YAG	2.94 μm	Diodes	Medical and dental; wavelength ideal due to water absorption in tissue
Ho^{3+}	YAG, YLF	Several lines near 2 μm	Diodes at 794 nm with transfer from Tm^{3+} or lamps	Bone surgery at powers ~100 W
Tm^{3+}	YAG, YLF	Several lines near 2 μm	Diodes at 794 nm	Not as useful as Ho^{3+} at comparable wavelengths
Cr^{3+}	Al_2O_3 (ruby) Alexandrite	694 nm (ruby) 700–800 nm Alexandrite		Pulsed holography; scientific. Tunable laser in alexandrite
Ti^{3+}	Al_2O_3 (Ti:sapphire)	650 nm–950 nm (tunable)	Nd:YAG + SHG cw argon ion	Scientific, ultrafast

lasing in both Nd:YAG and Nd:glass, the two lasers have little else in common. Even the wavelengths of the two lasers are slightly different (because the electric fields in the two hosts shift the energy levels differently). Nd:glass is much better at storing energy than Nd:YAG, and glass can be made in much larger sizes than crystals. For example, a typical Nd:YAG laser rod is 6 mm in diameter and cut from a synthetic single crystal that is perhaps 10 cm in diameter, whereas Nd:glass plates that are more than 40 cm in diameter are used in high-energy, Q-switched applications such as laser fusion. Large Nd:glass lasers can produce pulses of 100 J and greater, whereas Nd:YAG is typically limited to about 1 J from a single, Q-switched oscillator.

On the other hand, glass has much lower thermal conductivity than YAG, so YAG lasers are preferred for high-average-power applications. Thermal conductivity is important because it is the only way to remove waste heat from the interior of the laser material. A coolant then removes the heat from the rod surface. If the heat is not removed fast enough from the interior of the crystal, the host overheats and distorts the optical quality of the crystal, and with enough heat input can fracture the crystal. A single Nd:YAG diode-pumped oscillator, containing several Nd:YAG rods lined up in series, can produce an average output power of 1 kW or more.

Note that the potential for combining some of the size advantages of glass lasers with the excellent thermal conductivity of YAG can now be obtained with ceramic materials that are made by compressing under high temperature a powder of nanocrystals of doped YAG into large plates or other shapes. The resultant ceramic material can have lower bulk absorption than rods or slabs cut from Czochralski-grown single crystals. These new ceramic materials are finding their way into a variety of applications and novel laser material configurations, at both low power and high power. Another material advance, of interest in certain applications, is the bonding of doped and undoped materials together; for example, Nd:YAG and YAG. This finds use in end-pumped lasers, discussed below, as a means to ameliorate some of the aberrations produced by nonuniform heating of a material.

Let us look at some other ion/host combinations listed in Table 15.1. The triply charged erbium ion (Er^{3+}) can be added to YAG, glass, and other hosts. In YAG, the lasing wavelength of 2.9 μm is strongly absorbed by tissue, making the Er^{3+}:YAG laser useful in medical applications like bone or dental surgery. In glass, Er^{3+} can lase at 1.5 μm, an eye-safe wavelength with minimal loss in optical fibers. These lasers are the core elements of the fiberoptic telecom industry. Erbium-doped fiber amplifiers (EDFAs), glass fibers with Er^{3+} ions doped into the glass, are excited by semiconductor lasers. They have been widely deployed in the global telecommunication system, in which they directly boost signals in the 1.55 μm band carried on fiberoptic cables. Before EDFAs were integrated into the fiberoptic-based telephone system, the optical signal in the fiber cables had to be converted to an electrical signal for amplification, and then back to optical for further transmission. Fiber lasers are discussed in detail in Chapter 16.

The thulium (Tm^{3+}), erbium (Er^{3+}), and holmium (Ho^{3+}) ions can lase at various wavelengths between 2 and 3 μm. These wavelengths are strongly absorbed by the water in soft tissue and bone, so these lasers have found extensive use in surgery and other medical applications.

Sometimes, a second ion may be added to the host because it absorbs the pump light from lamps more efficiently than the lasing ion. For example, a Ho:YAG laser might be codoped with chromium (Cr^{3+}) or thulium (Tm^{3+}) ions, which strongly absorb the pump light and transfer the energy to the lasing Ho^{3+} ion. Likewise, ytterbium is added to increase absorption in erbium-fiber lasers and to efficiently transfer energy to the lasing erbium ion.

Another host listed in the Table 15.1, YLF (short for lithium yttrium fluoride, $LiYF_4$), has thermal and optical properties attractive for some applications. YLF is naturally birefringent, which offsets the effects of thermal stress (described later in this chapter) and avoids depolarization of beams observed in YAG and other hosts.

For completeness, Table 15.1 lists two unusual solid-state lasers: chromium (Cr^{3+}) in the host alexandrite, called the alexandrite laser, and the titanium (Ti^{3+}) ion in Al_2O_3, the Ti:sapphire laser. These lasers have a very wide bandwidth and can produce subpicosecond pulses or be tuned in wavelength across their output bands. These two unusual lasers, and others like them, are discussed in more detail in Chapter 20.

15.2 DIODE-PUMPED SOLID-STATE LASERS

Historically, lamp-pumped lasers found practical application decades before their diode-pumped counterparts could be commercialized, but we will reverse that order in our discussion. Diode pumping is intrinsically simpler than lamp pumping, and is becoming commercially dominant at low and moderate powers. Although powerful semiconductor diode arrays are costly, diode pumping is often preferable because it produces much less heat in the laser medium and has a significantly greater overall efficiency. The higher efficiency pays off in lower cost of power supplies and a simpler thermal design of the laser itself.

There is ambiguity in this terminology. The diode in diode pumping is a laser diode, never a light-emitting diode. As you will see later, the emission of a light-emitting diode is insufficiently "bright"—intense and focusable—to pump a solid-state laser well. Diode lasers are more intense and can be focused to produce sufficient gain.

We will explain diode pumping using the Nd:YAG laser as an example. The same approach can be applied to most of the lasers in Table 15.1, as long as a diode can be found to match a suitable absorption line of the active ion. Nd^{3+} is a four-level system, as you can see by comparing the energy-level diagram in Figure 15.1 with Figure 7.10. Each of the four levels is designated with the appropriate initials on the left side of Figure 15.1. These levels are not the only energy levels in the Nd ion, and some of the other levels are also shown. Some of these levels can serve as the lower laser level of secondary lasing transitions, at 1338 and 946 nm. (You can force the laser to lase at one of the secondary transitions by maximizing the feedback at that wavelength and minimizing the feedback at competing laser wavelengths.)

The ground state of Nd^{3+} is actually a set of five energy levels, all of which have some population at room temperature. The 1064 nm laser transition terminates on a

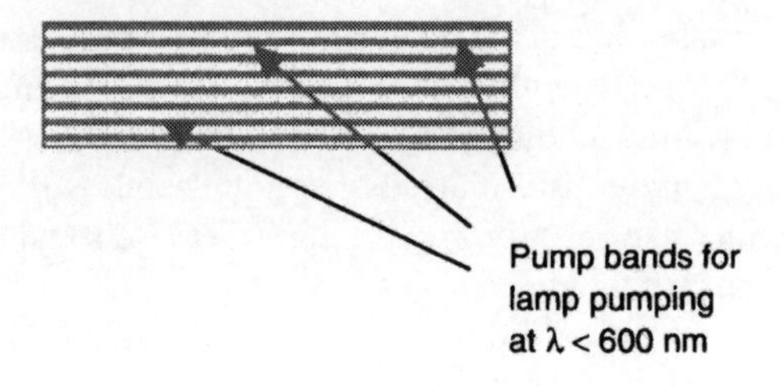

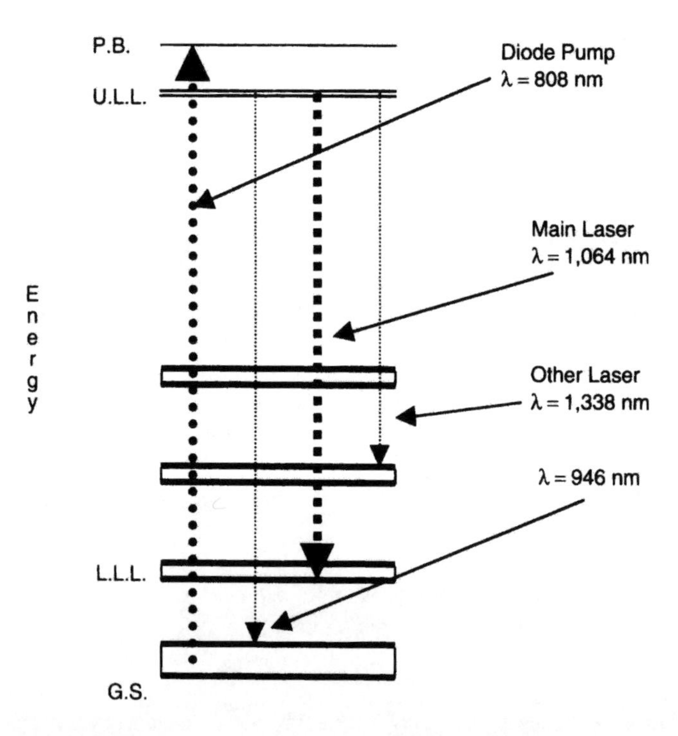

Figure 15.1 Energy levels for the Nd laser. Nd is a four-level laser system, as discussed in Chapter 7. The ground state is actually a set of five energy levels, all of which have some population at room temperature. The 1064-nm laser transition terminates in a state that has negligible population, even at the elevated rod temperatures of lamp pumping. In diode pumping, Nd ions are excited from the ground state (G.S.) to the pump band (P.B.) by obsorbing diode-pump photons at 808 nm. (The diode emits continuously, or in pulses of several hundred microseconds.) The population inversion is created between the upper laser level (U.L.L.) and the lower laser level (L.L.L.). If a population inversion is crreated between other energy levels, and if the resonator provides feedback at the appropriate wavelegth, lasing can occur on these lines as well.

state that has negligible population, even at the elevated rod temperatures of lamp pumping. In diode pumping, Nd ions are excited from the ground state (G.S.) to the pump band (P.B.) by absorbing diode-pump photons at 808 nm. (The diode emits continuously, or in pulses of several hundred microseconds.) The population inversion is created between the upper laser level (U.L.L.) and the lower laser level (L.L.L.). If a population inversion is created between other energy levels, and if the resonator provides feedback at the appropriate wavelength, lasing can occur on these lines as well.

Figure 15.1 also illustrates the advantage of diode pumping over lamp pumping. Because the diode laser's output is concentrated at the narrow wavelength at about 808 nm, it can be efficiently absorbed by the narrow energy level identified as the diode pump band (P.B.). Most of the broadband emission of a lamp falls outside this narrow absorption band and is wasted. The wasted light due to the poor overlap of the lamp spectrum with the absorbing ion spectrum leads to low efficiency. In addition, the primary excitation channel with a lamp is to levels higher than the upper laser level, as shown on the top of Figure 15.1, leading to lower quantum efficiency, the ratio of laser photon energy to the average pump photon energy. The result is that the power into a lamp also needs to be larger than the power into a diode or diode array, resulting in a more difficult cooling problem. The energy that the Nd ion releases in relaxing from the lamp pump band to the upper laser level is not only wasted, it seriously impairs the laser by heating the laser material. The laser material is further heated by all the wavelengths from the pump lamp that are not absorbed by the pump band. We will discuss these adverse thermal effects later in this chapter.

In Figure 15.2, we show typical efficiencies for each major step in the process of converting energy from electricity to light for the case of Nd:YAG. A typical pump diode laser converts about 50% of the input electrical energy to optical output at 808 nm directed toward the laser rod, chip, or slab. Depending on the design of the coupling optics from the diodes to the laser material, roughly 80% of the power can be put in the volume that the laser cavity will extract. This efficiency is a product of the

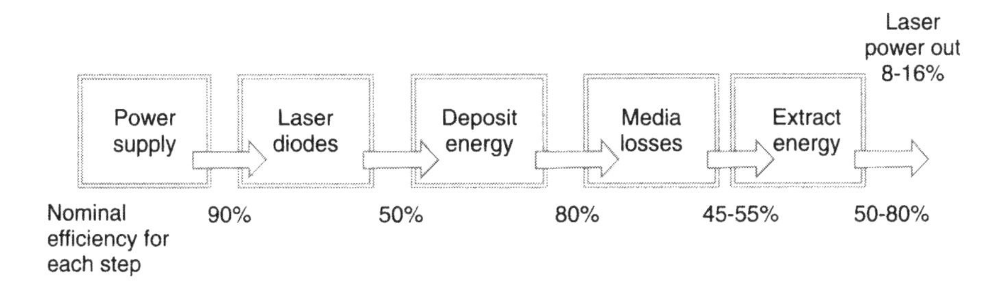

Figure 15.2 The major steps in transforming the energy from electricity to laser light are shown along with approximate efficiency for each step. This figure does not account for other power-consuming processes such as power needed to circulate coolant. The overall efficiency of a diode-pumped Nd:YAG laser can readily exceed 8%. Diode pumped Yb:YAG lasers, discussed later, can have efficiency of over 20%.

transport efficiency and the fill factor. (Lasers designed to emit a TEM_{00} mode typically do not have a "fill factor" efficiency this high.) Once the photons have arrived in the desired volumetric region, there are losses due to the quantum efficiency (76% for 808 nm diodes pumping Nd) and fluorescence decay and incomplete absorption of pump light. That leaves only about 50% of the stimulated-emission power available for the cavity to extract, dependent on the actual fraction of light absorbed. The extraction efficiency depends on the flux in the cavity, the cavity reflectivity, and the losses on mirrors and other surfaces in the cavity. Extraction efficiency typically is 50% to 80%. Put all these losses together, and the output power typically is 8% to 16% of the input electrical power. Efficiency may be lower if the pumped volume is considerably bigger than the extracted volume, if the pump light is incompletely absorbed, or if other losses occur. Power needed to move coolant or run the control system is not included in this analysis but is typically not a large decrement to efficiency.

Figure 15.3 is a similar diagram for a lamp-pumped Nd:YAG laser, which typically has overall efficiency around 1 or 2%. There are several reasons for the greater efficiency of diode pumping:

- As we explained above, the quantum efficiency of lamp pumping is significantly lower.
- The narrow bandwidth of diodes means that much more of their output can be spectrally matched to the laser medium's absorption.
- The collimated output from diodes can be coupled directly into the laser medium, without the lossy reflections of a lamp pump cavity.

This greater efficiency of diode pumping, along with lower-voltage operation and the incredibly long lifetime of diodes, has driven the expansion of diode pumping during the past two decades. As the cost of diodes decreases further in the future, this technology will be applied more widely, and at higher powers, to lasers in the commercial world.

Many other ion/host combinations listed in Table 15.1 can also be diode pumped. Nd in other hosts behaves similarly to the Nd:YAG case discussed previously. The Yb:YAG laser has the YAG crystalline host doped with Yb^{3+} ions and is an ideal candidate for high-power applications because the small separation between its

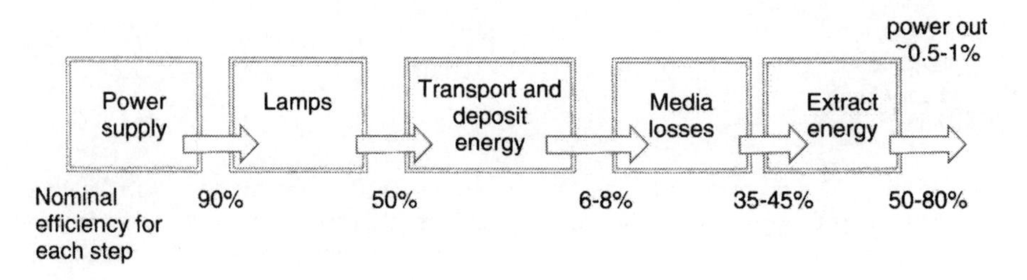

Figure 15.3 Overall efficiency of a lamp-pumped Nd:YAG laser rarely exceeds 2%.

pump band and laser wavelength gives very high quantum efficiency. Yb^{3+} doped into glass is the underlying material for high-power, 1 μm fiber lasers used for welding and similar applications. Yb-fiber lasers have high pump efficiency as well as high quantum efficiency, and their wall-plug efficiency can exceed 25%.

Ho lasers, with output in the 2–3 μm region, are often co-doped with Tm. The Tm ions efficiently absorb the pumping radiation from a laser diode at 785 nm and transfer the energy to the lasing Ho ion. The laser diodes that supply this 785 nm pump lamp are similar to those that pump Nd:YAG at 808 nm. Er:glass lasers can likewise be codoped with ytterbium (Yb), which absorbs pump light at 970 nm and transfers the energy to the erbium.

However, not all ion/host combinations have pump bands or internal kinetics that are suitable for diode pumping. In the case of Ti:sapphire, both factors play a role. Its pump band is in the green, where current diode lasers do not operate efficiently. Moreover, Ti^{3+} has a radiative lifetime of only about 4 μs in sapphire, so producing a population inversion requires pump intensities much higher than current short-wavelength diode lasers can generate. Instead, Ti:sapphire is normally pumped by green lasers such as frequency-doubled Nd:YAG.

15.2.1 Diode-Pumping Geometry

Several pump geometries are possible for diode-pumped solid-state lasers. We will start with side pumping of a traditional simple rod of solid-state material, a case analogous to most lamp-pumped lasers. Figure 15.4 shows a typical end view of the laser rod and the diodes and side pumping optics. In this case, the laser rod is surrounded by pump diodes that excite the laser crystal through a cooling jacket; the diode power supplies and cooling equipment are not shown. Water flows between the jacket and the rod, removing the heat that gets to the surface. Lenses focus the pump light through the coolant onto the laser rod, aligned perpendicular to the page. The entire laser resonator includes one or more laser heads, any polarization or Q-switching optics required to produce a short pulse, and the resonator mirrors. The fully packaged laser includes the resonator package, plus the power supplies and timing electronics for the diodes and Q-switch, and the coolant circulation system with its heat exchanger.

15.2.2 Pump Diodes, Pulsing, and Packaging

The diodes are a major cost driver in diode-pumped solid-state lasers. Single emitters are used for low-power lasers, diode bars for moderate-power lasers, or stacks of these bars (see Figure 14.9) for high-power applications. Pump diodes may run continuously, or be pulsed on and off at intervals on the order of the storage time of the lasing ion, often called quasi-cw pumping. A commercial, diode-pumped, continuously running, continuous-wave (cw), Nd:YAG laser can produce 100 W or more in the infrared, absent power-limiting items in the cavity such as a Q-switch. Diode-pumped lasers in the multikilowatt range are now commercial products, and 100 kW laboratory lasers have been reported. Quasi-cw pumping tends to be used

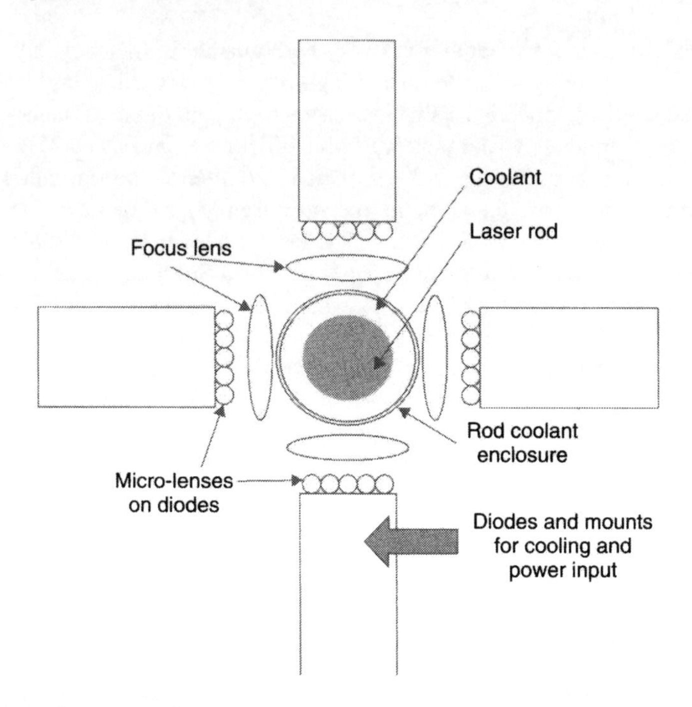

Figure 15.4 Schematic of a side-pumped diode-array-pumped solid-state rod laser. Side pumping is often used for moderate-power Nd:YAG lasers. This end view shows a laser rod at the center of the drawing. It is typically 5 to 8 mm in diameter and 5 to 10 cm long. It is surrounded by coolant that flows between an outer glass tube and the laser rod. The pump light from the diode arrays is coupled into the rod by coupling optics. CW diodes that are run continuously, and quasi-cw diodes that are pulsed on for a time, comparable with the upper-state lifetime of the laser medium (e.g., about 240 microseconds in Nd:YAG), have both been used in this configuration.

for high-energy, low-pulse-rate, Q-switched applications. Due to the smaller market and less efficient use of the diode materials and processing costs, the quasi-cw diodes tend to be more expensive and are not used as widely.

Diode-pumped lasers can be Q-switched (Chapter 11). When a Q-switch is added to a cw-pumped laser, a pulse-repetition frequency of tens of kilohertz can be achieved. Due to the limited power handling capability of the Q-switch, these lasers typically produce less than 100 W of average power. With quasi-cw pumping, Nd:YAG lasers can produce 1–2 joule, 10 ns pulses at rates of 100 Hz or more and average powers over 100 W. They usually are run in master oscillator/power amplifier mode to avoid the average-power-handling limits of the Q-switch. Nonlinear optics (Chapter 13) can convert the infrared output of these lasers to many different wavelengths in the visible and even in the ultraviolet. Because the duty factor of these lasers tends to be low (e.g., 250 μs pump pulses but 10 ms between pulses), the diodes are not as cost-effective as they are in a cw systems.

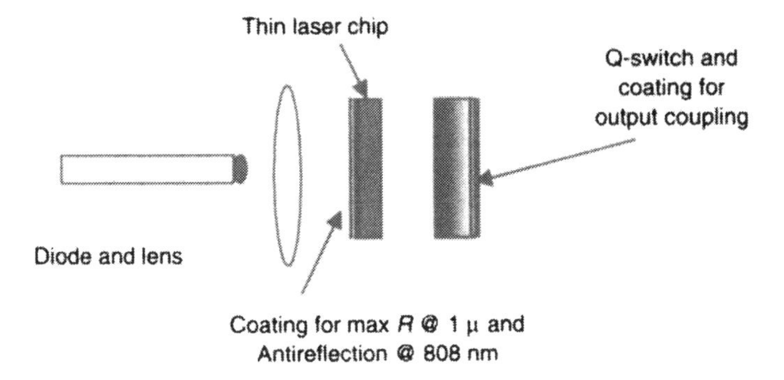

Figure 15.5 End-pumped microlaser. This simple configuration is used widely in low-power applications of diode-pumped Nd lasers. It can be run cw, or a Q-switch can be added for pulsed output. The increasingly ubiquitous green laser pointers utilize this scheme (without the Q-switch).

The diodes themselves come in a variety of packages. The quasi-cw diodes are often stacked to form high-intensity pumps for higher energy outputs. Typically, these stacks produce 100 mJ/cm^2 of stack surface area and comprise roughly 10 individual diode arrays. A Q-switched Nd:YAG laser producing about 0.5 J requires roughly 3 J of diode-pump energy. Moderate power (perhaps 10 W), cw Nd:YAG lasers can be pumped with a few cw, single-bar arrays that provide about 40 W. Many commercial products fiber-couple the diode to the laser to separate the diodes from the laser head. This removes a major heat source from the laser head and also facilitates replacement of diodes when necessary.

An important variety of the diode-pumped solid-state laser is the end-pumped microlaser, illustrated in Figure 15.5. The simple, compact design allows average powers of 1 W or more in the infrared, and a nonlinear crystal (Chapter 13) can be included to produce visible output. The laser mirrors are applied directly to the YAG or other host material, so the laser can be quite small, perhaps 1 cm long. Advanced designs use a passive Q-switch fabricated from YAG doped with the Cr^{4+} ion, which can be directly bonded to the laser medium. In this case, the entire laser resonator, Q-switch and all, is a single, monolithic structure.

15.3 LAMP PUMPING

We first encountered lamp-pumped solid-state lasers in Chapter 7. Lamp pumping is well established, but fading as diode pumping becomes practical at higher powers. Today, lamp-pumped lasers are found mostly in high-power, industrial applications.

Because lamp pumping is less efficient than diode pumping, lamp-pumped lasers require heavier power supplies and larger cooling systems to remove waste heat than diode-pumped lasers. Lamps must be replaced more frequently than diodes, although the cost of replacement lamps is a small fraction of the cost of replacement diodes. The reliability and lowered costs of diodes and diode arrays are limiting the market for lamp-pumped lasers.

As we explained in the discussion of Figure 15.3, the efficiency of lamp pumping is much lower that that of diode pumping. One reason for that is spectral mismatch between the lamp's emission and the laser medium's absorption. Another reason is the loss inherent in collecting lamps' nondirectional emission and delivering it to the laser medium. Typically, pump lamps are placed parallel to a laser rod inside a cylindrical reflector with an elliptical crossection, shown in Figures 7.12 and 15.6. If the lamp and rod are placed at the two foci of the ellipse, the reflector ideally would focus all the rays the lamp emitted onto the rod, if the light was emitted perpendicular to its surface, in the plane of the ellipse. However, the lamp also radiates at other angles, increasing losses.

The lamps that pump solid-state lasers are usually tubular, as shown in Figure 15.6, but other shapes, such as a helix, are sometimes used. Helical lamps are often found in high-energy ruby lasers but are not as conducive to water cooling as the linear flash tubes used in higher-average-power Nd lasers. High-energy Nd:glass lasers use linear flash tubes, sometimes longer than a meter.

Once the energy reaches the upper laser level, the laser resonator can extract it at about the same efficiency as in the diode-pumped case. The major advantage of diode pumping is in the directionality and narrow bandwidth of the diode and the improved quantum efficiency.

Let us examine the Cr:ruby laser as an example of lamp pumping. Because Cr:ruby absorbs strongly across a broad band in the blue and green, it is not suited for narrow-bandwidth diode pumping. We learned in Chapter 7 that Cr:ruby is a three-level system, and therefore requires intense pumping to achieve a population inversion. (Why? Because the population inversion is created between the upper laser level and the heavily populated ground state, not between the upper laser level and an unpopulated lower laser level.) The energy levels of Cr in the Cr:ruby laser are shown in Figure 15.7. The lamps that excite a ruby laser have a gas fill and pressure chosen to enhance emission in the blue and green where ruby has a broad absorption, so they can load more energy into Cr:ruby than is necessary in Nd:YAG. In fact, it takes so much energy to create a population inversion in Cr:ruby that the laser cannot normally operate in a continuous-wave mode; only flash-pumped, pulsed operation is possible. Nd:YAG, by contrast, can be held in a steady-state population inversion with a continuous-wave lamp or laser diodes.

The Tm, Ho, Er, and Yb lasers all are quasi-three-level systems (discussed in Chapter 7), as is the 946 nm transition in Nd. These lasers can be lamp pumped by codoping with Cr, making use of that ion's broad spectral absorption in the green and blue. Then a population inversion can be created despite the thermal population of the lower laser level. Chilling the laser depopulates the lower laser level, and can make these lasers quite efficient. For example, the Ho^{3+} laser, emitting at 2 μm, can be readily pumped with a simple tungsten-filament lamp (rather than an arc lamp) and can be quite efficient when it is chilled near liquid nitrogen temperature. These quasi-three-level lasers, with output between 2 and 3 μm, wavelengths that are readily absorbed by the water in living tissue, are well suited for medical applications. Although quasi-three-level lasers are frequently lamp pumped, they are more efficient when diode pumped.

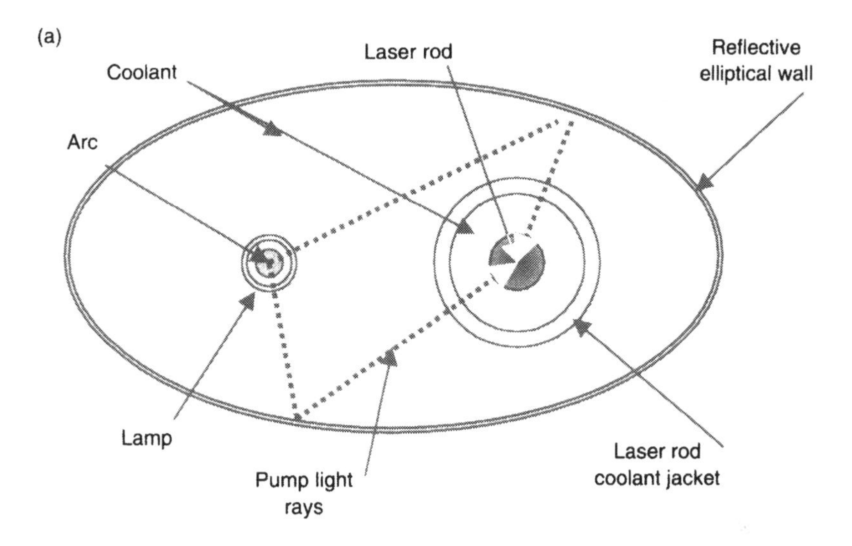

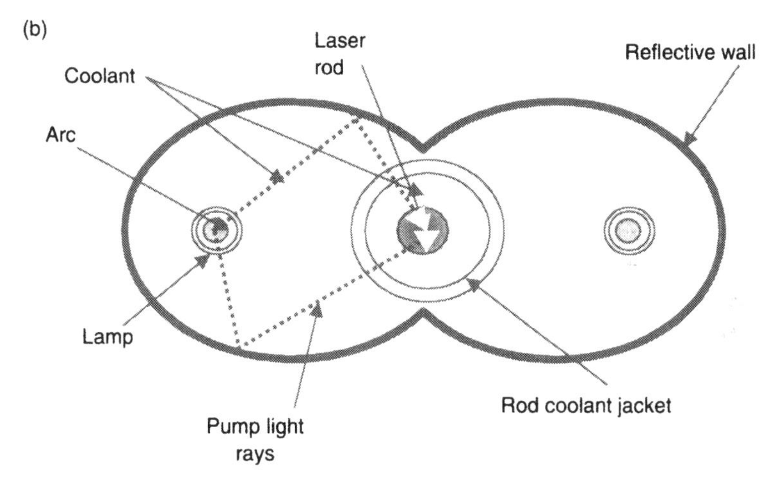

Figure 15.6 (a) Simple elliptical pump cavity for single-lamp, solid-state laser rod. In this schematic, we see the major components end-on. A lamp, which is a glass tube containing an arc discharge into a noble gas, is at one focus of an ellipse formed by the coolant flow chamber used to cool the lamp. The walls are coated with a reflector, typically gold or silver. The laser rod is typically 5 to 8 mm in diameter, and is positioned at the other focus of the ellipse. In practice, the lamps, walls, and laser rod are squeezed together more tightly than this schematic shows. Often, a coolant jacket for the laser rod has material doped into the glass to remove ultraviolet and infrared light from the arc to minimize rod heating. (b) A two-lamp configuration can provide twice the power loading. In this approach, the laser rod is excited from two sides and the outer chamber is machined to resemble two ellipses joined together, shown here not to scale. Multiple lamp designs are often needed for multikilowatt power levels.

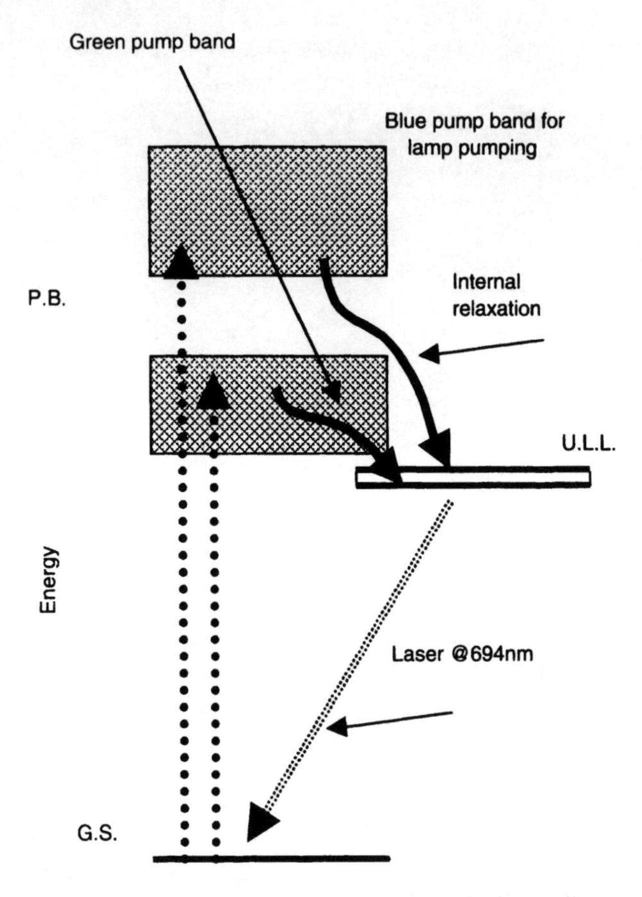

Figure 15.7 Energy-level structure of a Cr:ruby laser. The ruby laser relies on very intense absorption in the blue and green to excite Cr^{3+} ions to the pump bands (P.B.). The ions then relax spontaneously to the upper laser level (U.L.L.), which has a spontaneous lifetime of several milliseconds. Once more than 50% of the Cr^{3+} ions are in the upper laser level, an inversion is created. When the population inversion is sufficiently large so that the intracavity round-trip gain exceeds the round-trip losses, the laser begins to lase. However, as soon as half the ions are back in the ground state (G.S.), the population inversion has vanished and the laser is extinguished.

One laser that cannot be pumped at all by lamps, the 1.03-μm Yb:YAG laser, is excited efficiently by diodes, and laboratory versions have been scaled to produce average powers of 20 kW or more, suitable for industrial applications such as welding. An important advantage of Yb is the long spontaneous lifetime, about a millisecond, of the upper laser level. Because it can store energy four times longer than the upper laser level in Nd, Yb:YAG requires less-intense diode pumping than Nd:YAG. At the time of this edition, the Yb laser, in both YAG hosts and in glass fibers, is making increasing inroads into applications formerly served by lamp-pump Nd lasers.

15.4 THERMAL ISSUES IN SOLID-STATE LASERS

Whether a solid-state laser is diode pumped or lamp pumped, the pumping process leaves excess heat in the laser medium, which must be removed to keep thermal stress from distorting and ultimately breaking the solid. In a high-power gas laser or liquid laser, the heat is removed by flowing the gas or liquid out of the laser and then discarding the heat in a heat exchanger. In a solid-state laser, the heat must first be conducted to the edge of the laser medium, and then be transferred to whatever is touching the edges of the laser medium. In a low-power, diode-pumped laser, the edges of the laser medium can be wedged tightly against a metal block (called a heat sink) that transfers the heat to fins or some other cooling mechanism. High-power diode pumped lasers require more careful design and analysis. With lamp pumping or diode pumping at higher power, flowing water or other coolants are generally used to remove the heat.

Water is the most common coolant, but there are problems with any flowing liquid. Additional power is needed to operate the pump, diminishing the overall efficiency of the laser. Fittings must be of high quality and frequently maintained to prevent leaks. Corrosion is always a problem. And the liquid itself must be kept clean. Indeed, algal growth in the coolant is a common problem with many high-power, solid-state lasers.

The coolant flow rates for a kilowatt-class solid-state laser can be prodigious. As mentioned earlier, a lamp-pumped Nd:YAG laser is hard pressed to achieve 2% efficiency. Thus, a 2 kW Nd:YAG laser used for welding will need perhaps 100 kW of electric power, most of which must be removed from the lamps and the laser heads. If we allow the water to heat up to only 20°C as it flows through the entire laser head (minimal change in the water's temperature is good for certain optical considerations, which we discuss subsequently), the flow rate needed will be about 1 l/s or 15 gal/min. This flow must be directed over both the flashlamps and the laser rods, typically flowing in separate channels. Imagine the excitement when the hoses or connections spring a leak!

Heating causes thermal expansion of solids, creating two linked thermal-optical problems: thermal lensing and birefringence. Laser operation creates a temperature gradient in the laser rod, with the center of the rod hottest and the surface cooler. For example, a high-power, lamp-pumped, Nd:YAG welding laser with 1 kW of output power is likely to have four laser rods arranged in series inside the laser head, each producing about 250 W. If the efficiency is 2%, each rod must be pumped by 12.5 kW. As indicated in Figure 15.3, about 5% of this wall-plug input must be removed from the laser rod. (In Figure 15.3, 8% of the electrical energy reaches the rod and 2% is extracted as laser power. Another 1% of the original energy is lost to spontaneous emission and intracavity optical losses; this 1% is not shown in Figure 15.3. Eight percent minus 2% minus 1% equals the 5% heat to be removed from the rod.) That is, some 625 W of waste heat must be removed from the laser rod.

Of course, much more heat must be removed from the pump cavity and lamp itself: 4 kW from the pump cavity and 7.5 kW from the lamps, according to Figure

15.3. Still, the 625 W deposited in the rod can have serious effects. For one thing, YAG expands when it is heated, and because the center of the rod is hotter than the edges, the center expands more than the edges. If the center expands too much, the resulting stress will crack the laser rod—an unfortunate event that we have witnessed more than once. As a rule of thumb, thermal loading of an Nd:YAG rod should be limited to about 120 W/cm of rod length to avoid disaster. Thus, for the laser we are discussing here, each rod should be at least 6 cm long. Ten-centimeter rods would allow a more comfortable margin for error.

Typically, the center of the rod might be 60°C hotter than the edges, and the edges might be 40°C hotter than the water flowing past. This thermal gradient in the rod has three detrimental effects on the laser: (1) because the rod's refractive index is temperature dependent, the refractive index at the center of the rod is greater than the refractive index at the edges; (2) the center of the laser rod physically expands more than the edges; and (3) the stress of this uneven expansion induces a birefringence in the laser rod.

The first two effects both contribute to an effect called thermal lensing of the laser rod. The heated rod acts like a thick lens inside the resonator, distorting the shape of the intracavity beam. If the heated rod were a good, uniform lens, the effect could be compensated for (at least for a given pump power) by adjusting the curvature of the resonator mirrors to take the lensing into account or by machining a curve into the ends of the rod themselves. Unfortunately, the thermal gradient in the rod is generally uneven, so it is rarely possible to completely compensate for the thermal lensing. Hence, the output power and/or beam quality is reduced.

Thermal effects also can induce birefringence, a more subtle effect that seriously diminishes the power available from a polarized laser, for example one with a Q-switch in the cavity. A birefringent rod acts somewhat like a waveplate (Chapter 3) and couples light in one plane polarization to the other polarization. The difference between a hot rod (a hot *laser* rod, that is) and a waveplate is that the birefringence in the rod is irregular and cannot be compensated with a simple waveplate. If the resonator includes a Brewster plate or another polarizing element to force oscillation in a single polarization, the rod's thermal birefringence constantly converts light from the favored polarization to the polarization that is rejected by the polarizing element. This conversion of light from the favored polarization is an intracavity loss (Chapter 8) and reduces the output power.

One technique for improving the laser power output and optical beam quality at higher power for a lamp-pumped laser, is to utilize a slab configuration. These types of "thin laser" configurations are discussed in greater detail in the following section on diode-pumped, high-power, solid-state lasers. However, the slab configuration which was first examined as a lamp-pumped laser is illustrated below in Figure 15.8.

In a typical slab, the power is extracted by having the light trace out a zigzag path through the medium, as shown in Figure 15.8. If you imagine Figure 15.8 rotated 90 degrees, you see that in this configuration the thermal gradients are perpendicular to the zigzagging light. It is these gradients that give rise to the thermal lensing and birefringence problems. Making the beam traverse the aberrated slab in a zigzag path rather than go straight through it can minimize the lensing effect and compensate birefringence, improving beam quality

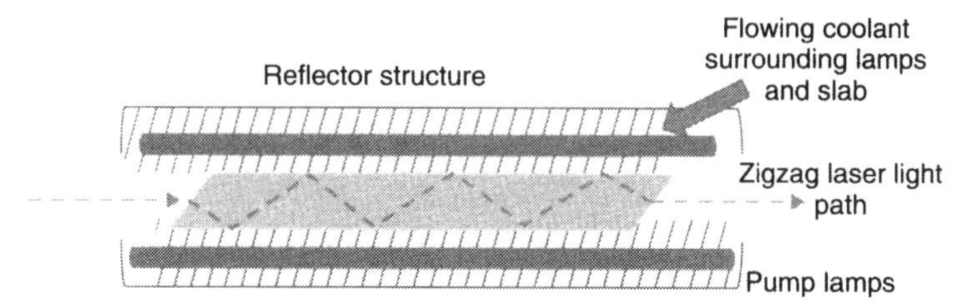

Figure 15.8 A sketch of the layout of a lamp-pumped slab, solid-state laser, as viewed from the top. The laser material is structured to be a thin slab rather than a rod, with a thickness in the range of 3 to 8 mm and a height that is perhaps 3 to 10 times the thickness.

The average power, with good beam quality, of a slab can be three to ten times greater than that of comparable rod but a major drawback is cost. The cost of a suitably shaped and polished, large-volume slab cut from a melt-grown boule, say for a 200 W Nd:YAG laser, is much more than that for a comparable rod. As a result, lamp-pumped slabs are not found as often as lamp-pumped rods in commercial use, even for high-average-power applications. In certain low-energy and low-power applications, simple slabs are used because the cost issues are not crucial and the slab geometry permits other subtle features to be optimized. Diode pumping makes the best use of the slab geometry at moderate and high powers.

15.5 SCALING DIODE-PUMPED LASERS TO HIGH POWER

Several approaches now make it possible to scale solid-state laser power to 10 kW and well beyond for industrial and military applications. This section describes these techniques in roughly historical order.

The overriding issue in scaling solid-state lasers is the removal of heat from the laser material and related optical problems. Heat removal requires maintaining a temperature gradient from the hot part of the solid to another, cooler portion in contact with the heat sink. The basic equation ruling this phenomenon, simplified here, relates the temperature gradient in the material, ΔT, from the hottest point in the solid to the cooling wall; the thickness of the material, t; the thermal conductivity of the material, K; and the heat load per unit volume Q:

$$\Delta T = Q(t^2/8K)$$

That is, the higher the heat load Q, the greater the temperature gradient needed to remove the heat. The thicker the distance to the coolant wall, the higher the temperature gradient will be. The better the thermal conductivity, the lesser the ΔT needed to reject the heat. Let us discuss each of these parameters in turn to see where the sensitive directions may be.

The thermal conductivity constant K is typically in the range 0.01 W/cm^3 for a solid medium like glass and 0.1 W/cm^3 for a good crystalline conductor like YAG.

With greater thermal conductivity, YAG lasers could produce higher outputs per unit volume. Glass lasers, however, can be produced in much larger sizes, but the tenfold smaller thermal conductivity limits the power density. Since K is in the denominator of this equation, the higher K is, the lower the ΔT will be. The lower ΔT, the less the thermal-optical issues like thermal lenses and other aberrations that relate to the differences in optical path due to temperature variation of the index of refraction. Also if ΔT is too large, stress is greater and both fracture and birefringence become important.

These effects often have straightforward solutions in 100 W rod lasers. For example, a simple spherical focus in a rod laser can be corrected by a simple lens. But in the 10 kW range more creative solutions are needed and that drives one to other geometries.

The other term in the heat conductivity equation is the heat load per unit volume, Q. This can only lowered by lowering the gain (not a good choice) or by increasing the quantum efficiency of the system. This is a place where excitation by diode arrays has a large impact, for several reasons. For starters, they can channel their output to a much lower pump level, as indicated in the example of Nd:YAG in Figure 15.1. This dramatically reduces the amount of waste heat that has to be removed. Also, because their output is collimated, they can be placed at a farther distance from the laser medium. Finally, diode pumping permits novel material choices not available in lamp pumped systems, such as Yb^{3+}-based lasers.

Finally, there is the thickness term. The temperature gradient goes down, for a given Q, as the thickness decreases. And it is not a linear term but rather gets better as $1/t^2$. So making the thickness or distance to the wall smaller is critical to high power. Slab lasers with lamp pumping, as discussed above, were the first approach to the use of thin material. Whether with lamp or diode pumping, the slab style geometry enables a thin dimension for heat rejection. Figure 15.9 and its caption describe this thermal property in more detail.

An interesting comparison can be made between a fairly standard 6 mm diameter cylindrical rod used in many commercial Nd:YAG lasers and a 3 mm thick slab. The details of this are a bit beyond the scope of this discussion, but for the same volume of material, a thin slab of dimensions of 3 mm × 100 mm can handle roughly four times the power as a 6 mm diameter rod of similar volume. This is a simple reflection of the fact that the cooled wall is half the distance away in the slab. So a 50 W rod laser requires about as much material as a 200 W slab laser. In practice, the power difference may be only a factor of three due to end effects. Note also that for efficient use of the diode pump light, the diodes must have a narrow spectral spread. Multiple passes (two or more) of the pump light through the medium also improves diode pump utilization in slabs.

A variant on the slab is the thin-disk or active-mirror design. It is a plate or thin disc that has a large diameter and is cooled on the back side, with appropriate coatings to allow the lamp or diode pump power to excite the medium and to reflect the laser beam itself. A single plate in this configuration can produce multi-kW power levels, with further scaling in power with multiple plates in line in the laser cavity. This thin-disk concept is sketched in Figure 15.10.

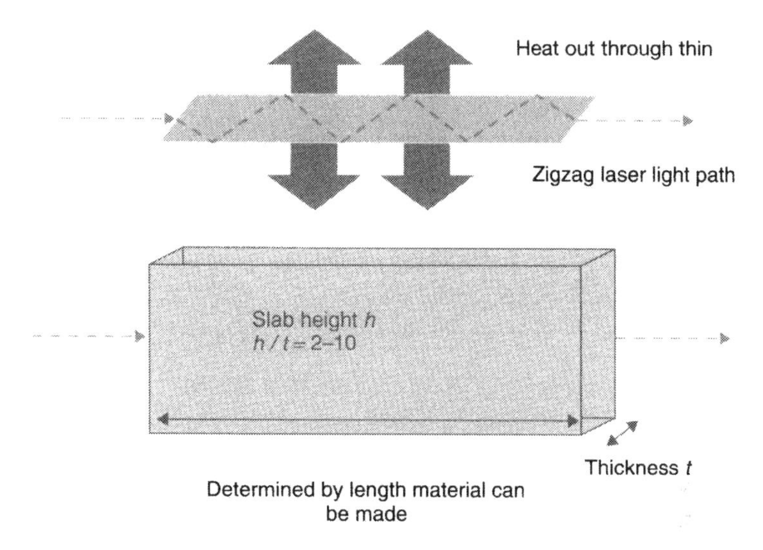

Figure 15.9 Concept of a slab laser for a zigzag path for a laser beam through the slab (upper panel, top view). The heat is transferred to coolant or cooled heat sinks through the thin dimension. The slab is typically a parallelepiped with Brewster angles at the ends where the beam enters and exits. Pump lamps or diodes are not shown but the pump light often enters through the cooling surface (the large area of height h and length L), or the pump radiation can be piped in thru the ends for diode pumping. The bottom panel is a side view.

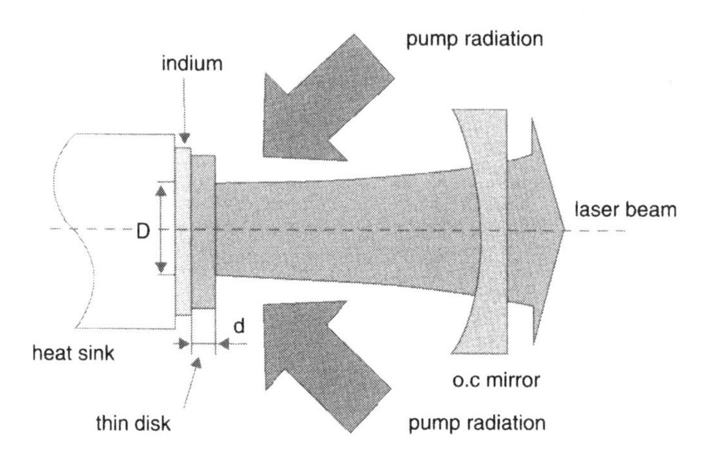

Figure 15.10 Schematic of thin-disk laser. The disc is cooled on the back side of the active mirror, which also reflects the laser beam. Lasers can be designed with a single disk, as shown here, or with multiple disks in a single resonator. Pump lamps or diodes are not shown. This approach works well for power scaling low-conductivity materials that can be fabricated into large discs, such as Nd:glass.

Though these thin-dimension concepts predated the era of affordable diode arrays and were originally lamp pumped, these concepts came into greater usage as the price of diode arrays dropped. Moreover, as diodes further developed in the range of wavelength capability, other media could be excited. The Yb:YAG disc laser is a key commercial example of this application. Since Yb^{3+} ions doped into hosts such as YAG or glass offer considerably higher quantum efficiency than Nd^{3+} lasers in the same host, the heat load of Yb:YAG is one-third that of a Nd:YAG laser. Lower heat load per unit power input means that more power can be pumped into the disc before deleterious effects of lensing and birefringence set in. With harder pumping, greater average power outputs result for the same stress in the host medium.

The thin-disk Yb:YAG laser stands out as the first industrial laser to use such high-power technology, with power outputs of over 20 kW possible. Unique properties of the Yb^{3+} ion doped in YAG permit the use of very thin discs (200 μm) and novel diode-pumping approaches. A discussion of the fundamental properties Yb:YAG is best started by reference to the energy-level diagram, Figure 15.11. Note high-power fiber lasers use Yb^{3+} in a glass medium and have a similar but not identical energy level structure, and are pumped at a slightly different wavelength than for the YAG host. Yb is a quasi-three-level laser. To create gain, one needs to

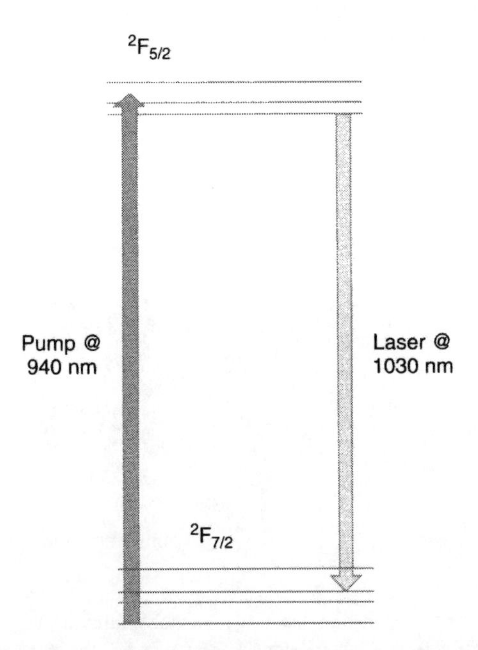

Figure 15.11 Very sparse energy level diagram for the Yb^{3+} ion in YAG. Although the sparseness of the number of energy levels makes it very difficult to pump with a lamp, it is ideal for diode-laser pumping. The laser transition terminates in a state that is thermally populated at room temperature and, therefore, these lasers must be pumped intensely or be cooled to below room temperature for efficient operation.

pump hard enough to both make a sufficient number of excited states but to also deplete the lower laser level to assure sufficient inversion. Note that the sparseness of energy levels makes Yb:YAG a poor candidate for lamp pumping.

The beauty of the quasi-three-level Yb laser is that it has very high quantum efficiency. When one pumps at 940 nm and the laser emits at 1030 nm, the quantum efficiency is 91%, much better than the diode-pumped Nd:YAG's quantum efficiency of 76%. Looked at from a heat-generation point of view, the Yb laser will nominally convert 9% of the pump radiation into heat, compared to 24% for Nd:YAG. Measurements suggest that the ratio of heat production is more like 3:1, rather than the 2.7:1 ratio suggested by the relative quantum efficiencies, due to some other losses in Nd:YAG.

Another interesting point is that Yb can be doped into the host YAG crystal at much higher doping density than Nd. This is because the ionic radii of Yb^{3+} is comparable to that of Y^{3+} ions of the host crystal. Thus, one can make the doping density high enough that a very thin disc of Yb:YAG will absorb virtually all of the pump radiation when multipassed. As discussed above, *thin* means good conductive heat transfer, which means the potential for high power. But because the low-lying lower laser level is thermally populated, one needs to pump the medium very strongly to empty most of the population out of the lower laser level. One can achieve this with high pump intensity from the diodes, but as the population of the lower level shrinks, there are fewer and fewer ions left to absorb the pump photons. This is not most efficient use of expensive diode-pump photons. To obtain high efficiency for absorbing the pump light while simultaneously depleting the lower laser level, efficient Yb:YAG designs pass the pump light through the

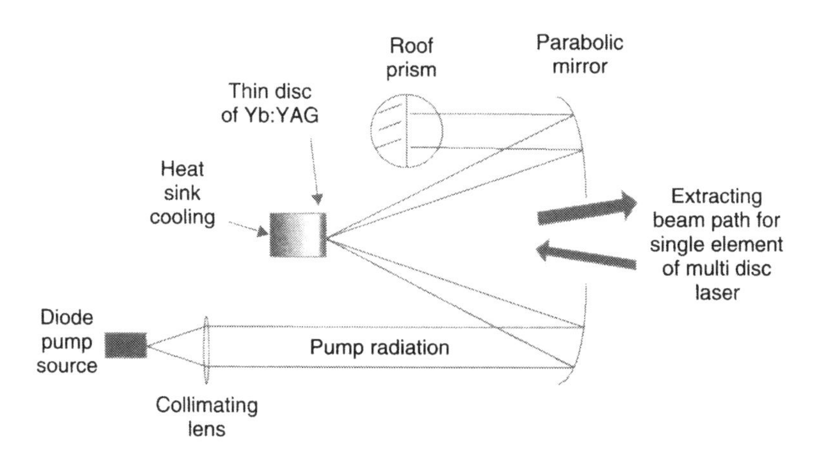

Figure 15.12 A thin disc of Yb:YAG is multipassed by collimating the diode pump light and focusing it on an the thin disc. The refected light, after traversing the Yb:YAG, is then redirected through a prism assembly to another region on the mirror to repeat the process. Very high pump fluxes are thus maintained in the laser gain medium and the very thin disc efficeintly absorbs the pump light. Note that Figure 15.10 illustrated a thin-disk laser with a single disk, while this figure shows a single gain medium in a multidisk resonator.

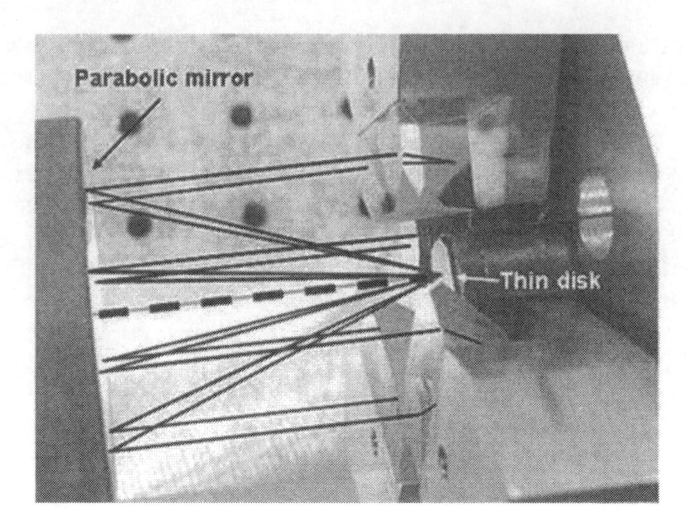

Figure 15.13 This illustration uses solid lines to show how the (infrared) pump light passes multiple times though the thin disk of laser material. The intracavity laser beam is represented by the heavy broken line.

Yb:YAG multiple times. Commercial Yb:YAG lasers take this multipass concept to a novel extreme, passing the collimated diode pump light 20 times or more through a very thin disk, typically 200 microns thick. A one-dimensional sketch of a typical multipass arrangement is shown in Figure 15.12, and Figure 15.13 is a photograph illustrating how the pump beam zigzags in three dimensions to make multiple passes through the gain medium.

Figure 15.14 A photo of three thin disc-laser heads that, when ganged in a single resonator, produce 15 kW of output.

Thin-disk lasers can generate multiple kilowatts from a single disk, and by ganging together multiple disks in a single resonator, even higher outputs can be obtained. Because the pump beam is relatively large at the face of the disk, these lasers cannot easily operate in the TEM_{00} mode. But even though the output beam is far from diffraction limited, its divergence is sufficiently small for many materials-processing applications such as welding and cutting. Figure 15.14 shows a Trumpf commercial laser, with three thin-disk gain modules in its resonator.

Chapter 16

Fiber Lasers

During the first decade of the twenty-first century, fiber lasers exploded from the laboratory into thousands of applications all around the world. Although they are solid-state lasers in the literal sense, fiber lasers are different in many ways from the conventional, optically pumped solid-state lasers discussed in the previous chapter. Because of these differences, fiber lasers are almost always have a higher electrical efficiency than conventional solid-state lasers with similar output power, and their beams are of higher optical quality.

An optical fiber is a thin strand of glass, several micrometers to over a millimeter in diameter, and meters long, or, in some cases, hundreds of kilometers long. A simple fiber consists of two layers, a core that guides the light, and a cladding of lower refractive index that surrounds the core. Light is guided through the core by the phenomenon of total internal reflection described in Chapter 3. If a ray of light B enters the fiber core at an angle nearly aligned with the length of the fiber, as shown in Figure 16.1, it will experience total internal reflection when it reaches the cladding, and be trapped or guided along the length of the core. However, a ray entering at a larger angle A will leak out, as shown in Figure 16.1.

16.1 ACCEPTANCE ANGLE AND NUMERICAL APERTURE

To be trapped inside the core, an incoming ray of light must be within the core's acceptance angle θ, and that angle can be visualized as an imaginary cone at the end of the fiber, as pictured in Figure 16.2.

Another important fiber-optic concept is the *numerical aperture* of a fiber, which is defined as the trigonometric sine of the core acceptance angle θ, as shown in Figure 16.2. Although it is not obvious, some complicated math involving Snell's law shows that the numerical aperture is a function of the refractive index of the fiber core and the surrounding cladding:

$$\text{Numerical aperture} = \text{NA} = \sqrt{n_{\text{core}}^2 \quad n_{\text{cladding}}^2}$$

It is important to realize that the physical aperture of the fiber, that is, the core diameter, has nothing to do with the numerical aperture. The numerical aperture depends only on the refractive indexes of the core and the cladding.

Introduction to Laser Technology, Fourth edition. C. Breck Hitz, J. Ewing, and J. Hecht

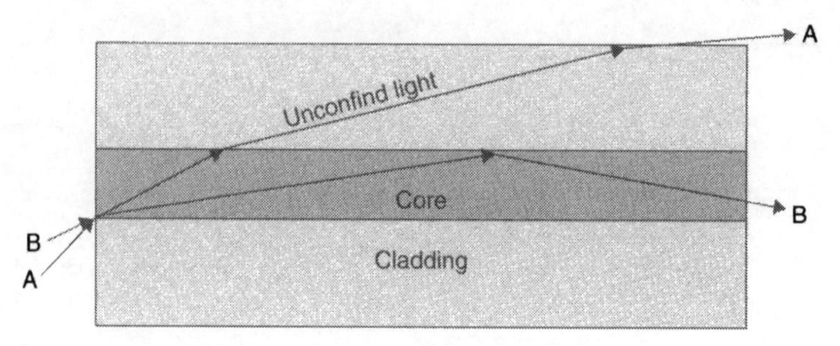

Figure 16.1 Ray A enters the core at the end of the optical fiber and is refracted out at the edge because its angle of incidence is less than the critical angle for total internal reflection. But ray B enters the fiber core at an angle greater than the critical angle and is totally internally reflected. Ray B will travel down the entire length of the fiber without escaping. It is important to realize that this figure is not drawn to scale; in a typical fiber, the core will be tens of micrometers in diameter, and the fiber will be many meters long.

16.2 DOPING OPTICAL FIBERS

The standard optical fibers used in fiber-optic communications or for beam delivery in materials working are passive devices with pure glass cores that do not generate light. Fiber lasers, and the fiber amplifiers used in optical communications systems, are made from fibers in which the core is doped with an impurity that lases, just as a conventional solid-state laser is doped with a lasing impurity. The most common dopant in conventional lasers is probably neodymium, whose strongest laser line is at 1.06 μm, and neodymium is also a dopant in fiber lasers. More common in fiber lasers, however, is ytterbium, with a lasing wavelength tunable between 1.02 and 1.10 μm. Other common dopants for fiber lasers include erbium, with laser lines at 1.5 and 2.9 μm, and holmium, with lines at 2.1 and 2.9 μm.

Fiber lasers have mirror-like structures at both ends, which form a resonant cavity in which the laser oscillates. It is also possible to use fibers as amplifiers, as in the

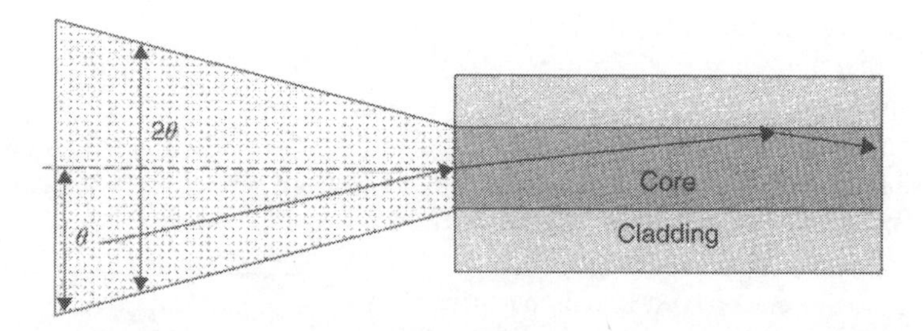

Figure 16.2 The acceptance angle can be visualized as a cone extending from the end of the core, which contains rays that are trapped inside the fiber by total internal reflection. Rays outside the acceptance angle are not trapped. The numerical aperture is the sine of the acceptance angle θ.

MOPA scheme discussed in Chapter 8, Section 8.6. The most common widespread application of fiber amplifiers is in fiber-optic communication systems, where erbium-doped fiber amplifiers, or EDFAs, amplify weak signals in the 1.5 μm band without requiring the optical signals to be converted to electrical ones.

What distinguishes fiber lasers from conventional solid-state lasers is their unusual geometry—very long and very skinny. Because the fiber is so long and skinny, its ratio of surface area to volume is much greater than in a laser rod. All that surface area makes it much easier to remove waste heat and, of course, heating is a fundamental limitation of most solid-state lasers. In a long, skinny fiber laser, the problem of heat removal becomes much easier to deal with.

The long, thin shape of a fiber laser is also responsible for a fundamental limitation. Its power is generated in a thin core, producing very high power densities that facilitate deleterious nonlinear effects, like stimulated Brillouin scattering, stimulated Raman scattering, and four-wave mixing, which convert laser light to other wavelengths, and hence diminish the circulating power. The magnitude of these nonlinear effects increases with distance, and becomes very large in a long fiber. Thus, the fundamental limitation on fiber lasers is usually not heat removal but the loss of intracavity power to nonlinear processes. The issue is especially acute in pulsed lasers with high peak power.

16.3 PUMPING FIBER LASERS

Fiber lasers, like other solid-state lasers, are optically pumped. But unlike many other solid-state lasers, fiber lasers are rarely pumped from the side. Instead, the pump radiation for a fiber laser is coupled into the fiber from one or sometimes both ends, and is trapped inside the fiber, where it can be absorbed over the fiber's entire length. Fiber amplifiers are pumped in the same way.

For end pumping to work, the pump radiation must be coupled into the fiber, which requires two things. First, the incoming light must fit into the light-guiding core. Second, it must enter the core within the fiber's acceptance angle. That requirement eliminates lamps and LEDs as pumps, because those sources are too large or emit into too wide an angle, so only a tiny portion of their power would ever get into the fiber. Fiber lasers are almost always pumped with diode lasers, which, as you learned in Chapter 14, are efficient and emit from a small area. However, many doped fibers have cores are so small that coupling even a diode laser into them is not easy.

In Section 9.2, we discussed transverse modes of conventional laser resonators, and fiber lasers also have transverse modes. They are not quite the same modes, because total internal reflection provides sideways, or lateral, confinement in the core of a fiber laser. (There is no lateral confinement in a conventional laser resonator. The lateral dimension of the intracavity beam is controlled by intracavity focusing elements, like the laser mirrors and thermal focusing in a solid-state laser rod.) But the transverse modes of a fiber laser are similar in many ways to those of a conventional resonator. In particular, the lowest-order mode produces the highest-quality (lowest-divergence) output beam, and the laser can be forced to operate in the lowest-order mode by confining it to a narrow aperture. In practice, a single-mode fiber laser—one operating in the lowest transverse mode—has a tiny core, usually ten micrometers or less in diameter, and single-mode fibers are preferred for many applications.

Pump diodes can deliver enough light to the core of a single-mode fiber for most fiber amplifiers and for some low-power fiber lasers. However, it's hard to squeeze high pump power into a single-mode core. Focusing pump light with a strong enough lens can squeeze it into a very small focal spot. But remember from Section 5.2 that the tighter you focus light, the more divergent it becomes on each side of the focus. Thus, if you focus the light from a diode laser down onto the 10-micrometer core of a single-mode fiber, much of the light will be outside the fiber's acceptance angle. If you relax the focusing to keep the pump light is within the acceptance angle, a lot of it misses the core altogether.

The ingenious solution to this problem is creating an additional layer in the fiber that serves as an outer core or inner cladding, as shown in Figure 16.3. Only the tiny inner core is doped in these *double-clad* fibers. It is surrounded by an inner cladding (or outer core) having a refractive index intermediate between the inner core and the outer cladding.

Pump light is coupled into the inner cladding, a much bigger target than the core itself. Total internal reflection traps the pump light inside the inner cladding, and as it travels down the fiber it bounces back and forth through the inner core, where it is absorbed by the core's dopants. But the inner core confines the laser light, which, therefore, emerges from the end of the fiber in a tiny, high-quality, single-mode beam.

A potential drawback to double-clad lasers is weaker absorption of the pump light. In a core-pumped fiber laser, the pump light is always in the core, and hence can be absorbed in a relatively short length of fiber. In a cladding-pumped laser, the pump light is inside the core only part of the time and, hence, must travel a longer distance in the fiber before it is absorbed. However, in practice, fiber lasers can be designed with fiber long enough so this factor is not a problem.

16.4 FABRICATING OPTICAL FIBERS

There are several techniques for fabricating optical fibers, but the most common is to draw them from the top of a tower. The process begins with a *preform,* a large-

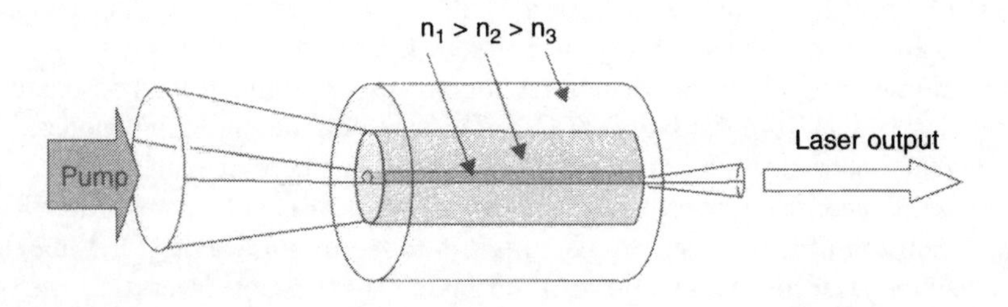

Figure 16.3 In a double-clad fiber, the pump light is confined by total internal reflection in the inner cladding, while the laser light is confined in the inner core. Reminder: This drawing is ridiculously not to scale. A typical double-clad fiber might be 125 micrometers in diameter and ten meters long.

scale replica of what the final fiber will look like: a rod of high-index, doped glass surrounded by the glass cladding layers. (A preform might actually resemble the object drawn in Figure 16.3.)

The preform is placed at the top of a drawing tower and heated to the softening point, when it begins to flow. As gravity pulls downward on the softened glass, it flows into a narrow strand descending from the bottom of the preform. This strand exactly mimics the structure of the preform, but on a much smaller scale. As the strand is pulled downward by gravity, it solidifies and is flexible enough to be wound on a spool at the bottom of the tower, as depicted in Figure 16.4.

16.5 FEEDBACK FOR FIBER LASERS

Like any laser, a fiber laser requires feedback to sustain oscillation. A pair of normal laser mirrors could be placed at the two ends of the fiber, but this would require careful alignment of the external mirrors, which would be subject to misalignment

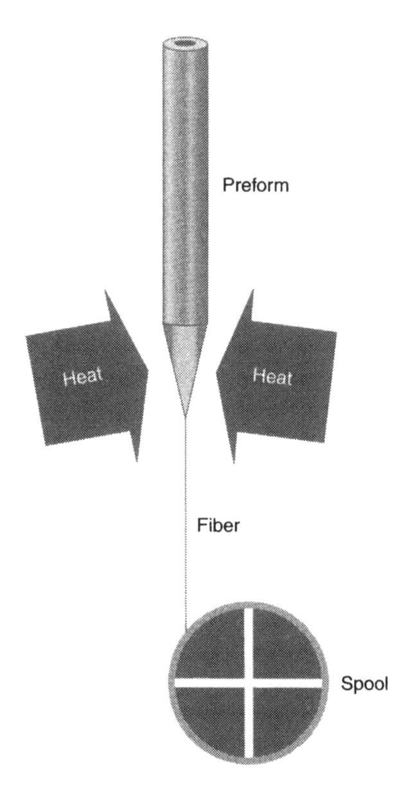

Figure 16.4 Optical fibers can be fabricated by softening a preform at the top of a drawing tower. As the glass softens, it flows into a narrow strand—the optical fiber—descending from the preform. The strand solidifies as it falls, and at the bottom of the tower it is flexible enough to be wound on a spool. This drawing is not to scale: the preform might be inches or feet in length, and the tower is typically several yards tall.

resulting from thermal changes, physical bumps, or other disturbances. Also, the ends of the fiber would have to be antireflection coated, just as the ends of a Nd:YAG rod are, to prevent intracavity reflective losses.

An alternative is to fabricate the mirror *inside* the optical fiber, which would automatically align the mirror and avoid the need to coat the end of the fiber to reduce intracavity loss. A mirror that is fabricated inside the fiber is a *fiber Bragg grating,* or an FBG.

An FBG, pictured in Figure 16.5, works exactly like the laser mirrors discussed in Section 8.8. The stripes in the core have a different refractive index than the rest of the core, and light is reflected from them. If the spacing between the stripes, Λ in Figure 16.5, is exactly equal to half the wavelength of light in the glass, then the light reflected from any stripe will be in phase with the light reflected from all the other stripes. The result is a very high reflectivity from the FBG at the selected wavelength.

An FBG is written in a special fiber whose core is photosensitive, usually as the result of germanium doping. The fiber is placed in an area illuminated by two overlapping ultraviolet beams that interfere, creating a fringe pattern, similar to the fringes created in Young's experiment (Section 4.3), which "write" the grating into the core.

FBGs cannot easily be created in the rare-earth doped core of a laser fiber so, usually, FBGs are fabricated in special photosensitive fibers that are spliced onto the ends of the laser fiber.

16.6 HIGH-POWER FIBER LASERS

A high-power fiber laser for industrial applications such as metal welding or marking would resemble the drawing in Figure 16.6. The output beams from multiple diode lasers are combined in a multimode coupler, and coupled into the inner cladding of the double-clad fiber. The double-clad fiber, usually several meters or longer in length, is coiled so the whole device fits into a conveniently sized container. FBGs at either end of the fiber provide the feedback necessary for laser action, and output power can reach kilowatts, limited by the nonlinear effects mentioned earlier or by optical damage.

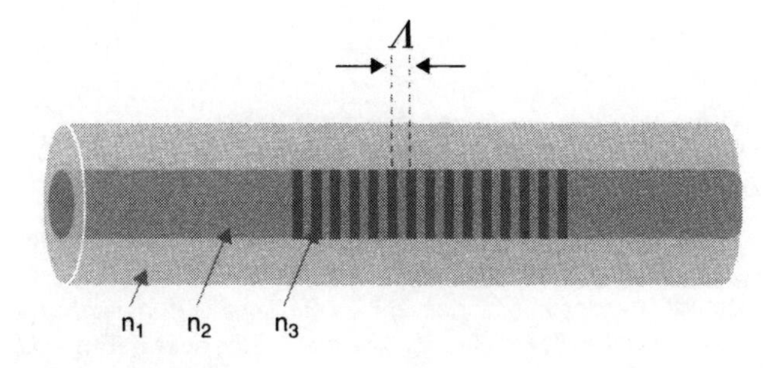

Figure 16.5 A fiber Bragg grating (FBG) is an optical fiber with a periodic grating written into its core. In a fiber laser, an FBG can serve as the laser mirror.

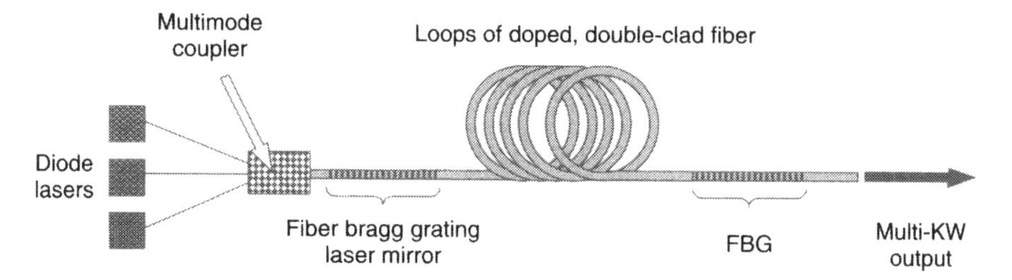

Figure 16.6 A high-power fiber laser configuration, showing multiple diode-laser pumps and laser feedback provided by fiber Bragg gratings.

16.7 LARGE-MODE-AREA FIBERS

The obvious way to reduce the high power density in the core of a fiber is to increase the diameter of the core. That is precisely what large-mode-area (LMA) fibers do, but in Section 16.3 we said that a fiber laser needed to have a small core (that is, a small mode area) in order to be single mode. So the question is, How do you make an LMA fiber oscillate in the lowest-order transverse mode?

One successful approach question has been to employ bending loss. When a fiber is bent or coiled, some light traveling down the core no longer intersects the core-cladding interface at an angle greater than the critical angle, so it can escape from the core. The tighter the coil, the greater the bending loss. More important, bending losses are greater for high-order modes than for the low-order modes, so a fiber can be coiled at a radius chosen so that losses high-order modes are high enough to allow only the lowest-order mode to oscillate in the laser.

As the output of the fiber laser increases, the laser must be coiled ever more tightly to keep high-order modes below threshold. Eventually, the coil must be so

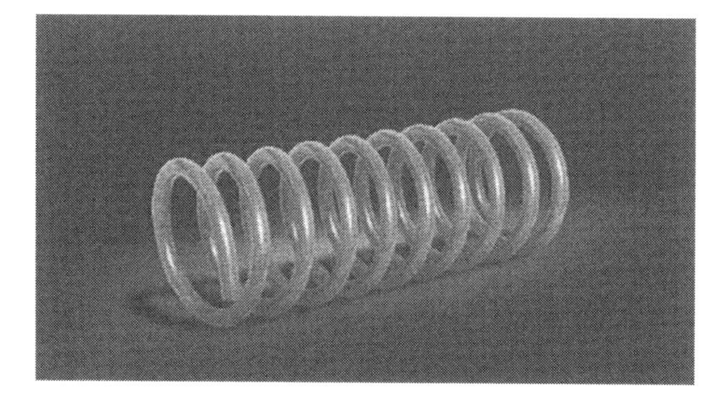

Figure 16.7 The high-order modes in an optical fiber experience greater bending loss in a coiled fiber than low-order modes do. If a fiber laser is coiled with exactly the right amount of bend, only the lowest-order mode will oscillate.

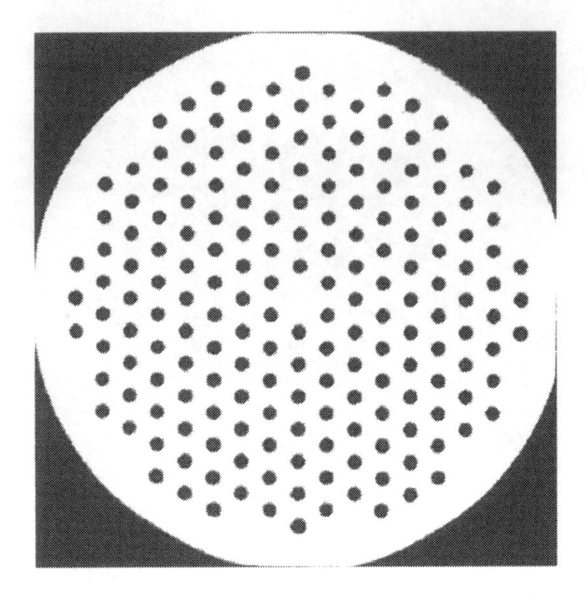

Figure 16.8 This holey fiber, approximately 125 μm in diameter, has a lattice of tiny airholes running its entire length. The part of the fiber with the airholes is the cladding, and the core is the single "defect," or lack of an airhole, at the center.

tight that the fiber breaks. Helical-core fibers were developed to avoid this pitfall. A helical-core fiber is fabricated from a preform with an off-center core. Then, as the fiber is fabricated in a drawing tower, as illustrated in Figure 16.4, the preform is rotated so that the core forms a helix in the perfectly straight fiber. A bend in the core of a helical-core fiber can be much sharper than the bend achieved by coiling a conventional fiber.

16.8 HOLEY FIBERS

Whoever first used the term "holey fiber" probably intended the humorous double entendre, but the name is apt. Holey fibers are precisely that: fibers with a lot of holes in them, as illustrated in Figure 16.8. The fiber core is the solid region in the center, surrounded by a cladding made of the same glass, but full of regularly spaced holes that reduce its refractive index, making it act as a cladding. Also called microstructured fibers or photonic crystal fibers, holey fibers have unusual properties that make them very useful in fiber lasers.

An important attraction of holey fibers is that they can be designed to transmit only a single mode no matter how large the core becomes. This odd behavior can be intuitively understood by thinking of the fiber as a sieve for the modes traveling in the core. As shown in Figure 16.9a, the lowest-order mode has a single intensity maximum that is too big to fit between the air holes, so it cannot leak out of the core. Figure 16.9b, on the other hand, shows a higher-order mode in the same holey fiber, which has four intensity maxima, each of them small enough to

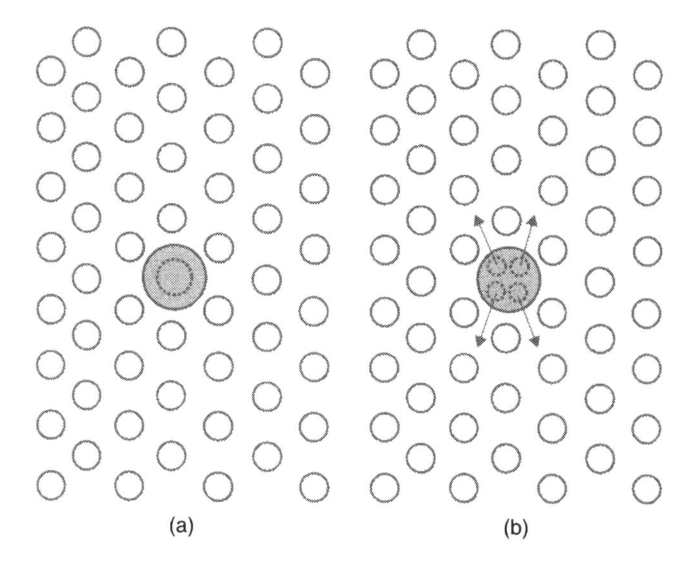

Figure 16.9 The lowest-order mode shown in the core of the holey fiber (a) is too large to fit between the airholes in the cladding of the fiber and, thus, is confined to the core. The higher-order mode shown in the core of (b) consists of several smaller-intensity maxima that can escape between the airholes.

fit through the spaces between the air holes, allowing it to leak out of the core as it propagates.

However, large-core singlemode holey fibers also have other properties that make them less attractive for fiber lasers, such as severe sensitivity to bending losses, which makes it impossible to coil them.

QUESTIONS

1. Suppose the fiber pictured in Figure 16.3 has a cladding diameter of 125 μm and a core diameter of 15 μm. The index of the cladding is 1.47, and the core index is 1.51. What is the numerical aperture of this fiber?

   Suppose another fiber has the same core and cladding indexes, but the aperture of the core is 30 μm. What effect does this have on the numerical aperture?
2. Using the Internet, find the highest-power, single-mode fiber lasers that are commercially available today. (Hint: Don't miss the websites of companies like IPG and Nufern.)
3. Calculate the surface-to-volume ratio of a four-inch-long Nd:YAG rod with a quarter-inch diameter. Compare that to the surface-to-volume ratio of a 10-m-long fiber with a 10 μm core.
4. A fiber Bragg grating (FBG) is fabricated in an optical fiber whose nominal refractive index is 1.5. What must the spacing between the stripes (Λ in Figure 16.5) be if the FBG is to reflect light at 1.1 μm?

Chapter 17

Gas Lasers: Helium–Neon and Ion

Laser action is possible in a wide variety of gases. The basic requirements are gain and feedback: a suitable excitation mechanism, a gas that can be excited to produce a population inversion, a tube to contain the laser gas, and an optical resonator to generate the beam.

The physics of atoms and molecules in the gaseous state was well understood by the mid-1950s, when the laser concept was developed. The first proposals for making lasers envisioned using a gas as the active medium. However, pulsed solid-state lasers excited by powerful flashlamps proved easier to make, as described in Chapter 15. The first continuous-wave gas laser followed only seven months later and was demonstrated in December 1960.

Gas lasers proliferated quickly once specialists figured out the principles of laser physics. You can fill tubes with different types of laser gas much faster than you can grow crystals of different compositions for solid-state lasers. Researchers have demonstrated laser action on thousands of transitions in a wide range of gases, although only a few ever proved practical and found wide use.

For many years, gas lasers were the stalwarts of the field. Semiconductor and solid-state lasers have eclipsed them in many applications, but gas lasers remain versatile and important tools. There are three important families of gas lasers today. One—relatively low-power gas lasers such as the helium–neon (He–Ne) and ion lasers described in this chapter—remain common sources of visible beams at powers ranging from under a milliwatt to tens of watts. A second family, the carbon dioxide lasers described in Chapter 18, generate high infrared powers for industrial applications. A third family, the excimer lasers covered in Chapter 19, produce pulsed UV beams used in semiconductor manufacture, refractive surgery, research, and other applications. A few other types of gas lasers exist but their applications are quite limited.

These gas lasers offer combinations of important features not readily available from semiconductor or solid-state types. Gas lasers offer better beam quality and coherence than semiconductor lasers, as well as higher power at wavelengths toward the short end of the visible spectrum and into the ultraviolet. They offer wavelengths not readily available from solid-state lasers. Sometimes, they are used as replacement parts or in older equipment to avoid the need for redesign to use with diode or solid-state lasers. Advances in semiconductor and solid-state lasers are eroding these advantages, but gas lasers remain a significant part of the laser world.

17.1 GAS-LASER TRANSITIONS

The fundamental distinction among gas lasers is in the type of energy-level transition that generates the laser light. Gas molecules have three types of energy levels, as shown in Figure 17.1. As we explained in Chapter 6, molecules, like atoms, can store energy in electronic energy levels. Unlike atoms, however, molecules can also store energy in vibrational and rotational levels. Rotational energy levels are closely spaced, so laser transitions typically have energies that correspond to wavelengths longer than about 100 μm. Rotational transitions also can occur at microwave wavelengths, where they produce cosmic masers. Lasers can operate on purely rotational transitions, but they are of little practical use because they produce photons of such limited energy.

Vibrational levels are farther apart than rotational levels and, typically, transitions between vibrational states correspond to infrared wavelengths between about 1 and 100 μm. In practice, molecules change their rotational state at the same time they change their vibrational state. The change in rotational energy can add or subtract from the change in vibrational energy on the laser transition. The most important vibrational-level laser is the carbon dioxide laser; Chapter 18 explains how these transitions work.

Changes in the energy states of electrons in gas molecules or atoms produce transitions that generally have higher energies, usually corresponding to wavelengths shorter than about 2 μm. Electronic transitions affect the bonds between atoms making up a molecule and often cause dissociation of the molecule. The excimer lasers described in Chapter 19 are electronic-transition molecular lasers, but the molecules are unstable. Most electronic transition gas lasers are based on atoms, such as the He–Ne and rare-gas ion lasers described in this chapter.

The electrons making the transitions normally are the outer (or "valence") electrons, which also are involved in chemical bonding. These transitions typically generate light at visible, near-infrared, and near-UV wavelengths, between about 200 and 2000 nm. Electrons can make lower-energy transitions if suitable energy levels are

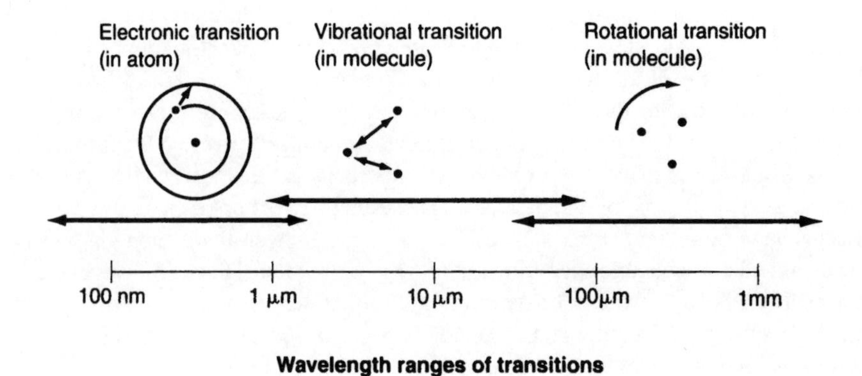

Figure 17.1 Types of laser transitions.

closely spaced, but these transitions have found little practical use in lasers. Electronic transitions can release much more energy—generating X rays—if the electrons drop into vacancies in the innermost energy levels close to the nucleus, but this normally requires removing most of the atom's electrons and is done only in the laboratory.

Table 17.1 lists the principal types of electronic-transition lasers in practical use. This chapter covers the most important atomic types: He–Ne, argon (Ar) ion, and krypton (Kr) ion.

Table 17.1 Wavelengths of important gas lasers

Type	Wavelength (nm)	Approximate power (W)[a], cw or average	Operation
Electronic transitions			
Molecular fluorine (F_2)	157	1–5	Pulsed
Argon-fluoride excimer	193	0.5–50 (avg.)	Pulsed
Krypton-fluoride excimer	249	1–100 (avg.)	Pulsed
Argon ion (deep UV)	275–305	0.001–1.6	Continuous
Xenon-chloride excimer	308	1–100 (avg.)	Pulsed
He–Cd (UV)	325	0.002–0.1	Continuous
Nitrogen	337	0.001–0.01 (avg.)	Pulsed
Argon ion (near UV)	333–364	0.001–7	Continuous
Krypton ion (UV)	335–360	0.001–2	Continuous
Xenon-fluoride excimer	351	0.5–30 (avg.)	Pulsed
He–Cd (UV)	353.6	0.001–0.02	Continuous
Krypton ion	406–416	0.001–3	Continuous
He–Cd	442	0.001–0.10	Continuous
Argon ion	488–514.5	0.002–30	Continuous
Copper vapor	510, 578	1–50 (avg.)	Pulsed
He–Ne	543	0.0001–0.003	Continuous
He–Ne	594.1	0.0001–0.006	Continuous
He–Ne	604.0	0.0001–0.001	Continuous
He–Ne	611.9	0.0001–0.003	Continuous
Gold vapor	628	1–10	Pulsed
He–Ne	632.8	0.0001–0.05	Continuous
Krypton ion	647[b]	0.001–7	Continuous
He–Ne	1153	0.001–0.015	Continuous
Iodine	1315		Pulsed
He–Ne	1523	0.0001–0.001	Continuous
Vibrational transitions (see Chapter 18)			
Hydrogen fluoride	2,600–3,000[c]	0.01–150	Pulsed or cw
Deuterium fluoride	3,600–4,000[c]	0.01–100	Pulsed or cw
Carbon monoxide	5,000–6,500[c]	0.1–40	Pulsed or cw
Carbon dioxide	9,000–11,000[c]	0.1–45,000	Pulsed or cw

[a]For typical commercial lasers.
[b]Other wavelengths are also available.
[c]There are many lines in this wavelength range.

17.2 GAS-LASER MEDIA AND TUBES

The active medium in a gas laser is a gas sealed in a tube to isolate it from the environment. The gas may be a single pure gas, such as Ar, or a mixture of two (or more) gases, such as He and Ne. Laser operation depends on the relative concentrations of the two gases and the total pressure. It also depends on the purity of the gas, because certain contaminants can poison some gas lasers, impairing their operation. Continuous-wave lasers are usually operated at pressures well below atmospheric pressure to sustain stable electric discharges.

He, Ne, Ar, Kr, and Xe, the family of rare gases, are the most common elements used in gas lasers. Under normal conditions, these elements are atomic gases. In lasers, they may be ionized by the removal of one electron, which alters their energy levels to create the common ion-laser transitions in Ar and Kr. Vapors of metals such as copper, cadmium, and gold also can be used in gas lasers; the metals are solids at room temperature, but are vaporized to make the laser operate.

The names used for lasers identify the active elements in the gas. In the He–Ne laser two elements are active: He collects energy from the electric discharge and transfers it to the Ne atoms that emit the laser light.

Figure 17.2 illustrates a generic gas-laser tube. The details of tube structures differ among gas lasers, but this generic diagram highlights the main features of tubes. The most obvious function is to separate the laser medium from air. Air is pumped out of the tube, the laser gas is pumped in, and then the tube is sealed. The longest-lived gas lasers are He–Ne, with operating lifetimes that can reach 50,000 h. Some sealed lasers, particularly Ar- and Kr-ion lasers, can be refurbished by opening the seal, cleaning the interior, replacing the gas, and resealing the tube. Other gas lasers are designed for repeated fillings (e.g., the excimer lasers covered in Chapter 19), or operation with flowing gas (some carbon dioxide lasers in Chapter 18).

An electric discharge passes along the length of the tube in Figure 17.2 between a pair of electrodes. The tube must accommodate conductors that conduct electricity into the gas without allowing gas to leak in or out. Differences in thermal expansion coefficients of the tube material and the conductor are a prime challenge. At least part of the tube must be made of insulating material.

A pair of mirrors must be located at opposite ends of the tube to form a resonator in which the laser beam oscillates, as described in Chapter 8. One mirror reflects all light reaching it from inside the resonator, while the other transmits a small fraction of the light and reflects the rest back into the resonator. As shown in Figure 17.2, the mirrors often are not in contact with the laser gas; transparent windows (often at Brewster's angle, although that configuration is not shown) can transmit light to the mirrors while containing the gas. This is desirable because high-energy ions present in many gas lasers can damage the optical coatings used on highly reflective mirrors. In many He–Ne lasers, on the other hand, the resonator mirrors are attached directly to the end of the discharge tube, replacing the windows.

All electronic-transition gas lasers convert only a small part of the input electrical power into laser light; the rest remains as heat, which must be removed. In low-power lasers, convective cooling is adequate to remove the waste heat. Higher-power

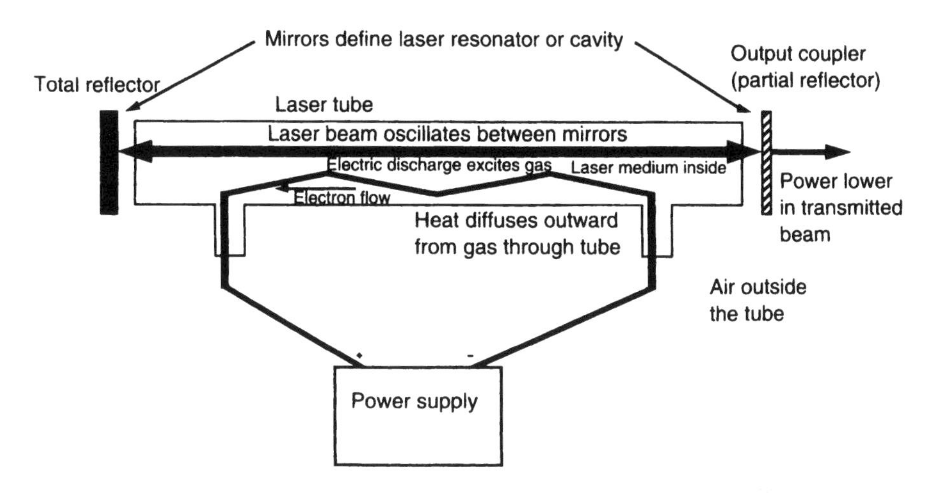

Figure 17.2 Generic gas-laser tube.

lasers require either forced-air or flowing-water cooling of the tube. Thermal requirements play an important role in the selection of tube material; low-power lasers often have glass tubes, but higher-power lasers require tubes with greater thermal conductivity, constructed of ceramics or metal and ceramic.

Like other lasers, gas lasers are packaged for safety and handling considerations. You may see a compact cylindrical gas laser head or a larger box that contains a laser tube, but unless someone opens the cover, you will not see the bare laser tube.

17.3 LASER EXCITATION

All commercial electronic-transition gas lasers are excited electrically. A power supply converts alternating current from the commercial power grid into higher-voltage direct current. A high voltage applied across the gas, usually longitudinally along the length of the tube, accelerates electrons, which collide with gas atoms and transfer some of their energy to the gas. Modulating the electrical input can modulate the optical output, but the response is too slow for most practical applications. An initial voltage spike ionizes the gas so that it can conduct a current, and then the applied voltage drops to a lower level while the laser operates continuous wave. Optical pumping is possible and occasionally employed in the laboratory, but it is not practical in commercial gas lasers. For example, the 1315 nm transition of atomic iodine can be pumped optically or pumped by energy transfer from chemically excited oxygen molecules. The chemical oxygen–iodine laser has been in laser weapon experiments, notably in the Airborne Laser and Advanced Tactical Laser programs.

Energy transfer is a complex process that differs among laser types. In lasers that contain two species, such as an He–Ne laser, the more abundant species (He) absorbs the electron energy, then transfers it to the other species (Ne), populating the upper laser level to produce a population inversion.

As in other lasers, depopulation of the lower laser level is as important as population of the upper laser level. There are, after all, two ways to increase the population inversion in a four-level system (see Chapter 7): you can add population to the upper laser level, or you can remove population from the lower laser level. In most gas lasers, the laser transition is far enough above the ground state that thermal population of the lower laser level is negligible. In fact, as you will see later in this chapter, electronic transition gas lasers typically are far above the ground state. Although this ensures that there is practically no thermal population of the lower laser level, it limits how efficiently the laser can convert input electrical power to laser light. For example, if the pump level is 20 eV above the ground state and the laser photon carries 1 eV of energy, the other 19 eV is lost, limiting efficiency to at most 5%. In practice, efficiencies are even lower. This means that most energy deposited in the laser gas winds up as heat, which must be dissipated.

17.4 OPTICAL CHARACTERISTICS

The gain per unit length of electronic-transition gas lasers typically is relatively low, so the resonator losses must be strictly minimized. The output mirror must reflect a large fraction of the beam back into the resonator, as shown in Figure 17.2. This means that the power circulating in the laser resonator is much higher than that in the output beam. Unstable resonators (Chapter 8) are rare in low-power gas lasers.

An intracavity window oriented at Brewster's angle (Chapter 3), about 57° for glass, can polarize the output beam. The Brewster window reflects part of the light polarized parallel to the face of the window, but transmits all of the orthogonal polarization. Thus, loss for the transmitted polarization is less, and the resulting laser beam is polarized in that direction.

The rear reflector in a gas laser typically is a long-radius spherical mirror, defining a stable resonator, as shown in Figure 17.3. This design generates a good-quality, diffraction-limited beam with low divergence, typically in the milliradian range.

17.5 WAVELENGTHS AND SPECTRAL WIDTH

Intracavity optics combine with characteristics of the laser gas to determine both the wavelengths and spectral widths of gas lasers. Gas lasers typically can produce gain on many transitions over a range of wavelengths. Each transition has its own characteristic gain. The laser mirrors select among them by being strongly reflective at the desired wavelength but having lower reflectivity on other transitions. For example, a He–Ne laser designed to oscillate on its green transition should have mirrors strongly reflective in the green, but with low reflectivity on the stronger red and infrared transitions. Because the round-trip gain must be greater than the round-trip loss (Chapter 8) for any lasing wavelength, the low mirror re-

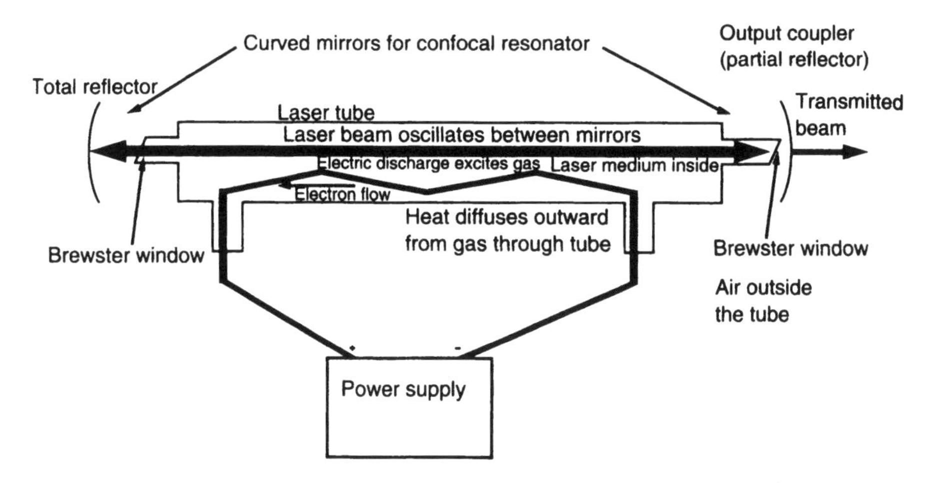

Figure 17.3 Gas laser with Brewster-angle windows and curved external mirrors forming a confocal resonator.

flectivity for the red and infrared wavelengths will prevent those transitions from reaching threshold.

Resonator mirrors reflective over a broad range of wavelengths allow lasers to simultaneously emit on several wavelengths. The strength of each line depends on the gain of the laser transition. Such mirrors are used in multiline Ar- and Kr-ion lasers, as well as in mixed gas lasers containing both Ar and Kr and oscillating on lines of both. Later in this chapter, you will see that resonator elements, similar to those discussed in Chapter 10, can be used to select a single transition in a multiline laser.

As you learned in Chapter 9, laser cavities oscillate at longitudinal modes, defined as wavelengths at which a resonator round-trip equals an integral number of wavelengths. Each of these longitudinal modes is quite narrow, only about 1 MHz wide. Their spacing is close enough that many of these modes fall under a gas laser's gain curve, so, normally, gas lasers oscillate in multiple longitudinal modes, as shown in Figure 9.15.

Individual gas-laser transitions have a well-defined wavelength, which is broadened by Doppler broadening (Chapter 10). For a typical He–Ne laser, the Doppler width (full width at half maximum) is about 1.4 GHz, or 0.0019 nm. This is small compared to the frequency of the laser's familiar 632.8 nm red transition (4.738×10^{14} Hz), but a thousand times broader than the 1 MHz bandwidth of one longitudinal mode in a typical resonator. If you look closely at a high-resolution spectrum of the output of an He–Ne laser, you will see a series of spikes, each one a separate resonator mode, spread out under the Doppler-broadened line (e.g., Figure 9.15).

Their relatively narrow linewidth gives gas lasers good coherence. With Doppler broadening, the coherence length is more than 10 cm, whereas limiting emission to a single longitudinal mode can increase the coherence length to over 100 m. Such long coherence is crucial for interferometry and holography.

17.6 He–Ne LASERS

The He–Ne laser has long been the most common gas laser. Ali Javan, William R. Bennett Jr., and Donald R. Herriott demonstrated the first He–Ne laser at Bell Labs in December, 1960. It was also the first gas laser and the first cw laser. (All three laser types demonstrated earlier that year were pulsed solid-state lasers.)

The first He–Ne emitted in the near infrared at 1152 nm. The familiar red line at 632.8 nm was discovered later and is not as strong, but it became extremely important because it is easily visible to the human eye. He–Ne lasers emitting a few milliwatts were the first lasers to be mass produced and found wide use in applications from supermarket scanners to surveying equipment. They were the most common lasers used for any application well into the 1980s and were the most common visible laser until inexpensive red diode lasers reached the market in the 1990s. He–Ne lasers continue to be used in various instruments, holography, and a number of other applications, as well as in classroom demonstrations. Red He–Ne lasers are better than red diode lasers for some applications because of their better coherence and beam quality, as well as their very narrow and stable linewidth when operated in a single longitudinal mode. Commercial versions can generate up to a few tens of milliwatts, although powers of a few milliwatts are more common.

In addition, He–Ne lasers can operate on several other lines in the visible and near infrared. The most important lines in the visible are as follows:

- 543.5 nm: green
- 594.1 nm: yellow
- 604.0 nm: orange
- 611.9 nm: orange
- 632.8 nm: red (primary visible line)

He–Ne lasers also can operate on near-infrared lines at 1152, 1523, and 3390 nm. The only one of much current practical importance is 1523 nm, which lies in the main band used for high-performance fiberoptic communication systems. Near-infrared He–Ne lasers are mainly used for instrumentation.

Decades of experience have made mass-produced He–Ne lasers simple, practical, durable, and inexpensive. Typical He–Ne lasers consist of a rectangular or cylindrical head containing the laser tube, and a separate power supply that generates the high voltage that drives the laser. About 10,000 V is needed to ionize the gas; after ionization, a couple thousand volts maintain the operating current of a few milliamperes. Tubes typically range from 10 to 30 cm long, except for the highest-power lasers, which are longer.

17.7 PRINCIPLES OF He–Ne LASERS

Figure 17.4 shows the major energy levels involved in He–Ne lasers. The laser gas typically consists of five parts He and one part Ne, with the more-abundant He

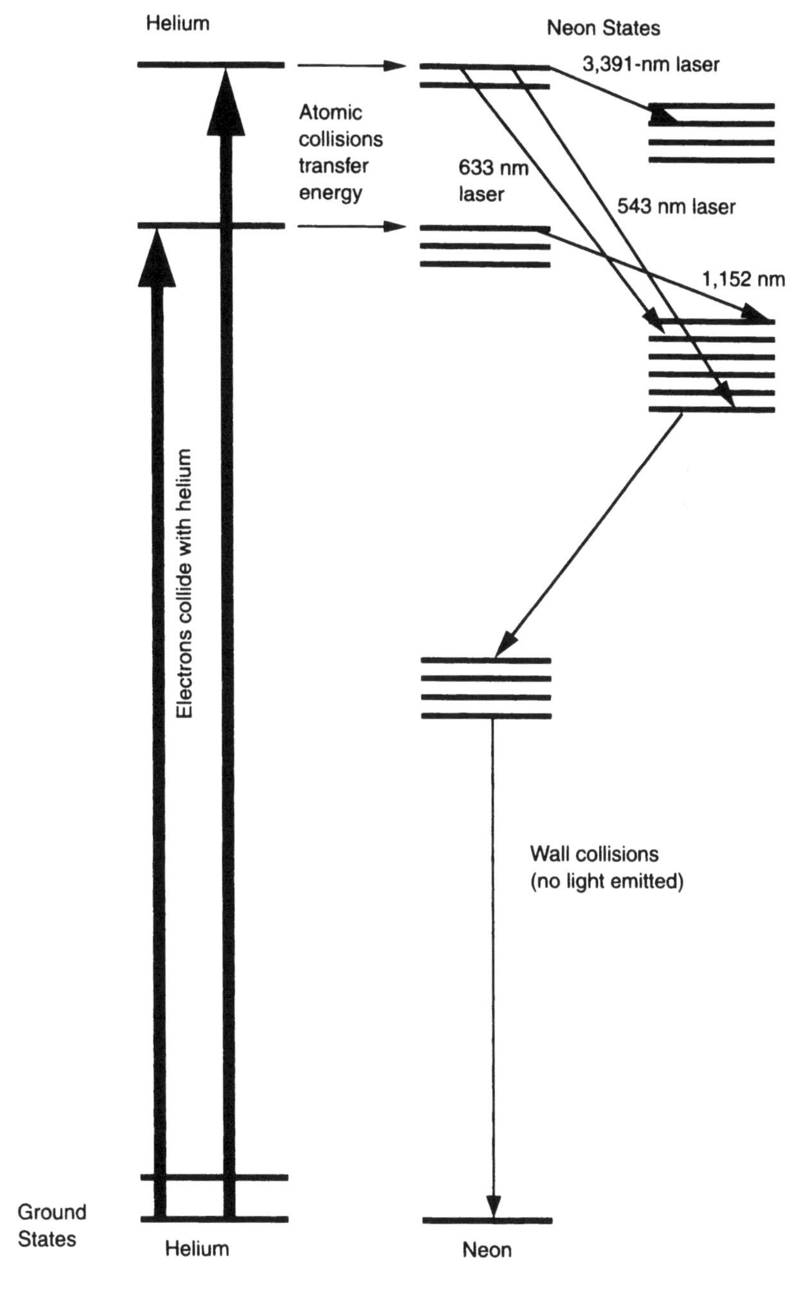

Figure 17.4 Energy levels in He–Ne laser; wide arrows show laser transitions (simplified and not to scale).

atoms collecting more energy from the electric discharge than the Ne. The He states excited by the electrons have nearly the same energy as two Ne levels, so when the atoms collide, the excited He atom readily transfers its energy to one of the two Ne levels. The Ne levels are metastable, so the atoms remain in them for a comparatively long time.

This excitation can produce population inversions on several transitions of Ne. Which transitions lase depends on the feedback provided by the resonator. The 1153 nm infrared line is particularly strong but rarely used because it has few applications. The 632.8 nm red line is the strongest visible line, generating powers to about 50 mW in commercial lasers. The green, yellow, and orange lines produce only milliwatts, but their colors are in commercial demand.

The lower laser levels are far above the ground state and depopulate quickly, dropping through a series of lower levels before they can be reexcited. This energy-level structure makes efficiency inherently low, only about 0.01–0.1% in standard He–Ne lasers.

17.8 STRUCTURE OF He–Ne LASERS

Figure 17.5 illustrates the internal components of a typical mass-produced He–Ne tube. The details differ significantly from the generic gas laser shown earlier. One important difference is that the discharge passes between a pair of electrodes at opposite ends of a narrow bore in the center of the tube, with internal diameter no more than a few millimeters. Confining the discharge this way raises excitation efficiency and helps control beam quality just as the aperture in a solid-state laser does. The rest of the tube serves as a reservoir for extra He and Ne, at a pressure of a few tenths of a percent of atmospheric pressure. Struts support the bore, keeping it at the center of the tube. The glass laser tube is contained in a metal housing providing structural support and protection, as well as isolating the user from the high voltage that powers the tube.

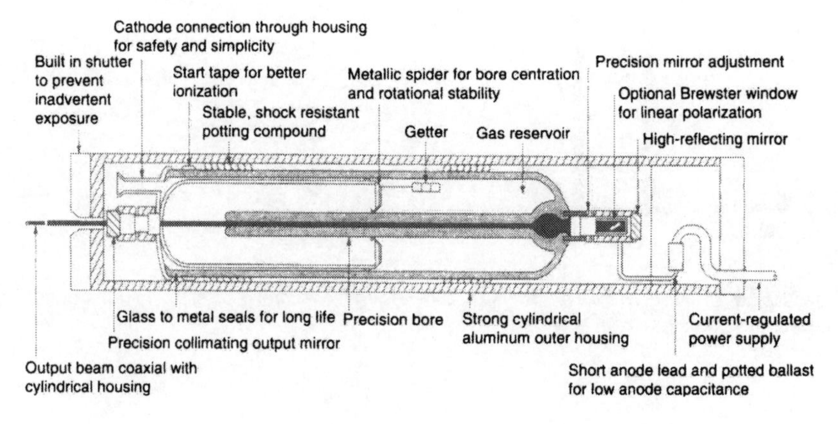

Figure 17.5 Internal structure of He–Ne laser (courtesy Melles Griot).

An interesting variant on the standard linear He–Ne laser is a *ring laser gyroscope* used to sense rotation. In this case, the tube forms a square or triangle, with mirrors at the corners that direct the oscillating internal beam from one arm of the tube to the next. When this ring rotates around its axis, the motion causes a phase shift between laser light going in opposite directions through the tube. Measuring this phase difference indicates the rotation rate. Such ring laser gyroscopes are used in some aircraft as alternatives to mechanical gyroscopes.

He–Ne lasers typically emit continuous TEM_{00} beams, with a diameter of about 1 mm and divergence of about 1 mrad. The beam is unpolarized unless an intracavity element such as a Brewster plate favors one polarization over the other. The output is monochromatic enough for many practical purposes. Typical Doppler-broadened width of 1.4 GHz for the red line corresponds to a coherence length of 20–30 cm, adequate for holography of small objects. Operation in a single longitudinal mode is possible with suitable optics, but such lasers are more expensive. Their 1 MHz bandwidth corresponds to a coherence length of 200–300 m.

The uses of He–Ne lasers are decreasing with the spread of visible semiconductor diode lasers, but diodes cannot meet all needs. The longer resonator of the He–Ne generates a cleaner beam in a single transverse mode, which is better for many applications in instrumentation and measurement in which the beam quality is at a premium. Examples include alignment of industrial equipment and biomedical instruments for cell counting and sorting. The good coherence of He–Ne lasers is important for applications in interferometry and holography.

17.9 Ar- AND Kr-ION LASERS

Ar- and Kr-ion lasers are another important type of electronic-transition gas laser. However, they are often called simply *ion lasers,* a term that is both vague and all-encompassing. Ar-ion and Kr-ion lasers resemble each other closely and can operate in the same tubes, but they oscillate at different wavelengths, the most important of which are listed in Table 17.2. Functionally, there are three important types: pure Ar, pure Kr, and lasers operating with a mixture of both gases. Ion lasers sometimes are classed according to their power levels, with the lower-power, air-cooled lasers (sometimes called small-frame) distinct from the higher-power, water-cooled (or large-frame) lasers.

Ar- and Kr-ion lasers are the most powerful cw gas lasers in the visible and UV. Ar-ion lasers can emit tens of watts on their most powerful green lines. Kr lasers can deliver several watts, and mixed-gas lasers emitting on lines of both gases can exceed 10 W. Smaller ion lasers also are available, with powers ranging from several milliwatts to hundreds of milliwatts.

The laser lines in Ar and Kr come from ions of the two gases. Visible wavelengths come from singly ionized atoms (Ar^+ or Kr^+), and UV wavelengths come from doubly ionized atoms (Ar^{+2} or Kr^{+2}).

The complex excitation kinetics start with an initial high-voltage pulse that ionizes the gas (at a pressure of about 0.001 atm), so it conducts a high current. Electrons in the current transfer some of their energy to gas atoms, ionizing them and

Table 17.2 The many wavelengths of rare-gas ion lasers

Argon (nm)	Krypton (nm)
275.4	337.4
300.3	350.7
302.4	356.4
305.5	406.7
334.0	413.1
351.1	415.4
363.8	468.0
454.6	476.2
457.9	482.5
465.8	520.8
472.7	530.9
476.5	568.2
488.0 (strong)	647.1 (strong)
496.5	676.4
501.7	752.5
514.5 (strong)	799.3
528.7	
1090.0	

leaving the ions in high-energy levels. These ions then drop into many metastable states, which are the upper levels of laser transitions. These excited ions can be stimulated to emit on many different laser transitions, dropping to a lower laser level, which is quickly depopulated, as shown in simplified form for argon in Figure 17.6. Ions dropping from the lower laser level emit light in the extreme UV (74 nm for Ar), creating harsh conditions within the laser.

It takes a large amount of energy to excite the visible laser transitions because they are far above the ground state of the singly ionized species. This contributes to their inherently low efficiency. The UV transitions are even less efficient because they occur in doubly ionized atoms, which must be raised even higher above the ground state of the atom.

Ionizing the laser gas causes its resistance to drop, so it can carry a large current, typically 10–70 A. That is more than a thousand times the current in an He–Ne laser, but the lower resistance also means the operating voltages are lower: 90–400 V. The overall energy deposited in the gas is much higher than in the He–Ne laser, reflecting the lower efficiency of ion lasers.

Ar, Kr, and mixed-gas lasers use essentially the same tube structures, such as shown in Figure 17.7; the difference is in the gas fill. Tube design and operation varies significantly with power level. Operating conditions can be quite difficult in high-power lasers, in which large quantities of heat must be dissipated. In addition, tube materials must be able to withstand the intense extreme UV light emitted by ions dropping from the lower laser level. Ceramic and metal–ceramic structures are common.

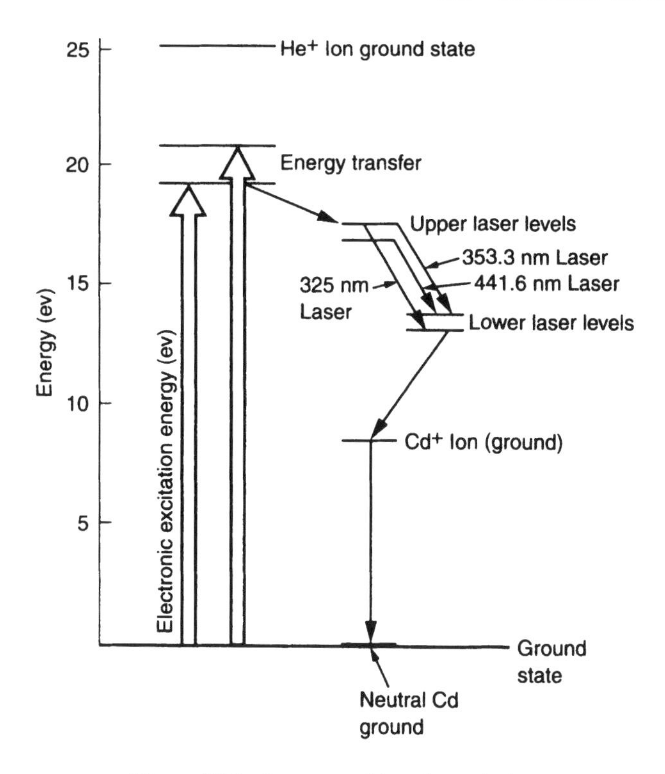

Figure 17.6 Laser transitions in an Ar laser.

As in other electronic transition lasers, the discharge typically is concentrated in a narrow central region. This may be a thin separate bore attached to a gas reservoir and fatter cathode region, or a series of large central holes in metal disks spaced along a larger ceramic tube. In both cases, the tube needs a return path for the positively charged ions that move toward the anode, although there is no metal to coat cool spots inside the tube. High-power Ar lasers may be up to 2 m long. With their low efficiency, Ar- and Kr-ion lasers need very high electrical in-

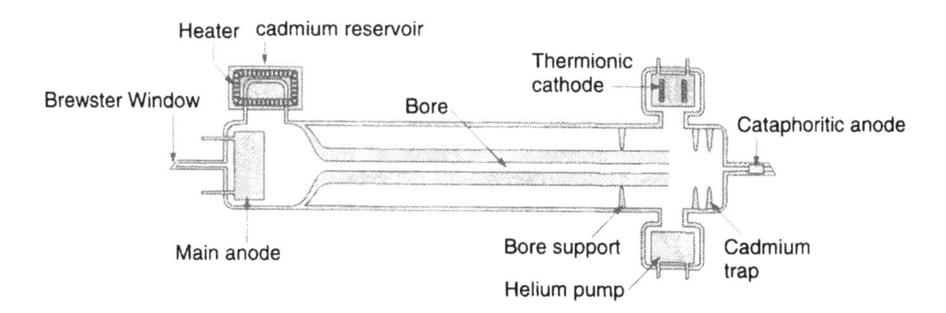

Figure 17.7 Structure of an Ar-ion laser.

put, and high-power models typically operate directly from 400 V, three-phase power.

Discharge conditions inside the tube are extreme, typically limiting tube lifetimes to 1000–10,000 h of operation, with the highest-power tubes the shortest lived. The tubes often can be reconditioned by cleaning and replacing worn components and adding fresh laser gas—a worthwhile undertaking because new tubes can be quite expensive.

Ar and Kr lasers have low gain, so resonator loss must be minimized, and the output-coupling mirror transmits only a small fraction of the intracavity beam. Some low-power ion lasers have internal mirrors, but others have Brewster-angle windows at the ends of the tube with external mirrors.

Resonator optics determine the wavelengths produced by Ar- and Kr-ion lasers. Only those transitions for which round-trip gain exceeds round-trip loss can achieve threshold. Mirrors with broadband reflectivity can allow simultaneous oscillation on several visible lines—all the way from blue to red in mixed-gas lasers. If sheer power is the goal, mirrors with limited band reflectivity can select the two strongest Ar lines at 488 and 514 nm, or the entire Ar band from 457 to 514 nm. Narrow-line lasers can be built that operate on single wavelengths. Adding tuning elements such as a prism to the resonator makes it possible to tune among several emission lines.

When operated on a single line, Ar- and Kr-ion lasers typically have a spectral linewidth of about 5 GHz, encompassing many longitudinal modes. Standard resonator optics produce TEM_{00} beams on single lines, but high-power lasers oscillate in multiple transverse modes. Beam quality is usually comparable with that of other electronic-transition gas lasers.

Ion lasers have a diverse range of applications, although low-power uses are endangered by steady progress in solid-state lasers. Air-cooled ion lasers have been used in high-speed printers and graphics systems, biomedical instrumentation, cell sorting, inspection systems, flow cytometry, and research. High-power water-cooled ion lasers have been used for optical-disk mastering, fabrication of fiber gratings and three-dimensional prototypes, photoluminescence studies, materials research, and semiconductor production. A major use of mixed-gas ion lasers has been for entertainment and displays, because they can simultaneously oscillate at several visible wavelengths from red to green and blue, generating multicolor displays.

Chapter 18

Carbon Dioxide and Other Vibrational Lasers

Gas lasers based on molecules emitting light on transitions between vibrational energy levels differ in important ways from the electronic-transition gas lasers described in Chapter 17. (The differences between vibrational and electronic transitions were reviewed in Chapter 6.) The most important functional differences are their longer wavelengths in the infrared and the higher efficiency of some vibrational transition lasers. The best known of this large family of lasers is the carbon dioxide (CO_2) laser, widely used in industry and medicine. Chemical lasers also are vibrational lasers, but their operating characteristics make them less suitable than CO_2 for industrial applications, and they have largely been developed as potential high-energy laser weapons.

The commercial importance of the CO_2 laser comes from its combination of high efficiency and high output power. Typically, 5–20% of input power emerges in the output beam, the highest of any gas laser, although lower than some semiconductor and fiber lasers. This high efficiency limits both power consumption and heat dissipation, so industrial CO_2 lasers can generate continuous powers from 1 W to 10 kW. Typical powers are below 1 kW.

In addition to being efficient, the CO_2 laser lends itself to efficient removal of waste heat left over from laser excitation. Flowing gas through the laser can remove waste heat efficiently, by transferring the heat to flowing air surrounding the tube, flowing water, or, in some cases, exhausting the laser gas directly to the atmosphere.

CO_2 lasers emit at wavelengths between 9 and 11 μm. The strongest emission is near 10.6 um, which is often listed as the nominal wavelength. These wavelengths are strongly absorbed by organic materials, ceramics, water, and tissue; therefore, tens of watts often suffice for applications such as cutting plastics or performing surgery. By contrast, most metals reflect strongly at 10 μm, so metal working requires higher powers at that wavelength than at shorter wavelengths. Years of engineering development have made the CO_2 laser a practical tool for moderate- to high-power commercial applications. The only gas lasers capable of delivering the 100 kW power levels required for laser weapons are the chemical lasers described at the end of this chapter.

Introduction to Laser Technology, Fourth edition. C. Breck Hitz, J. Ewing, and J. Hecht

18.1 VIBRATIONAL TRANSITIONS

Molecules have a large and varied family of vibrational modes, which are resonances that depend on the mass and bonding of the atoms in the molecule. The number of distinct modes increases with the number of atoms in the molecule. Each mode has its own precise excitation energy, and all the adjacent levels of a given mode are separated by that amount of energy.

Figure 18.1 shows the three principal vibrational modes of the CO_2 molecule: the symmetric stretching mode, ν_1; the bending mode, ν_2; and the asymmetric stretching mode, ν_3. Each vibrational mode has its own ladder of quantized energy levels, as shown at the bottom of Figure 18.1. In this case, the energy steps are smallest on the bending mode, roughly twice as large on the symmetric stretching mode, and even larger on the asymmetric stretching mode.

Transitions between pairs of these vibrational energy levels enable vibrational lasers. As in four-level, electronic-transition lasers, an external energy source populates a pump level. The molecule quickly drops from the pump level to a long-lived metastable state, which is the upper laser level. Then, stimulated emission drops the molecule to the lower laser level, which then decays to the ground state. All four levels are vibrational levels in the ground electronic level.

Vibrational transitions typically correspond to infrared wavelengths of 1–100 μm, so they require less excitation energy than electronic transitions. But remember that there are dozens of rotational levels associated with each vibrational level (see Chapter 6). And the rules of quantum mechanics require that a molecule change exactly one rotational level when it moves from one vibrational level to another. Take a minute to think about what this requirement means. If the molecule starts in the third rotational level of the upper vibrational level, it must end up in the second or fourth rotational level of the lower vibrational level. If it starts in the sixth rotational level, it must end up in the fifth or the seventh rotational level. The result is that each vibrational transition includes many possible rotational transitions. The overall effect is to make vibrational lasers emit at a broad range of wavelengths, corresponding to the energy of the vibrational transition plus or minus that of the rotational transition.

The CO_2 laser provides a good example of how this works. The first excited asymmetric stretching mode of CO_2 is the upper laser level for two laser transitions, as shown in Figure 18.1. If the excited molecule drops to the first excited state of the symmetric stretching mode, it releases a photon with a nominal wavelength of 10.5 μm. Alternatively, it can drop to the second excited level of the bending mode, releasing a photon at 9.6 μm. The 10.5-μm transition is somewhat stronger, but lasers can oscillate simultaneously on both bands.

As shown in Figure 18.2, each vibrational transition has its own family of rotational transitions. Each step away from the nominal center of the vibrational transition corresponds to a one-level step in rotational energy. Depending on the resonator optics, CO_2 lasers can oscillate on one or many lines simultaneously. Industrial lasers optimized for power delivery typically operate broadband, with a nominal wavelength of 10.6 μm, but actual emission is from 9 to 11 μm. Scientific or spe-

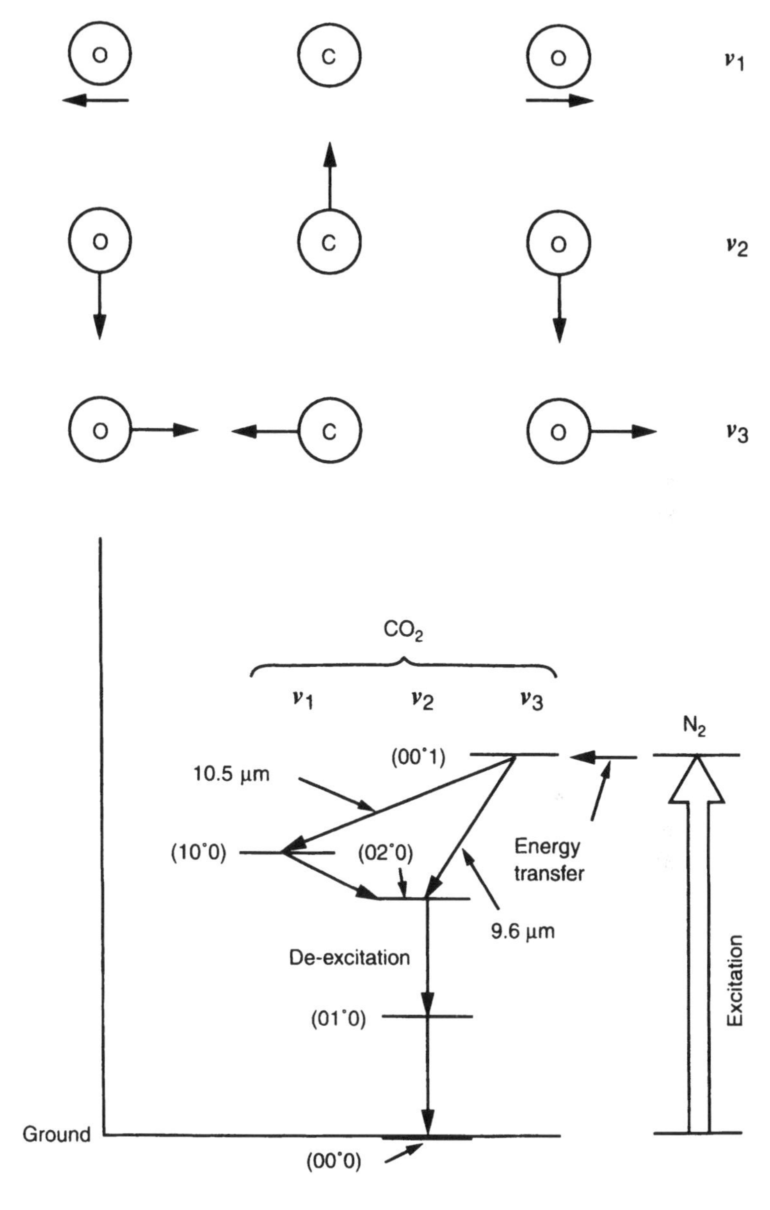

Figure 18.1 Major vibrational modes of CO_2 molecules (top) are symmetric stretching, ν_1; bending, ν_2; and asymmetric stretching, ν_3. The first excited level of the asymmetric stretching mode serves as the upper laser level for both 9.6 and 10.5-µm CO_2 transitions. (From Jeff Hecht, *Understanding Lasers: An Entry Level Guide,* 3rd ed. IEEE Press, Piscataway, NJ. Used with permission.)

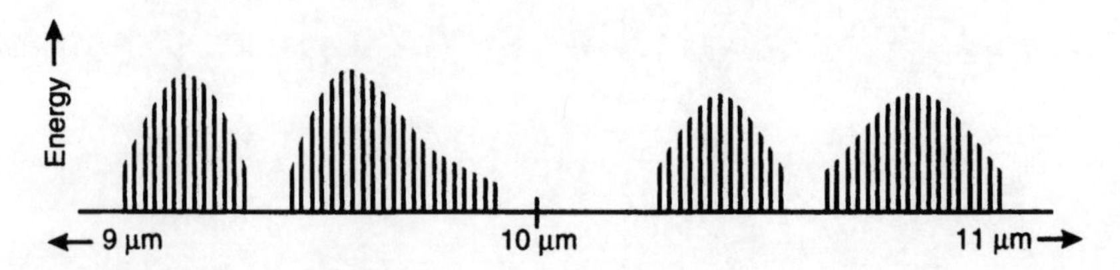

Figure 18.2 Families of rotational lines surround both the 9.6 and 10.5 μm vibrational lines of CO_2. (From Jeff Hecht, *Understanding Lasers: An Entry Level Guide,* 2nd ed., IEEE Press, Piscataway, NJ. Used with permission.)

cial-purpose lasers often are limited to oscillate on a smaller part of the band or on a single rotational line.

18.2 EXCITATION

The standard technique for creating a population inversion in CO_2 and most other vibrational lasers is to apply an electric discharge through the gas. Typical voltages are kilovolts or more, and as in electronic-transition lasers, a higher voltage may be needed to break down the gas and initiate the discharge. Total gas pressure must be kept below about one-tenth of an atmosphere to sustain a stable continuous discharge, and this is the standard operating mode for most commercial CO_2 lasers.

The addition of other gases aids in energy transfer to and from CO_2 molecules. Molecular nitrogen absorbs energy from the electric discharge more efficiently than CO_2 does. Because nitrogen's lowest vibrational mode has nearly the same energy as the upper laser level of CO_2, it readily transfers the energy to the CO_2 during a collision. Thus, adding nitrogen to the CO_2 in the laser discharge enhances the excitation process.

Helium is added to the gas mixture in a CO_2 laser because its thermal conductivity is much higher than CO_2's, so it can efficiently remove waste heat from the gas mixture. He also helps depopulate the lower laser level, thereby increasing the population inversion.

An alternative way of exciting CO_2 is by thermal expansion of a hot laser gas, in what is called a gas-dynamic laser. Very rapid expansion of hot gas at temperatures of about 1100°C and pressures above 10 atm through a fine nozzle into a near-vacuum produces a population inversion in the cool, low-pressure zone. This approach can generate high powers in the laboratory but has not found commercial application.

Chemical reactions also can create a population inversion in vibrational lasers. In this case, the energy comes from the chemical reaction itself, with the amount dependent on the reacting molecules. This type of excitation can produce the extremely high powers required for military applications, as described at the end of this chapter. (Chemical excitation is not utilized in CO_2 lasers, however.)

18.3 TYPES OF CO_2 LASERS

Several types of CO_2 lasers have been developed for particular applications. Like other gas lasers, commercial CO_2 lasers usually generate a continuous beam, but they also can be pulsed by modulating the discharge voltage.

The simplest type is a *sealed-tube CO_2 laser* with longitudinal discharge passing along the length of the tube, as shown in Figure 18.3. A radio-frequency-induced discharge also can excite the gas. Mirrors are placed on the ends of the tube to form a resonant cavity. Unlike He–Ne lasers, the discharge is not concentrated in a narrow bore but spreads through a larger volume.

Although this design is attractively simple, it has important limitations. The discharge dissociates CO_2 molecules, freeing oxygen and reducing CO_2 concentration below the levels needed for laser action within a matter of minutes. To regenerate CO_2, water or hydrogen may be added to the laser gas, which reacts with the carbon monoxide formed by the discharge. Alternatively, a metal cathode can catalyze gas regeneration. These measures can sustain CO_2 levels for thousands of hours of tube operation, after which the tube must be cleaned and refilled.

Another limitation is output power. The amount of energy a longitudinal discharge can transfer to the stationary gas is limited, imposing an output power limit of about 50 W/m of tube length, or roughly 100 W for the largest of these lasers. Although the helium in the gas mixture augments heat removal, dissipation of waste heat can still be a problem.

Waveguide CO_2 lasers are a type of sealed CO_2 lasers in which the discharge and laser gas are concentrated in a laser bore a few millimeters across, which functions as a waveguide for 10 μm light, as shown in Figure 18.4. The waveguide design reduces the diffraction losses that otherwise would occur in a laser with an output aperture a relatively small number of wavelengths across. Because the gas volume

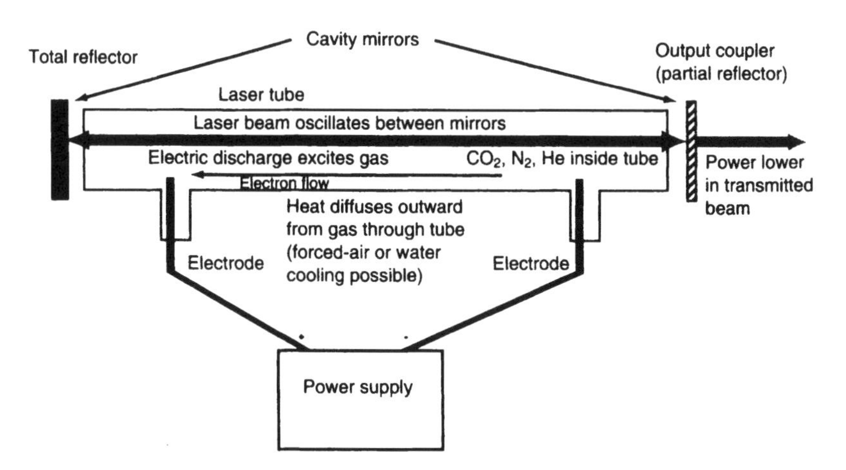

Figure 18.3 Sealed-tube CO_2 laser with longitudinal discharge.

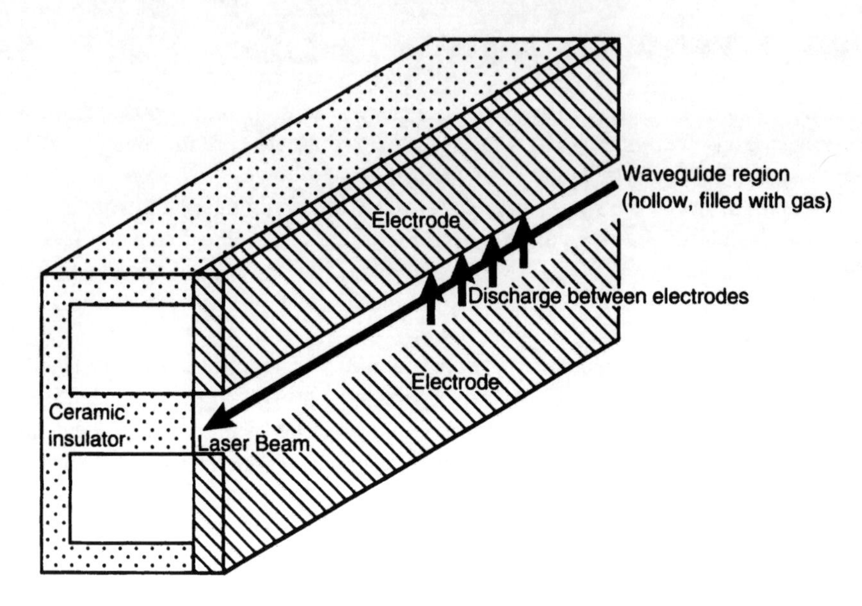

Figure 18.4 Cutaway view of a waveguide CO_2 laser, showing metal-insulator design in cross section; the right half of the insulator has been removed to expose the gas-filled waveguide. Cavity optics are not shown.

inside the waveguide is small, the sealed laser cavity includes a gas reservoir, and gas circulates through the waveguide and reservoir.

Typically a waveguide laser is excited by passing a radio-frequency or direct-current discharge transverse to the length of the waveguide. In the design shown in Figure 18.4, the laser tube is a metal–ceramic hybrid structure. Metal rods run the length of the tube at top and bottom; ceramic blocks separate them, sealing the waveguide and insulating the metal plates from each other. A radiofrequency discharge passes between the top and bottom metal plates, exciting laser gas passing through the waveguide.

The waveguide structure is attractive because of its compact size, high efficiency, and low cost. As in other sealed CO_2 lasers, it typically generates up to about 100 W, although folded tubes can deliver somewhat higher powers. The line between waveguide and conventional sealed CO_2 lasers can become hazy as waveguide diameters increase, and the term *waveguide* is not always used. Gas circulation typically is forced in waveguide lasers, unlike simple sealed CO_2 lasers.

Flowing-gas CO_2 lasers can provide higher powers than static sealed lasers because the gas is blown through the tube. The heated gas in the tube is constantly being replaced with new, cool gas.

The key variables in flowing-gas CO_2 lasers are the speed and direction of flow. Typically, flow is longitudinal, along the length of the laser, as shown in Figure 18.5. Pumps or turbines provide fast axial flow, improving heat transport and allowing higher power levels.

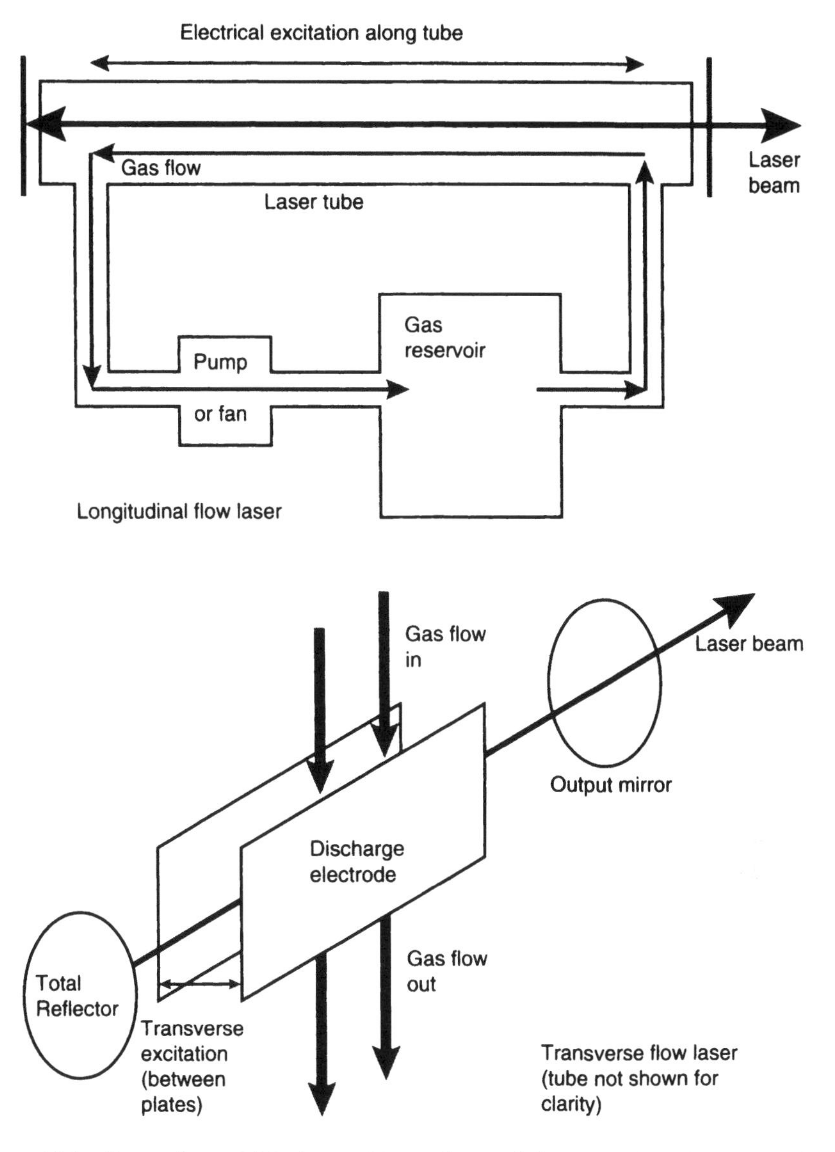

Figure 18.5 Comparison of CO_2 laser with gas flow and discharge along the length of the tube (at top) with one having gas flow and discharge transverse to the length of the tube and to each other (at bottom).

At the highest power levels used for industrial lasers, gas flow is transverse to the laser axis, or across the laser tube. In these high-power systems, also shown in Figure 18.5, the electrical excitation discharge is applied transverse to the length of the laser cavity. Because this gas flows through a wide aperture, it does not have to flow as fast as in a longitudinal-flow laser. Typically, the gas is recycled, with some fresh gas added.

Gas-dynamic CO_2 lasers are quite distinct from other CO_2 lasers because they rely on a different excitation mechanism. Hot CO_2 at high pressure is expanded through a small nozzle into a near-vacuum, a process that produces a population inversion as the expanding gas cools. When the expanding gas flows transversely through a resonant laser cavity, it can generate powers of 100 kW or more. Gas-dynamic CO_2 lasers have been used in high-energy military experiments but are delicate and have fallen out of favor except for research. Their high powers are not needed for civilian applications.

Transversely excited atmospheric (TEA) CO_2 lasers, unlike other types, are designed for pulsed operation at gas pressures of 1 atm or more. The electrodes are placed on opposite sides of the laser axis, as in other transversely excited lasers, but the discharge fires pulses into the gas lasting nanoseconds to about 1 μs, producing laser pulses of the same length. TEA CO_2 lasers cannot generate a continuous beam, but they are attractive sources of intense pulses of 40 ns to about 1 μs.

18.4 OPTICS FOR CO_2 LASERS

CO_2 lasers require different optics than used at visible and near-infrared wavelengths because conventional silicate glasses are not transparent at 10 μm. Solid metal mirrors are widely used because most metals are strongly reflective at 10 μm and metals are good thermal conductors for removing any excess heat. Some output-coupling mirrors are made of metal, with the transmitted part of the beam emerging through a hole and the rest reflected back into the laser cavity.

Windows and other transparent optics must be made of special infrared-transmitting materials such as zinc selenide. Many of these materials do not look transparent to the eye but do transmit the invisible 10 μm beam. No highly transparent optical fibers are available for 10 μm, but flexible hollow waveguides can carry the beam.

Special care is needed with CO_2 laser beams because the only indication that the laser is operating may be its emission indicator. Normal infrared viewers used with near-infrared, solid-state lasers respond to light in the 1 μm region, but not in the 10 μm band of CO_2 lasers. Military night-vision equipment does operate in the 10 μm band because that is the peak of thermal emission at room temperature, so soldiers and animals stand out as bright objects.

18.5 CHEMICAL LASERS

Chemical lasers rely on a chemical reaction to produce vibrationally excited molecules either directly as reaction products or indirectly by transferring energy from reaction products to another species. The concept is quite attractive because chemical reactions can generate large amounts of energy efficiently and can selectively produce the correct excited states to populate the upper laser level.

Figure 18.6 shows the structure of a simple chemical laser in which two reactants—hydrogen and fluorine—flow into a chamber where they react to produce vibrationally excited molecules of hydrogen fluoride (HF). The hot gas then flows through a laser cavity, which extracts energy in the form of a laser beam. Gas leav-

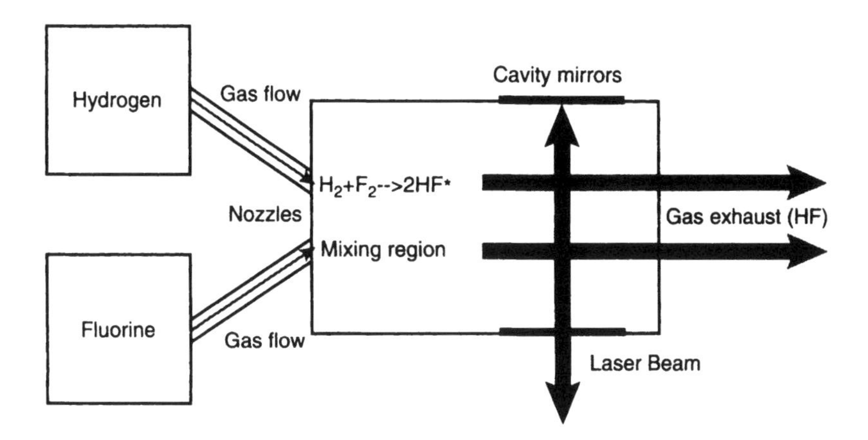

Figure 18.6 A simplified HF chemical laser, showing gas mixing and extraction of a laser beam downstream.

ing the laser cavity is pumped into a spent-gas container for treatment or disposal or is exhausted to the atmosphere.

Chemical lasers have generated little interest for commercial applications, but their high power has led to much interest for military development of high-energy laser weapons. This has led to the development of research lasers that resemble rocket engines with a laser cavity in the exhaust stream. Various hydrogen and fluorine compounds can fuel HF lasers, but it is simplest to view their operation as a chain reaction in which a fluorine atom reacts with a hydrogen molecule, producing an excited molecule designated HF^* and leaving a hydrogen atom. The hydrogen atom then reacts with molecular fluorine, again producing HF^*, but this time leaving a fluorine atom that can react with H_2 to produce another HF^*. The laser emits on vibrational lines between 2.6 and 3.0 μm when used with the abundant hydrogen-1 isotope. Using the rare, stable isotope deuterium (hydrogen-2) shifts the wavelengths to 3.6–4.0 μm, at which the atmosphere is more transparent. DF lasers have generated powers to 2 MW, but they did not prove practical for laser weapons.

For use in its two latest most recent airborne demonstration systems, the U.S. Air Force used a different type called the chemical oxygen–iodine laser or COIL, which emits at a wavelength of 1.315 μm. COIL is actually a hybrid of a chemical and electronic-transition laser. It draws its energy from the chemical reaction between chlorine gas and a mixture of hydrogen peroxide and potassium hydroxide, which generates excited oxygen molecules. The excited oxygen molecules then are mixed with iodine molecules, which dissociate to produce electrically excited iodine atoms, which emit a powerful beam. The 1.315 μm wavelength can be focused to a more intense spot than the longer HF wavelength, better able to destroy military targets.

Chapter 19

Excimer Lasers

Excimer lasers are important because they were the first lasers capable of producing high-power, ultraviolet (UV) output with good electrical efficiency. Remember that UV photons contain more energy than visible or infrared photons, so UV photons can often do things that less-energetic photons cannot. This fact makes excimer lasers useful in a wide range of applications. Though solid-state lasers have experienced significant improvement in power capability with the onset of less-expensive diode pump arrays, the conversion of a solid-state laser's near-infrared output to the ultraviolet requires nonlinear optical frequency shifting, a step that limits achievable power. In practice then, if one needs an average output power greater than about 5 W in the UV, or to access wavelengths where nonlinear materials severely limit the UV power, the excimer is the laser of choice.

The lasers discussed in Chapter 17 are gas lasers also capable of producing UV output, albeit less efficiently than excimers and at much lower power. Unlike these lasers, excimers operate at a high pressure. The laser itself is of a very different design, featuring discharge excitation that is transverse to the optical axis. Figure 19.1 is a schematic diagram of the layout of an excimer laser *transverse discharge* configuration. A transverse discharge differs from a *longitudinal* discharge found in He–Ne and rare-gas ion lasers, which tend to have very small bore tubes in which the discharge runs. The transverse discharge is required for a high-pressure laser. Higher pressure, of course, means more molecules in a given volume, and more molecules can provide a larger flux of laser photons. The peak power during a pulse is typically well over 1 MW compared to less than 1 W for other UV gas lasers. Average powers over 10 W are readily attainable by pulsing the gas mixture at 50 to 1000 pulses per second. Thus, for a reasonably sized (and reasonably priced) device, excimers are the lasers of preference for UV outputs above 1 W.

There is another difference between excimers and the low-pressure gas lasers discussed in Chapter 17. The electrical energy that creates the population inversion in an excimer laser is delivered in pulses, rather than continuously. That is because the gain coefficient of the excimer is much lower than that of an ion laser and the pump power during the discharge needs to be much higher. But such a high-energy discharge necessary in an excimer laser becomes unstable if it is on for more than roughly 50 ns. This requirement posed serious electrical challenges to the development of excimer lasers. Like the Q-switched, solid-state lasers discussed in

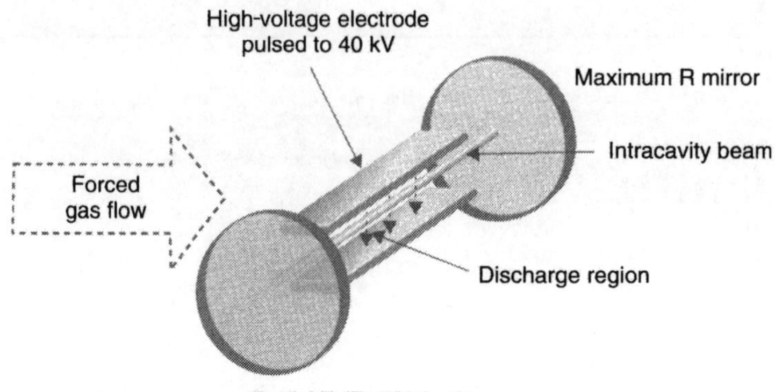

Figure 19.1 Unlike lower-power, cw ion lasers, the excimer laser runs at high pressure and with pulsed excitation. As such, the layout is of a transverse-discharge type. The laser optical axis runs perpendicular to the direction of current flow in the laser head. Since each discharge pulse both disrupts the optical properties of the gas and dissociates the excimer laser's precursor molecules, the gas is pushed through the laser head continuously with an external flow system. Typically, the gas flows in the third direction perpendicular to both the discharge direction and the optical axis.

Chapter 15, excimers produce short—perhaps 10 to 20 ns duration, pulses. But the excimers are not Q-switched; the short pulses are natural for excimer lasers.

Beam quality is yet another difference between excimers and most other gas lasers. In a low-gain laser, the photons must make many passes back and forth through the population inversion before they are amplified to the output level. That means that all the photons have to be well aligned with the axis of the resonator or they will go astray before completing all those round-trips. But in a high-gain laser, with a short pulse time dictated by the fast pulse electrical driver like an excimer, the photons only have to make a few round-trips to get out of the resonator. The output-coupling partial reflector does not have a very high reflectivity. Moreover, because the aperture of the excimer laser is usually quite large, a few square centimeters compared to less than a square millimeter for an ion laser, the photons can take a variety of paths to get out of the resonator, so they need not be so well aligned. These poorly aligned photons create a beam that is more divergent than the beam from a typical low-gain gas laser like the He–Ne laser. Many excimer lasers use an unstable resonator, as discussed in Chapter 8, to generate a beam with fairly low divergence that can be focused to a small spot.

The biggest difference between excimers and other lasers is the nature of the lasing medium itself. Chapter 7 discussed three- and four-level lasers, but those classifications do not exactly apply to excimer lasers. In terms of Figure 7.10, excimers are four-level lasers in which the lower laser level disappears very fast (in a picosecond or less). Because of the unique properties of these lasers, we describe them here starting from the basic physics and chemistry of the laser medium and working out to the excitation, gas processing, and optical systems.

Commercially, the most important excimers are the rare-gas halide* variety, such as krypton fluoride (KrF) with output at 248 nm, argon-fluoride at 193 nm, and xenon chloride at 308 nm. There are other, less common excimer lasers based on transitions in xenon fluoride (XeF) at 353 nm, and in fluorine (F_2, which is not a rare-gas halide excimer but uses similar lasing mechanisms) at 157 nm. There are also weaker, less-common excimer lasers using krypton chloride at 222 nm, xenon bromide at 282 nm, and the broadband transition in XeF at about 480 nm. Mercury halides, such as mercury bromide at 502 nm, exhibit transitions that can make very efficient excimer lasers, but these lasers must operate at such high temperatures that they find no practical application. Very early research into the excimer concept used relativistic electron beams to produce excimer outputs in the vacuum UV (wavelengths shorter than about 185 nm) from xenon (Xe_2), krypton (Kr_2), and argon (Ar_2) lasers, but such lasers have not found practical application.

19.1 EXCIMER MOLECULES

We talked about molecular energy levels in Chapter 6, so you know that molecules, like atoms, can exist in excited states. Ordinary molecules can start out in the ground state, absorb energy from an external source, and become excited. Once in an excited state, they can lose the energy and go back to the ground state. However, excimer molecules are different. They can exist as a form of stable molecule (i.e., they have some chemical binding energy relative to their parent atoms) only if they are in an excited state. If an excimer molecule loses its energy and falls back to the ground state, the molecule breaks apart into individual atoms in a few picoseconds.

KrF is an example of an excimer molecule. Kr is a rare gas, an inert element that does not normally form molecules because its outer shell of electrons is completely full. Normally, the closer an Kr and F atom get to each other, the harder they push each other away. They will never form a stable molecule. But suppose the Kr atom absorbs energy from an electric discharge and enters an excited state. If that excited Kr atom approaches an F atom, the force between them will be attractive rather than repulsive. The atoms will pull together to form a stable molecule of KrF in an excited state. In other words, the extra energy in the excited Kr atom will be used to make an excited state of the excimer molecule, above the energy of the ground state excimer molecule but stable relative to the atom's excited state.

The excited molecule will not be stable for long, though. Typically, the lifetimes of these excimer molecules are measured in nanoseconds. The excited excimer molecule loses its energy, mostly by emitting a photon, and decays to the ground state. But because the ground state is inherently unstable and cannot exist for more than a picosecond or so, the molecule flies apart. This property of lower-level dissociation provides a simple means to have a population inversion, one requirement for laser gain.

*The rare gases are the chemically inert elements (He, Ne, Ar, Kr, Xe, etc.) that appear along the right side of the periodic table. The halides are the chemically active elements (F, Cl, Br, etc.) appearing one column from the right of the periodic table.

That is the basic mechanism of an excimer laser. The photons are produced when the excimer molecule decays to its ground state and flies apart. Let us take a closer, step-by-step look at what happens in an excimer laser. The laser's physical structure is not like that of low-power gas lasers, but more akin to a very high-power CO_2 laser. There is a chamber containing the appropriate gases (a small amount of Kr and F_2 in the case of a KrF laser, plus a buffer gas such as He or Ne), and the population inversion is created by an electric current that runs from the high-voltage electrode through the gas to the other electrode at ground voltage. The flow of current is in a direction perpendicular to the optical axis and gas flow.

The cycle begins when an electron collides with a ground-state Kr atom. The electron can both excite the Kr to an excited state or, if it has enough energy, it can break an electron off the Kr, creating a Kr ion. These two reactions can be written as follows:

$$e + Kr \rightarrow Kr^* + e$$

or

$$e + Kr \rightarrow Kr^+ + 2e$$

In this shorthand notation, e is an electron, Kr is a ground-state Kr atom, Kr^* is an excited-state Kr atom, and Kr^+ is a Kr ion (i.e., a Kr atom from which an electron has been removed).

Also, some of the electrons collide with F_2 molecules. The collision of electrons with F_2 molecules breaks the molecules apart, and may ionize the individual atoms:

$$e + F_2 \rightarrow F + F^-$$

Note that the F^- ion is negatively charged—it has absorbed an extra electron. The Kr ion loses an electron and is positive.

Now there are all these excited-state Kr atoms, F ions, and Kr ions in the laser discharge region. They can combine in several ways to create excimer molecules, such as

$$Kr^+ \rightarrow F^- \text{ (+ a third body such as He)} - KrF^* \text{ (+ the third body)}$$

or by reaction of the excited state of the Kr atom, Kr^*:

$$Kr^* + F_2 \rightarrow KrF^* + F$$

One nice thing about excimer lasers is that it is easy to create a population inversion. There is no ground state, so a single excimer molecule corresponds to a condition of population inversion. All you have to do is add the appropriate feedback and you have what it takes to make a laser: a population inversion and a resonator. Of course, you must have enough gain to overcome the resonator's round-trip losses, and this is a difficult task. You need more than the inversion density caused by one excimer molecule being created. The inversion needed to overcome the round-trip losses is roughly 10,000 times more than that required for other kinds of gas lasers. Greater inversion needed for lasing implies greater power input to get that number of excited states. As a result, excimer lasers are pulsed because it is too difficult to maintain such a high inversion, and pour energy into the gas medium, over an extended period.

Unfortunately, there are other ways the excimer molecule can lose its energy without producing a laser photon. It can decay spontaneously, or it can give off the extra energy in the form of heat:

$$KrF^* + h\nu \rightarrow KrF + 2h\nu \qquad \text{stimulated emission (good)}$$

$$KrF^* \rightarrow KrF + h\nu \qquad \text{spontaneous emission (bad)}$$

$$KrF^* + F_2 \rightarrow Kr + 3F + \text{heat} \qquad \text{(bad)}$$

The F_2 collides with KrF* and is dissociated, and the KrF* is deactivated by the collision. Note that the lower laser level KrF noted in the reaction above falls apart in about a picosecond.

Here, $h\nu$ is the photon. If the energy in the excimer molecule is not claimed by stimulated emission within a few nanoseconds, it will be lost to one of the other mechanisms. To make sure that stimulated emission occurs before the energy is lost, there must be a sufficient flux of photons present to stimulate the excited molecule; that is, the photon flux in an excimer laser has to be large enough to ensure that a photon gets to the excimer in the few nanoseconds between its creation and its spontaneous decay. As a practical matter, that means that excimer lasers must sustain a photon flux of a few megawatts per square centimeter, thousands of times higher than the flux in other gas lasers.

We mentioned earlier that there is a buffer gas such as He, Ne, or Ar in a KrF laser. This other inert gas such as He serves much the same purpose it does in a CO_2 laser: its high thermal conductivity helps remove heat from inside the laser. The He also acts as a catalyst—a chemical agent that facilitates a particular reaction—in creating the KrF* molecules in the recombination process.

19.2 ELECTRICAL CONSIDERATIONS

Excimer lasers cannot be pumped continuously because of the high pressure needed for the kinetic processes discussed above and the resulting instability of the electrical discharge. They need the electricity that creates the population inversion to be provided in high-current pulses, because the inversion needs to be so large, and the discharge pumping so intense—that the electrical discharge typically becomes unstable if it is maintained for more than about 50 ns. (How long and hard an excimer laser can be pumped is a subtle trade-off of the degree and uniformity of preionization, the shape of the electrodes, and the rate at which the power supply can feed current into the discharge.) Thus, all the electrical energy must be deposited in the gas in a few tens of nanoseconds. This makes the electrical design of an excimer laser a challenging task, one that requires more expensive components than you would use in an ordinary gas laser.

The electrical circuits that work in cw discharge-pumped lasers, such as CO_2 and He–Ne, are far too slow to work with excimers. An entirely different approach is necessary. In most cases, the technique involves a two-stage power supply: the first stage to charge up an energy-storage system, and the second stage to switch the stored electrical energy to the laser head. The typical times for these two stages are roughly 1 μs for the charging stage and roughly 50 ns for the fast-discharge stage. The fast switch in the second stage is usually a thyratron in combination with a magnetic circuit. A magnetic circuit is a fast (and expensive) combination of a gas switch with sophisticated magnetic circuitry capable of handling high currents and high voltages, and ultimately providing a fast voltage and current pulse to the laser head. This doubling of the number of power-conditioning components contributes to the expense of building an excimer laser.

The laser head itself is one of the most important components of the discharge circuit. The placement and design of electrical components in the head are critical considerations. Figure 19.1 shows the general layout of a transverse discharge and Figure 19.2 shows the mechanical/electrical design of a typical excimer-laser head. The high-voltage electrode, which is typically pulse charged to 40 kV (40,000 V) or higher, is separated from the body of the laser by an insulator. Of course, this insulator must be made of material that will withstand the corrosive gases in the laser chamber and hold off the voltage pulse applied to the "hot" electrode. Ceramics or halogen-compatible plastics are often used. The engineering of the interface between the insulator and the electrode and the high-voltage feedthroughs is one of the more difficult aspects of designing a reliable excimer laser.

Another tricky problem that must be solved is the spurious side or corona discharge created by the rapid, high-voltage pulses applied to the electrode. This corona discharge creates dust, and the laser's gas-flow system must remove this dust before it can settle on the windows of the laser chamber. The dust can be from the insulator or the electrode or from the halogen molecules reacting with materials in the gas-flow system. High-reliability excimer lasers use a bleed flow of fresh gas over the windows to keep dust off the windows.

The discharge in an excimer laser would be very erratic and uneven if it were not for the preionizer and its electrical drive circuitry. The preionizer creates a small arc or discharge in the gas a few nanoseconds before the high-voltage pulse is applied. This preionization makes sure that there are seed electrons in the discharge volume that will multiply rapidly when the main pulse is applied. The preionizer often consists of a set of small, spark-plug-like electrodes beside the main electrode. Both the preionizer and main discharge electrodes produce dust that must be removed from the laser chamber before it settles on the windows. Some excimer laser designs use a corona discharge from the main electrode to preionize the gas. This approach is somewhat cleaner than the spark-plug approach, but the design of the electrode itself becomes much more difficult in this case. A very clean and uniform method to preionze is to use X-rays, an approach taken on experimental devices but not in commercial machines. Figure 19.3 shows some of the layouts used for preionization of an excimer laser.

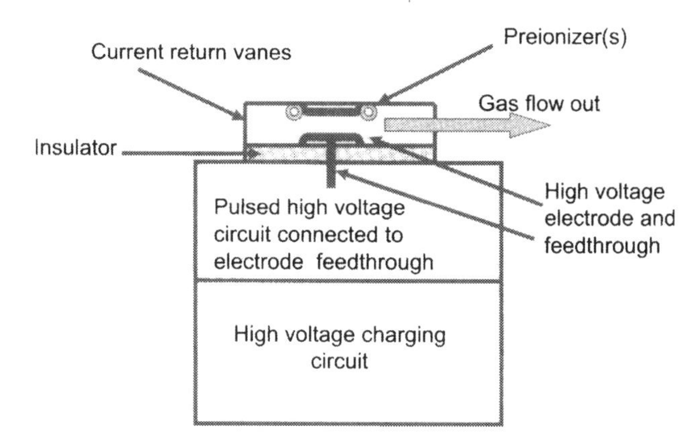

Figure 19.2 An excimer laser laser head will look something like this drawing. The action takes place in the discharge region, typically 50 cm–100 cm long and a few square centimeters in aperture. The discharge region is excited by an electrical pulse applied to the high-voltage electrode. When the gas breaks down, becoming an electrical conductor, the current flows across the gas to the ground electrode and then back through return vanes to the fast high-voltage power supply. Some form of preionization is needed, often provided by an array of sparks off to the side of the discharge. The laser light would come out of the plane of the paper. The gas is pushed through the laser by an external circulating fan for pulse rates of more than a few pulses per second. The current return path is typically a set of rods or vanes positioned a few centimeters away from the discharge region. They need to be close enough to allow the current to flow quickly, open enough to allow gas flow, and far enough away to keep the discharge from going straight to the current returns without going to the discharge region.

Finally, all the electrical components for the fast circuit must be laid out in a way that is conductive to the very fast, very high-current pulse that flows in the laser head. A number of circuits have been created that can achieve this, but the earliest circuits used are still used in the bulk of commercial product offerings. Figure 19.4 shows a typical fast circuit. The circuit is charged slowly, but the placement and values of the capacitors are chosen to allow the voltage to rise in ~10 ns and the current pulse to be less than ~50 ns after the laser medium has begun to conduct. Figure 19.4 shows a spark gap as the switch, but these short-lived start switches have been replaced by thyratrons with magnetic compression circuits, or, in some cases, with all-solid-state circuitry featuring magnetic-pulse compression and semiconductor switches.

19.3 HANDLING THE GASES

Although several of the lasers discussed in Chapter 17 and 18 (e.g., Ar-ion and CO_2) can produce tens of watts of output without changing the gas in the laser tube, excimer lasers do not work that way. In an excimer laser operating at more than several pulses per second, the gas inside the laser must be constantly flowing. Unlike the low-pressure gas lasers that have small diameter laser tubes that reject heat directly to the

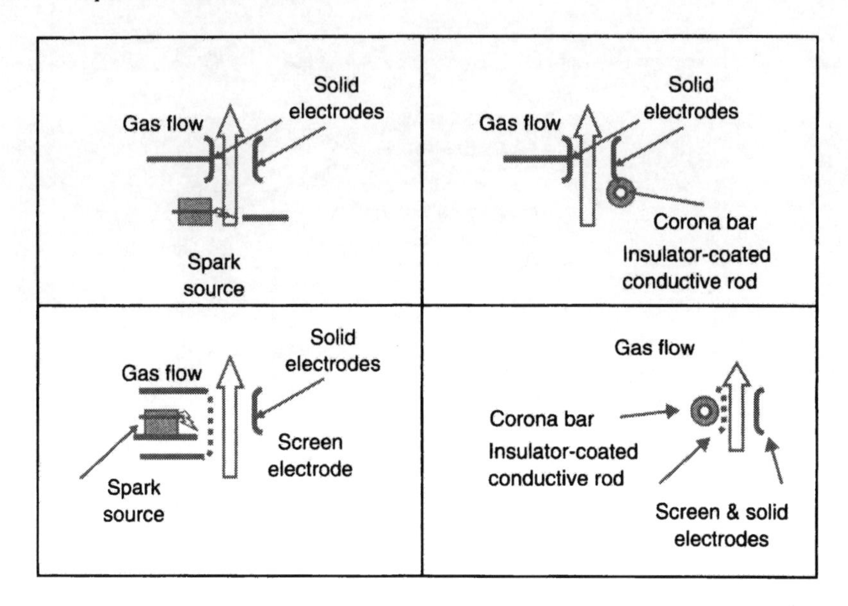

Figure 19.3 The goal of a preionizer is to create seed electrons in the discharge region that the main discharge pulse avalanches to the final discharge conditions. Absent a preionizer, the large volume discharge would become totally uncontrolled—an arc. Although a variety of preionization sources and configurations have been tested, two basic schemes and two basic layouts are the most common source of preionization. Sparks adjacent to the electrodes and in the gas flow are the most common means of preionization. Sparks produce very large densities of electrons. Reaction of the spark plug with the halogen gases and dust formed by the spark itself leads to dust in the laser resonator, requiring regular window cleaning. A corona preionizer is much cleaner relative to production of dust but produces a lower density of electrons in the discharge region. Either type of preionizer can be placed in the gas stream or behind one of the electrodes. If the preionizer is mounted behind an electrode, the electrode needs to be a screen or perforated metal electrode to allow the hard UV light produced by the preionizer to reach the discharge region. The electric circuits driving the preionizer are timed to produce the preionization pulse a few tens of nanoseconds before the main discharge pulse is turned on. In some circuits, the preionizer power is tapped off the main discharge voltage pulse.

wall, this heat rejection method does not work for high pressure and larger aperture size. The heat released during the creation and disintegration of the excimer molecules creates thermal aberrations that can greatly distort the beam in a few thousandths of a second. (An analogy is a mirage that you might see while driving across the desert, an optical aberration caused by localized heating.) The thermal distortion in the excimer laser is greatly minimized by flowing new, cool gas into the chamber.

Another reason for flowing the gas is the reaction that we saw earlier:

$$e + F_2 \rightarrow F^- + F$$

This reaction burns out the halogen fuel needed for the creation of excimer molecules. It takes a relatively long time, typically a tenth of a second, for the F atoms

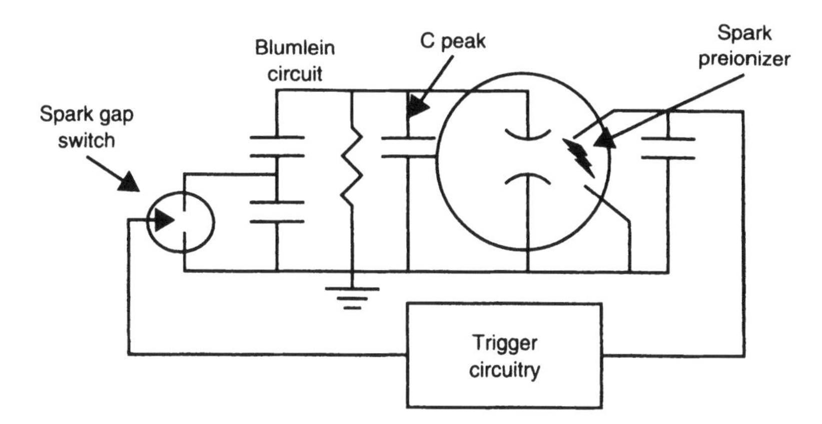

Figure 19.4 The simplest electrical circuit for exciting an excimer laser is one with a preionizer and a simple discharge circuit switched by a start switch, shown here as a spark gap. The spark gap has come to be replaced by a longer-lived gas switch, called a thyratron, in combination with a magnetic circuit to further sharpen the pulse applied. The magnetic component most importantly improves the life of the thyratron switch. The charging circuit, which applies about 20 to 40 kV to the capacitors in this fast circuit, is not shown. The voltage depends on the spacing of the electrodes and gas pressure. When the capacitors are charged, the start switch discharges the circuit into the laser head. The circuit shown here allows the voltage being applied to the laser head to be about two times the charging voltage. The capacitors and inductors in the circuit and any magnetic components must be chosen to provide a pulse that will rise and fall in less than roughly 50 ns while delivering about 1–10 J of electrical energy to the head.

that are created to recombine in the laser chamber. Yet another reason for flowing the gas is that any dust created by the electrical discharge and reaction of the halogen gases with walls and insulators must be removed before it settles on the windows at the ends of the laser chamber.

As a rule of thumb, the gas in the laser-discharge region must be changed three or more times between laser pulses. This ensures a good, fresh charge of gas in the main discharge region as well as in all the little nooks and crannies. In most excimers, the gas moves in a direction perpendicular to the direction of the laser axis and the discharge. A critical region near the laser discharge region is typically 5 cm wide, including the current return path, so the gas must move 3 × 5 cm = 15 cm between pulses. If the laser operates at 100 Hz, the gas must flow at approximately 15 m/s, or somewhat greater than 30 mph. Designing a flow system to move the gas that rapidly inside the laser requires careful if not sophisticated aerodynamic engineering. For many low-pulse-rate applications, the flow system is quite straightforward. For high-average-power excimer lasers operating at 1 kHz or faster, the gas must flow at hundreds of miles per hour inside the laser discharge region (though it is slowed down in other parts of the chamber). In these lasers, the laser chamber becomes a small wind tunnel, and the design of all the components is extremely sophisticated. The difficulty of designing for gas flows that are this fast limits most commercial excimers to roughly 500 Hz. The lasers built for photolithography, an application discussed in Section 19.4 can operate reliably at a much higher pulse rate, but these lasers command very high price premiums. A very schematic end-on view of how a wind-tunnel type of flow system can be laid out for an excimer laser is given in Figure 19.5. As with the preionizer, there are several ways to engineer a specific solution to this requirement for rapid gas flow.

Although conceptually simple, the engineering of a flow system for very high pulse rate can be complex. The flow is usually driven by a small fan that is magnetically coupled to an external motor. The bearings of this fan, which must operate in the hostile environment of halogen gas, are frequently a critical issue in the design of robust, commercial excimer lasers. In many excimer lasers, the problem of keeping dust off the inner surfaces of the windows is solved by constantly squirting a stream of clean gas at the surfaces. Part of the flowing gas is scrubbed in an external

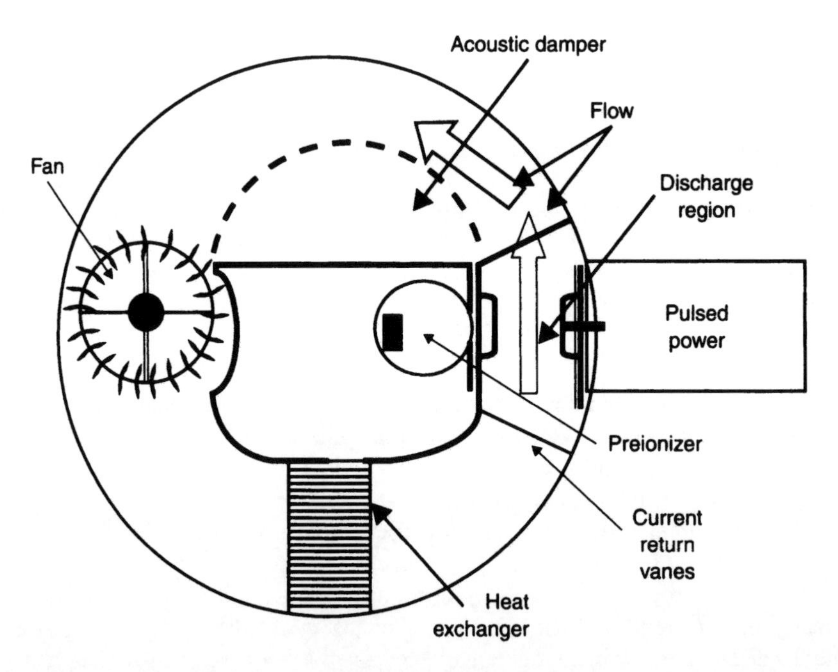

Figure 19.5 A high-power excimer laser can look like this when viewed from the end. As mentioned it is a small wind tunnel with the gas moving fastest through the discharge region. A fan pushes the gas through the discharge region and a heat exchanger removes the heat from the gas. In low-power lasers, the heat exchanger is simply the walls of the chamber itself. Only very sophisticated lasers use acoustic dampers to muffle the pressure waves that can be set up in the gas, especially at a very high pulse rate. The laser head has a high-voltage electrode surrounded by an insulator, a ground electrode, current returns, and a preionizer. For a 1 J per pulse class excimer laser, the discharge region has an aperture area of order 1.5 × 3 cm and is roughly 50 cm to 1 m long. Lower-energy lasers have smaller active regions. Lower-energy lasers with smaller dimensions require lower flow velocity to clear the spent gas out of the discharge region.

system, the mixture made right again to correct for any halogen reaction with walls, carefully filtered and injected directly over the windows. This usually protects the windows for hundreds of millions of pulses. (If you think hundreds of millions of pulses means forever, though, you should calculate how long it takes a laser operating at 500 Hz to reach 10^8 pulses.)

19.4 APPLICATIONS OF EXCIMER LASERS

Excimer lasers are useful primarily in three applications: research and development, material processing, and medical devices. Earlier in this chapter, we mentioned that excimer lasers can be excited by discharges or electron beams. All the commercial applications to be discussed here rely on discharge-pumped lasers. During the 1980s, there was much development of very large, electron-beam-pumped excimer lasers for military applications and inertial-confinement fusion.* Several lasers were built that could produce upward of 5000 J per pulse, and some of these lasers could be repetitively pulsed to produce extremely high average powers. Market demands for such lasers are not obvious. In fact, the market demands appear to be at the 1 kW level or less, where discharge pumping is appropriate.The excimer's utility in research and development stems mainly from its UV wavelength and the resulting high energy of individual photons. The energy of an UV photon is greater than the binding energy of many molecules; therefore, a single excimer-laser photon can break apart a molecule of the material illuminated by the excimer laser. This makes the excimer laser a very useful tool in chemical and biological research. It can be used as an extremely delicate crowbar to open a molecule for study, and to open it exactly the same way every time.

Frequently in chemical and biological studies, a wavelength-tunable source is desired to probe a chemical or biochemical phenomenon. When we discuss dye lasers in the next chapter, you will see that these useful devices can provide wavelength-tunable laser light over a wide spectral range, but they must be optically pumped by a source whose wavelength is shorter than their own. Thus, an excimer laser is an excellent pump source for a tunable dye laser whose output is in the long-UV and short-visible range. The dye laser can subsequently be frequency doubled (see Chapter 13) to produce a tunable output in the UV. Like many things in laser technology however, the excimer pumped dye has lost market share to solid-state lasers such as the Ti:sapphire laser discussed in Chapter 20.

The materials-processing applications of excimer lasers entail jobs as mundane, but profitable, as repetitive hole-drilling in plastics and ceramics, and tasks as sophisticated as fabricating state-of-the-art semiconductor memories and integrated circuits. When it comes to drilling tiny, deep holes in plastic, diamond, and metal,

*In inertial-confinement fusion (ICF) energy is applied, usually by laser, to a cryocooled pellet of hydrogen. The energy is applied so quickly that the atoms' own inertia holds them together as their nuclei fuse to form He. The fusion reaction is similar to that in a hydrogen bomb, and by studying ICF, physicists can better understand the nuclear dynamics of weapons. Also, ICF may be harnessed in the future as a source of civilian electrical power, though the advance of diode pumped solid-state lasers may well render the excimer solution moot.

the excimer laser has few rivals. Such holes, in metal, can be used in precision fuel injection nozzles. When drilled in ceramic or plastic electrical components, the holes can provide precise clearances for wires.

Excimer lasers have become an invaluable tool in the fabrication of new, more powerful semiconductors. To understand why, you have to know a little about how computer memories and integrated circuits are made. The intricate patterns in these devices are created by a process called photolithography, in which the desired pattern is created optically on a layer of photoresist. The light causes a chemical change in the photoresist, and a chemical reaction can then etch the pattern into the photoresist and the underlying semiconductor. But the minimum size of features created by photolithography is proportional to the wavelength of the light used. Thus, as features become ever smaller, the wavelength must also shrink.

If short wavelength alone were adequate for fabrication of very small features in semiconductors, then UV arc lamps could be used, and for many years they were. But recall that in Chapter 3 we discussed dispersion and said that the refractive index of any material depends on the wavelength of the light being refracted. Because of dispersion, every lens suffers chromatic aberration: different wavelengths of light passing through the lens are focused at different distances from the lens. A single, sharp focus is possible only for monochromatic light. Thus, a lens is able to focus the broadband light from an arc lamp less sharply than the narrow-band light from an excimer laser. That is why excimer lasers succeed and UV lamps fail in the fabrication of tiny features by photolithography. The relentless drive for faster computers holding more memory or having higher speed has seen feature sizes decrease by a factor of 10 as memory capability has increased a thousand-fold. Computer chip speeds have increased as well, but not as fast. This market demand has driven semiconductor manufacturers to use the short wavelength of excimer lasers to produce chips. The chip manufacturers tend to introduce a new generation of chips roughly every 2 to 3 years with correspondingly smaller minimum feature sizes needed. Ultimately, optically based lithography using UV or vacuum UV excimer sources may yield to X-ray sources for the production of chips with even finer features.

The excimer lasers used in photolithography are different from the excimer lasers used in other material-processing or medical applications. They must have a narrow bandwidth to minimize the effects of chromatic aberration, the blurring of the focal point for different wavelengths. The natural bandwidth of an excimer though better than that of a lamp, is usually too great, and it must be reduced with prisms, etalons, and other techniques discussed in Chapter 10. Surprisingly, a photolithography laser need not have great beam quality. In fact, a multimode, divergent beam is better, because the beam is more uniform in a multimode beam, and beam uniformity is crucial for smooth, uniform illumination of the exposed photoresist. The reliability of a photolithography laser is also crucial, because its unexpected failure can shut down a billion-dollar chip factory.

The state of the art in lithography devices has changed dramatically over the last 10 years in such a way that the excimer laser (first KrF and then primarily ArF) has replaced lamps in new photolithography equipment. As the feature sizes on chips have decreased, the ArF wavelength (193 nm) has become more prevalent than KrF (248 nm). Other optical methods, such as immersing the chip being irradiated in water, have been adapted that have also allowed the excimer based technology to make yet finer features with 193 nm exposure. At one time there were hopes that the F_2 laser at 157 nm would replace the ArF devices. However, the damage that these very-short-wavelength, relatively high-intensity photons can have on optical windows has worked against the adaptation of yet shorter wavelengths. In a classic example of technical evolution, the future now appears to be in the use of extreme UV (EUV) wavelengths. These EUV wavelengths can be made from discharges or laser produced plasmas. The lasers for producing the plasma that emits the EUV light can be excimers, solid-state lasers, or CO_2 lasers. With these laser options appearing to have the current edge in technology, one can expect the use of excimer lasers for lithography to decrease in the future.

The ArF laser enables one of the most publicly acclaimed applications of excimer lasers: eye surgery. Although other lasers, especially CO_2 and Nd:YAG, frequently find application in medical devices, there is a fundamental difference between these lasers and excimer lasers. YAG and CO_2 lasers, with output in the infrared or visible, are thermal lasers: they deposit heat in the tissue until the desired effect is achieved. A beneficial side effect associated with Nd:YAG and CO_2 lasers is that an incision is cauterized as it is made, thereby reducing the amount of hemorrhaging and the danger of infection. The UV light from an excimer does not deposit heat in the tissue, but its energetic photons disrupt the chemical bonds that hold the molecules together. Thus, an excimer laser can ablate tissue with minimal thermal damage to adjacent tissue. Accordingly, excimers are often called "cold" lasers.

The cold excimer laser has proven quite effective in ophthalmic surgery, in which it ablates tissue from the cornea. At 193 nm, the photons from the ArF laser penetrate only a few micrometers into the cornea. In corneal sculpting, the corrective prescription can be ground into the cornea by UV laser light; your glasses are permanently etched into the surface of your eyeball. In one approach, the first certified by the Food and Drug Administration and called PRK, the front surface of the cornea is ablated in a prescribed way to correct a person's eyesight. This approach does have a drawback: the thin membrane covering the cornea must grow back over the laser-treated cornea.

PRK has been displaced by a much more "medically clean" approach to corneal sculpting, termed LASIK. In this method, the patient's eyeglass prescription is literally carved inside the cornea with the beam of the ArF excimer laser. A small flap of the cornea is first removed with a precision knife (keratome) and an inner portion of the cornea is exposed to the excimer laser. After the prescription is carved, the corneal flap that was opened is then put back into place over the ablated cornea. The corneal material heals itself with fewer side effects than were seen with PRK. Myopia as well as astigmatism and farsightedness can be addressed with this technique now. Because a large portion of the world's population is myopic, millions of these operations are performed every year. However as with the photolithography story related above, no laser application stays the same forever. Research into medical applications of fiber-based ultrafast lasers, described in the following chapter, may yet displace the excimer for ophthalmic surgery.

Chapter 20

Tunable and Ultrafast Lasers

Fixed-frequency lasers—those discussed in other chapters—are suitable for many practical applications, but sometimes a laser whose wavelength can be varied is far more useful. In a variety of scientific applications, for example, it is necessary to find a frequency of light that exactly matches an atomic or molecular resonance. In this chapter, we discuss these wavelength-tunable lasers and also cover ultrafast lasers, which are closely related to tunable lasers and are expanding greatly in their usage and applications.

In Chapter 10, we discussed laser bandwidth and some of the techniques for reducing the bandwidth of a laser. Whereas most lasers have relatively narrow natural bandwidths, the bandwidths of the lasers discussed in this chapter are enormous. A good measure of the narrowness of a laser's bandwidth is the ratio of the bandwidth to the absolute wavelength. For example, Table 20.1 shows that this ratio has a value of about 10^{-5} in an Nd:YAG laser. Compare that with the ratio of about 10^{-1} for a Ti:sapphire laser and you see what we mean by enormous: the relative bandwidth of Ti:sapphire is nearly 10,000 times greater than that of Nd:YAG.

The tunable lasers we discuss here start with an enormous bandwidth and utilize many of the techniques discussed in Chapter 10, not only to reduce the bandwidth but to tune it to the exact frequency desired. These narrowing techniques almost always involve manipulating the feedback of the laser resonator, as explained in Chapter 10. However, if you start with a laser with an enormous bandwidth, you can do something other than make a tunable laser out of it. You can make an ultrafast laser. The term ultrafast usually means pulses less than a picosecond (10^{-12} s). The reason that you need a broad-bandwidth laser to make an ultrafast laser derives from a fundamental principle of physics. The Heisenberg uncertainty principle defines a relationship between the duration of an optical pulse and the bandwidth of the source. Mathematically, it looks like this:

$$\delta f \times T \approx 1$$

in which δf, is the laser's bandwidth and T is the duration of its pulse. The product of these two must be a constant. Thus, for a short pulse duration you need a broad bandwidth.

As a brief aside, the converse is also true. A narrow-bandwidth laser (like those discussed in Chapter 10) is incapable of producing short pulses. Another converse to this principle is that a very narrow line-width laser must have very long and fre-

Introduction to Laser Technology, Fourth edition. C. Breck Hitz, J. Ewing, and J. Hecht

Table 20.1 Tuning ranges for different classes of lasers

Type of laser	Example	Line width (typical) $\delta\lambda$ (nm)	Wavelength λ (nm)	$\delta\lambda/\lambda$	Comment
Gas or ion laser at low pressure	He:Ne	0,00006	633	10^{-6}	Narrow Doppler line-width is typically a factor of 10^{-6} of the laser wavelength λ.
Solid-state ion in crystalline host	Nd:YAG	0.5	1064	5×10^{-5}	ND^{3+} ions interact with the atoms of the host to broaden lines; varies a bit with host and ion.
Excimer	ArF, KrF	~2	193 (ArF) 248 (KrF)	$\sim 10^{-2}$	Line width depends on the shape of the lower level of the excimer transition and some excimers have much brader line widths.
Solid-state ion in glass host	Nd:Glass bulk or fiber	20	1054	$\sim 2 \times 10^{-2}$	Note that the tuning range for other ions such as Yb in glass fiber exceeds that for Nd:Glass as a bulk laser or fiber.
Solid-state Er^{3+} in glass fiber	Yb:Er:Glass	~50 nm	1540	3×10^{-2}	The broad bandwidth of these fiber lasers have made them the essential for telcom applications.
High-pressure gas	Co_2 with 10 atm buffer	1.6 μm	10600 (10.6 μm)	0.15	At very high pressure, the individual lines of the CO_2 laser merge into a broad band due to collisions. Though this has a huge tuning range, the difficulties of high-pressure-discharge technology make this laser very rare.
Tunable solid	Ti:Sapphire $Ti{:}Al_2O_3$	~180	~790	0.22	This laser has become the workhorse for ultrafast applications due to the ability to achieve pulses of 0.01 psec. Note that some ultrafast practical applications only need psec pulses and, as such, Yb:fiber lasers work fine.

quency-stable pulse duration. The extreme frequency-stability capability of moderate- and low-power diode pumped solid-state lasers makes these lasers the laser of choice for the ultrastable lasers used in the search for gravity waves in instruments like the Large Interferometric Gravity Observatory (LIGO) and similar systems being used by the astrophysics community.

Numerically, the previous equation implies that if you want a 1 ps (10^{-12} s) pulse, you need a laser bandwidth of at least a terahertz (10^{12} Hz). This is a greater bandwidth than can usually be achieved in, for example, an Nd:YAG laser. The 20 or 30 ps pulse typically obtained from a modelocked (Chapter 12) Nd:YAG laser is said to be bandwidth limited—shorter pulses cannot be obtained because the bandwidth is not sufficiently large. (Although the modelocking modulator does increase the bandwidth of the laser by coupling energy into outlying longitudinal modes that otherwise would be below threshold, the increase is not great enough to enable a 1 ps pulse.)

However, as already stated, ultrashort lasers have pulses of less than 1 ps (10^{-12} s), so they must have an enormous bandwidth of a terahertz (10^{12} Hz) or greater. Indeed, the lasers in the lower half of Table 20.1 achieve these bandwidths. Ti:sapphire lasers, for example, with a bandwidth greater than 10^{14} Hz are capable of producing pulses of several femtoseconds (10^{-15} s).

Yet another commercially important application of broad bandwidths is in fiberoptic communications, in which the high bandwidth of erbium-doped fiber amplifiers enables them to boost the signal at dozens of different wavelength channels spanning a significant bandwidth. The Er:glass laser is modestly tunable, but it is tunable enough to put 50 or more discrete communication channels under the top of the spectral band.

The dye laser has been the historical workhorse of tunable and ultrafast laser technology. However, drawbacks of these lasers are that the organic dye molecules degrade and decompose with time (some even while sitting on the shelf), that the solvents used are unattractive and often flammable, and that the lasers themselves are physically frail. In the past decade or two, tunable solid-state lasers that avoid the drawbacks of flammable solvents and degradation have largely supplanted dye lasers.

Nonlinear techniques such as frequency doubling and optical parametric oscillation (Chapter 13) are sometimes combined with the tunability of a laser to provide output from the infrared to the UV from a single system.

Table 20.1 also shows one other laser with extremely high bandwidth, the very high-pressure CO_2 laser. Because of the quite different nature of the long wavelength technology, the very large bandwidth available in high-pressure CO_2 does not readily translate into useful ultrafast or tunable technology.

20.1 DYE LASERS

Although they have been replaced by tunable solid-state lasers in many applications, dye lasers still find many uses. Flashlamp-pumped dye lasers, for example, are less expensive than solid-state lasers and provide the high energies needed in many medical applications in which a specific wavelength is required to match the absorption of a particular molecule.

A functional diagram of a tunable laser is presented in Figure 20.1. Many kinds of pump sources have been used to create the population inversion in a tunable laser, including frequency-doubled Nd:YAG, flashlamps, excimer lasers, diode lasers, nitrogen lasers, and ion lasers. A dye laser is used to show a practical example of this concept in Figure 20.2. As shown, an amplifier often is used in conjunction with the tunable oscillator. It is more efficient to generate a carefully tuned signal in the oscillator and then boost that signal in the amplifier than to produce high-energy output in

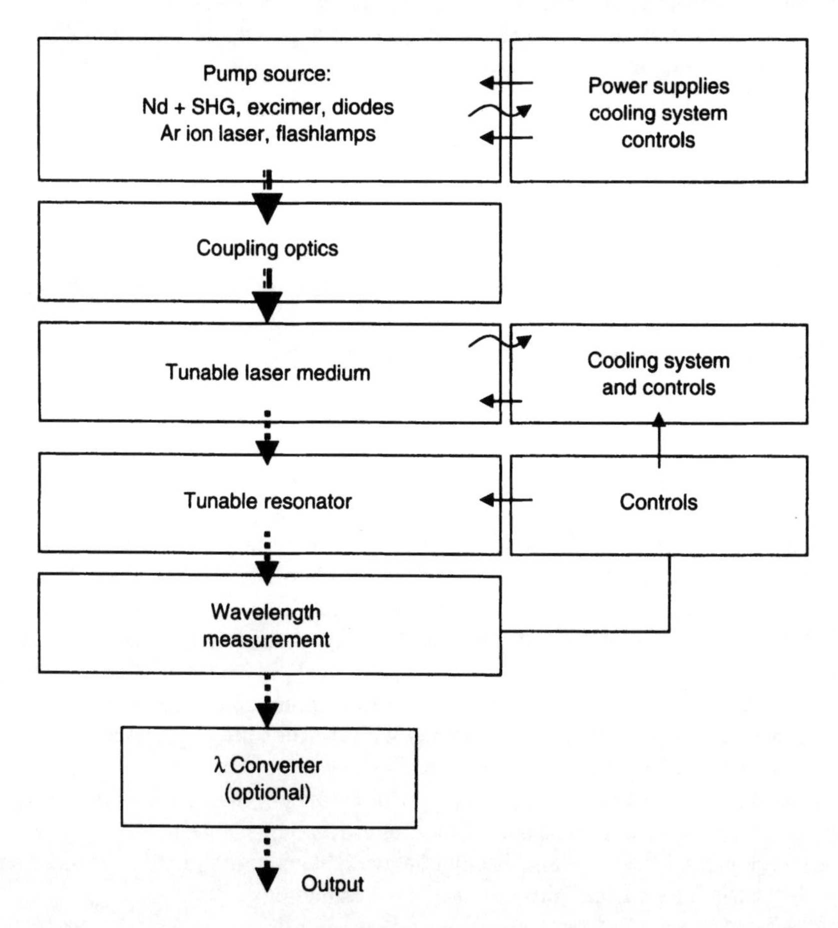

Figure 20.1 Functional or block diagram of a tunable laser. From a functional poiont of view, the medium can be a solid-state material such as Ti:sapphire, or an organic dye. The resonator allows the user to tune the wavelength of its feedback, thereby narrowing the output to the desired wavelength and bandwidth. Pump sources can be pulsed or cw lasers or lamps. A wavelength-control mechanism measures the laser's wavelength and bandwidth and can provide a feedback signal to electronically tunable elements in the resonator. Often, a wavelength converter (nonlinear optics) is added to extend the spectral range of the output.

a single step from an oscillator alone. The generic optical resonator configurations used with dye lasers have been adopted and modified for tunable solid-state lasers.

Because the dye is an efficient optical absorber, only a small volume of it needs to be exposed to the pump radiation. Transverse pumping, in which the pump illumination is transverse (perpendicular) to the laser's optical axis, is illustrated in Figure 20.2. (Longitudinal pumping would be utilized if the dye were a less efficient absorber.) Literally dozens of different dyes are available to provide wavelengths across the visible spectrum. A given dye can typically be tuned over a wavelength range of tens of nanometers. Rhodamine 6G, probably the most versatile dye, can be tuned across nearly 80 nm in the orange–red region of the spectrum. Figure 20.2 shows both a diffraction grating and an etalon, both of which are discussed in Chapter 10, as the narrowing and tuning elements.

Dye lasers operate as four-level laser systems, but the energy levels are so dense in an organic dye molecule that it is often difficult to distinguish one from another. This density of energy levels can contribute to lowered gain by way of an effect called excited-state absorption (ESA). ESA occurs when an energy level is located above the upper laser level by the exact energy of a laser photon. In most lasers, when a laser photon encounters an atom (molecule) in the upper laser level, only one thing is possible: stimulated emission. But if there just happens to be an energy level that the excited atom, ion, or molecule can reach by absorbing the laser photon, that process will compete with stimulated emission and act as an inherent loss mechanism to the laser. Depending on the specific dye used, ESA can be an important problem in dye lasers and leads to significant variation in the conversion efficiency of different dye/pump laser combinations.

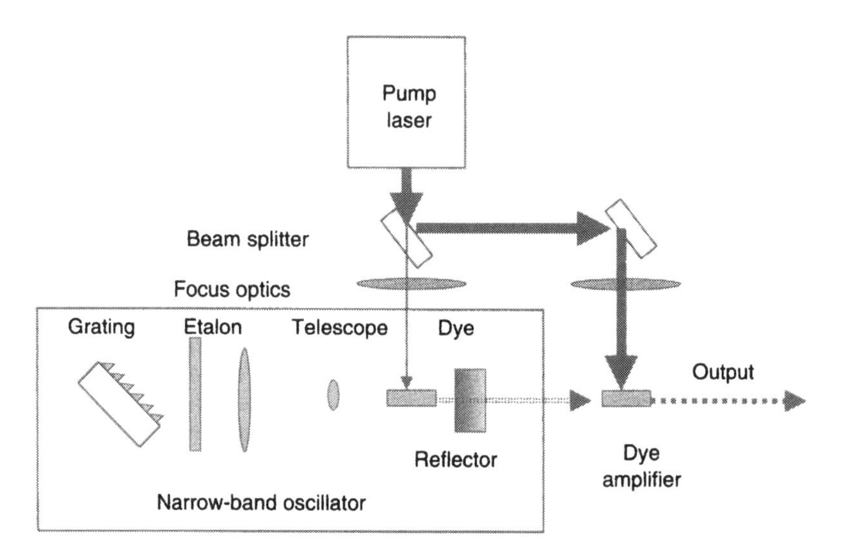

Figure 20.2 In this tunable dye laser, about 10% of the energy from the pump laser goes to the oscillator and the rest goes to the amplifier. In general, the resonator techniques employed to reduce and tune the bandwidth of dye lasers also work in tunable solid-state lasers.

The radiative lifetime of the upper laser level in dye lasers is very short, typically on the order of nanoseconds. Thus, Q-switching a dye laser is not practical. The dye laser output pulse will typically follow the time history of the pump laser. A 10-ns pump pulse from a frequency-doubled Nd:YAG laser or an excimer laser will result in a dye laser pulse of about 10 ns. A microsecond flashlamp-pumping pulse will yield a nominal microsecond dye laser pulse.

20.2 TUNABLE SOLID-STATE LASERS

Solid-state lasers have matured to the point that they can provide as broad a range of wavelengths as dye lasers, especially when nonlinear devices are used to extend their tuning range. Table 20.2 provides an overview of the types of tunable solid-state lasers and their tuning ranges. Of the lasers listed, only Ti:sapphire and Cr:alexandrite have found significant commercial applications.

As shown in Figure 19.1, there is more to the laser than the primary gain elements and, often, the pump laser with its controls and cooling is the most expensive part of a tunable laser. In addition to the pump laser and whatever nonlinear wavelength extender is present, the tunable system has other components. In practice, there are wavelength-measuring and control components, as well as the tunable cavity itself with the mechanically precise optical mounts and the software to run the tunable laser from a computer. The user also often has computerized control systems for these nonlinear optics. Thus, the solid-state tunable laser is one of the more expensive and complex laser systems imaginable, and typically finds application only in well-funded scientific research.

Ti:sapphire, probably the most versatile of the tunable solid-state lasers, is based on triply ionized titanium (Ti^{3+}) in a host lattice of sapphire (Al_2O_3). In a typical laser crystal, about 0.1% of the aluminum atoms in the sapphire lattice are replaced by titanium. Ti:sapphire absorbs in the green spectral region and emits with almost 100% efficiency in a broad band between 700 and 900 nm. Population inversions can be created across the entire fluorescence bandwidth and with sufficiently hard pumping and suppression of the higher gain gain region near the peak of the spectrum well into its wings, from about 650 nm to beyond 1000 nm. Ti:sapphire lasers can be classed into three distinct operating modes: purely cw lasers, pulsed lasers with substantial energy and output pulses on the order of 10 ns, and ultrafast lasers with subpicosecond outputs. The pure cw lasers are pumped by cw green sources (Ar-ion lasers or frequency-doubled, diode-pumped cw Nd lasers). The 10 ns variant typically is pumped with a Q-switched Nd laser. The ultrafast lasers usually have cw pumping but a specialized cavity to enhance the ultrafast operation. On rare occasions, one may find hybrids of these basically different configurations used together.

Figure 20.3 is a schematic of a pulsed Ti:sapphire laser. The green light from the pump laser enters the resonator through a dichroic mirror that reflects the laser wavelength but transmits green. The Ti:sapphire rod is end pumped to maximize the overlap of the pump laser light with the tunable laser's extracted volume.

An oscillator–amplifier configuration was shown earlier in Figure 20.2, with the laser-pumped dye laser as an example. The same configuration could alternatively

Table 20.2 Tunable solid-state lasers

Medium	λ peak	Tuning range, $\delta\lambda$	Pump	Comment
Ti:sapphire, Ti^{3+} doped in Al_2O_3	800 nm	~650 nm to ~1 μm	Green or blue λ: Nd + SHG Ar ion flashlamps	This the is the workhorse tunable solid-state laser. The short upper-state radiative lifetime of ~4 μs makes lamp pumping and diode pumping less attractive.
Alexandrite Cr^{3+} in $BeAl_2O_4$	Depends on temperature, ~750 nm	~700 nm to ~820 nm	Flashlamps, red diodes at λ ~670 nm	Long radiative lifetime, ~260 μs, makes flashlamp pumping quite practical. It is used extensively in place of the Nd laser + tunable laser for certain applications.
Cr:LiSAF Cr^{3+} in lithium strontium aluminum fluoride	~850 nm	~760 nm to ~940 nm	Flashlamps, red diodes at λ ~720	Longer radiative lifetime, ~60 μs, than Ti:sapphire make diode pumping practical, though short λ diodes are needed.
Ce: LiSAF	~300 nm	280 to 320 nm	UV laser, Nd + 4HG*	Very short radiative lifetime makes Q-switched laser pumping essential.
Cr^{4+}: YAG	~1.3 μm		1 μm	This passive Q-switch for Nd lasers is also a laser-pumped laser with ultrafast capability.
Co:MgF_2	~1.3 μm	1.8 to 2.4 μm	1.3 μm Nd laser	Competes with the Nd laser driving an OPO for tunable applications.

*4HG = the fourth harmonic of the Nd laser at λ ~266 nm.

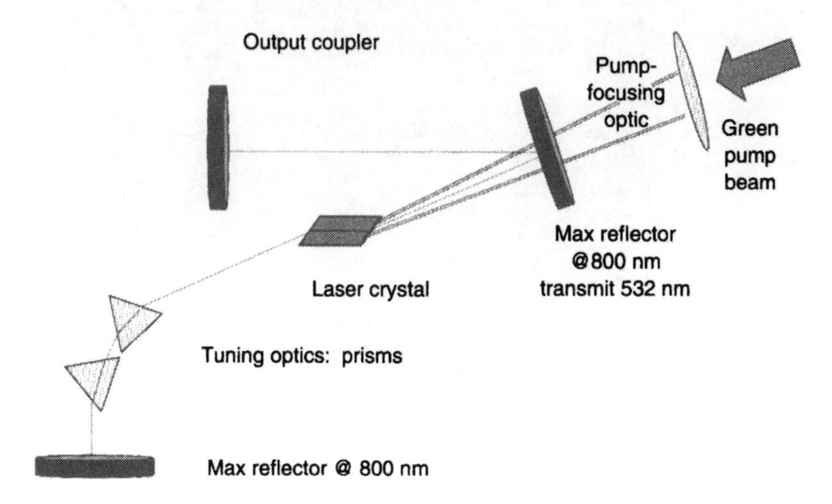

Figure 20.3 A tunable resonator for a Ti:sapphire laser has many similarities to a dye laser. In this case, the laser crystal is pumped from one end by a focused laser beam in the green or blue. Longitudinal pumping (end pumping) is typically used, as the absorption of the Ti ions is weaker than the absorption for dyes in a dye laser. A dichroic reflector is provided that transmits the green pump radiation but reflects the laser wavelength. Tuning optics, such as a set of prisms, as shown here, or a diffraction grating, are used to narrow the frequency to the desired output. The laser wavelength is controlled by adjusting the angle of the prisms (or diffraction grating). Oscillator–amplifier configurations can yield as much as 100 mJ outputs.

be used with Ti:sapphire as the active element for ~10 ns high-energy pulses. Pumping a Ti:sapphire laser with the green output from a Q-switched, frequency-doubled Nd:YAG laser, this system would produce tunable output pulses of about 100 mJ output in the region near 800 nm.

Ti:sapphire lasers have several advantages over dye lasers, in addition to their chemical stability and lack of obnoxious solvents. It is a messy and tedious procedure to change the dye in a dye laser, yet a given dye can be tuned across only tens of nanometers of spectrum. A Ti:sapphire laser, by contrast, can be tuned across its entire bandwidth from about 650 nm to beyond 900 nm, without changing the laser medium. (However, three sets of resonator optics are typically needed to cover the entire range.) Moreover, the solid-state laser material avoids all the messiness inherent in a flowing-liquid system, as these typically low-average-power lasers can be simply heat sunk to a metal holder, which is cooled by room air.

The other commercially important, tunable solid-state laser is Cr:alexandrite, based on triply ionized chromium ions (Cr^{3+}) in an alexandrite ($BeAl_2O_4$) host lattice. We discussed the fixed-wavelength Cr:ruby laser in Chapter 15, but the bandwidth of the chromium ion is much larger in an alexandrite host as a result of interactions between the ion and the surrounding lattice. Nevertheless, alexandrite's natural tuning range is a fraction of Ti:sapphire's. Another disadvantage of alexandrite is its relatively long pulse duration compared with that of Ti:sapphire, which makes wavelength conversion with nonlinear optics more difficult in alexandrite (because the peak power is lower in a longer pulse). Moreover, alexandrite's operating range is limited by ESA, as described earlier for dye lasers. However, alexandrite does efficiently absorb the broad bandwidth pump radiation from flashlamps, so it does not need the more expensive Nd:YAG pump laser usually required by Ti:sapphire. And it can readily produce multijoule outputs, a more expensive proposition for Ti:sapphire. Another advantage of alexandrite is that it operates at an elevated temperature, making heat removal relatively simple. Alexandrite's relatively short wavelength penetrates further into human skin than the 1 micron radiation from an Nd:YAG laser. This, and its low cost compared with that for Ti:sapphire, make it an excellent instrument for hair removal, a cosmetic medical application of lasers.

The short upper-laser-level lifetime of Ti:sapphire (several microseconds) makes it a poor candidate for Q-switching. In general, the pulse duration from Ti:sapphire is as short as one would obtain with Q-switching in any event. Alexandrite, on the other hand, with an effective lifetime of 200 μs—only slightly shorter than for Nd:YAG—can be Q-switched.

Other Cr^{3+} lasers have been developed in the laboratory, the most important being Cr:LiSAF. This material is of interest for both its broad tuning width (and potential for very short pulse duration) and its ability to be pumped by diode lasers. Despite great early promise in its development cycle, it has not found significant commercial application.

20.3 NONLINEAR CONVERTERS

Nonlinear optics, introduced in Chapter 13 as a technique to extend the wavelength range of fixed-frequency lasers, can greatly enhance the tuning range of tunable lasers. With a Ti:sapphire laser, it is possible to mix two photons from the laser (second-harmonic generation), or three photons, or even four photons. It is also possible to mix a photon from the Ti:sapphire laser with a 1.06 micron photon from the Nd:YAG laser that pumps the Ti:sapphire laser. In all these cases, the resultant photon has the combined energy of the two photons that are mixed to produce it. The wavelengths that can be obtained from a Ti:sapphire laser using these techniques are shown schematically in Figure 20.4

Most of these nonlinear techniques work only with pulsed lasers, because the high peak power is necessary to obtain reasonable nonlinear conversion efficiencies. (For clarification, see Figure 13.7.) And because we are dealing with nanosecond pulses, mixing the light from the Nd:YAG pump laser with the output of the Ti:sapphire laser presents an interesting timing problem. The green light from the Nd:YAG laser creates a population inversion in the Ti:sapphire laser, but it can take tens of nanoseconds for the circulating power inside the Ti:sapphire resonator to build up from spontaneous noise to create the output pulse. What do you do with the 1.06 pm pulse—which emerged from the Nd:YAG laser along with the green pump pulse, while that is happening? You have to send it on a long trip while the Ti:sapphire pulse builds up inside its resonator. If it takes tens of nanoseconds for the pulse to build up, then the delay line for the 1.06 pm pulse has to be tens of feet

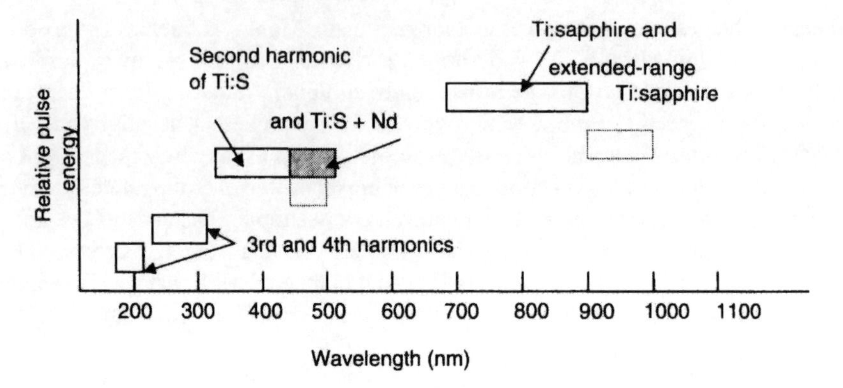

Figure 20.4 Harmonic generation and mixing greatly extend the wavelength range of the Ti:sapphire laser. In the ultraviolet, output on the order of 10 mJ can be obtained from 100 mJ Ti:sapphire pulses. The shaded box corresponds to the wavelength range available if the Ti:sapphire is equipped with special resonator reflectors to extend the tuning range beyond 900 nm.

long. Although conceptually straightforward, this task often proves to be awkward in practice.

One solution is to have an oscillator that has a much shorter buildup time for the laser output. This can be practically achieved by using an optical parametric oscillator (OPO) as the oscillator for the tunable system, in a cavity very similar to that used for a dye laser or Ti:sapphire oscillator. This OPO method also enables the gaps in the frequency tuning range shown in Figure 20.4 to be filled in. Figure 20.5 shows a configuration that is often used to extend the wavelength range of a Ti:sapphire laser and avoid delay lines. The same configuration could alternatively extend the tuning range

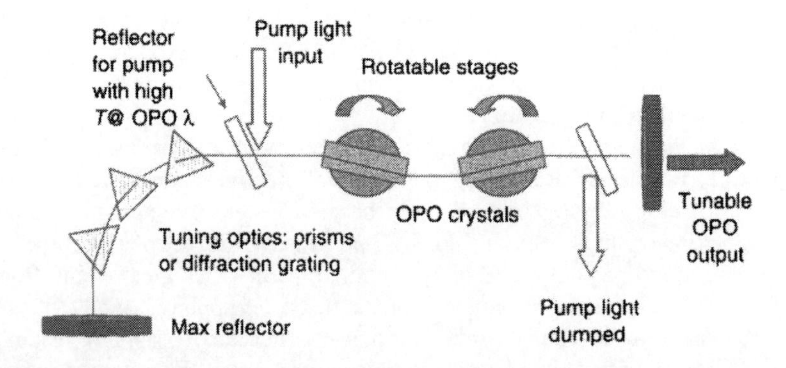

Figure 20.5 This OPO resonator could extend the wavelength range of a Ti:sapphire laser or be a tunable laser pumped by an Nd laser. To enhance the pumping efficiency, the second dichroic reflector (the element immediately to the left of the output mirror) could be turned orthogonal to the resonator axis, so the pump pulse from the Ti:sapphire or Nd laser makes a second pass through the nonlinear crystals.

of a frequency-doubled or -tripled Nd:YAG laser. The pump pulse from the pump laser, Ti:sapphire or Nd:YAG, enters the OPO from above and reflects off a dichroic reflector to pass through a pair of nonlinear crystals. (The dichroic reflector transmits the OPO wavelength.) Another dichroic reflector on the other side of the crystal extracts the spent pump beam from the OPO resonator. As explained in Chapter 13, one technique of phase matching in an OPO is to adjust the angle of the crystals, and that is the technique used here. The crystals are mounted on rotatable stages so that the phase-matching angle can be adjusted for the desired wavelength. The three prisms reduce the bandwidth of the resonator's feedback exactly as the prism in Figure 10.8 did. These prisms must be adjusted simultaneously with the phase-matching angle of the crystals. In many commercial systems, this complex simultaneous adjustment is accomplished under computer control. (In the case of noncritical phase matching, the angle of the nonlinear crystals does not have to be adjusted. It is obviously preferable to avoid the complexity of twisting the crystals, so the engineers who design these systems try to operate in the noncritical phase-matching region.)

These OPOs seem like (and are) very complex optical systems. But the payoff for all this complexity is shown in Table 20.3. All the additional wavelengths that can be obtained when an OPO is combined with a standard laser are given. Note that some of the wavelength regions require a 2- or 3-micron pump to cover the infrared. These pump lasers could be Nd lasers operating on long-wavelength transitions, or other lasers such as Ho:YAG.

A final point about OPOs is that the output of the device can come at two different wavelengths: the signal and the idler. For example, when converting the 1.06 μm

Table 20.3 Typical OPO materials

Nonlinear crystal	Typical pumping laser	Wavelength range
β-Barium borate, lithium borate	Frequency-tripled Nd:YAG, other Nd lasers, and XeCl eximer	~400 nm to ~2.5 μm
KTP and KTA	Frequency-doubled Nd: YAG, other frequency-doubled Nd lasers	~700 nm to ~4 μm
KTP and KTA	Nd:YAG, other Nd lasers, Yb	~1.5 μ to ~4 λm; shortest λ depends on pump λ and KTA versus KTP
$LiNbO_3$, lithium niobate	Nd:YAG, other Nd lasers, Yb	~1.4 to ~4 μm
Lithium niobate	Frequency-doubled Nd: YAG, other frequency-doubled Nd lasers	~700 nm to ~4 μm; can experience damage with green pumping
Periodically poled lithium niobate	~800-nm diodes, cw Nd:YAG; very high effective gain per unit power	~1.3 to ~5 μm; used with low-energy pulsed or cw lasers
Zinc germanium phosphide	2-μm Tm and Ho lasers, 3-μm Er laser	~3 to ~8 μm

output of an Nd:YAG laser to the eye-safe spectral region using potassium titanyl phosphate (KTP) or potassium titanyl arsenate (KTA), the 1.5 μm output is accompanied by a second beam at about 3 μm (see Figure 13.18). The second wavelength can be a major advantage in certain applications because it effectively broadens the range over which the tunable output can be produced. The specific ranges that are accessible depend on the OPO crystal itself and the pump wavelength. However, the fact that two wavelengths are emitted has enabled certain systems to cover virtually all of the visible and UV with one specific device.

20.4 ULTRAFAST LASERS

Ultrafast lasers are a burgeoning laser technology that extends the modelocking principles discussed in Chapter 12 to new ranges of very-short-duration pulses and makes use of newer media such as the extremely broadband Ti:Sapphire laser and the extremely compact and flexible fiber laser technology. Using nonlinear optical conversion techniques, these lasers offer the potential for very high intensity when developed into oscillator amplifier configurations as well as ultrafast sources for a wide variety of wavelengths including X-ray sources. The high peak intensity possible with an ultrafast laser makes wavelength conversion with parametric techniques discussed above straightforward and also provides very broad wavelength coverage. Though the industrial applications are just beginning to emerge, the fiber-based variants of the ultrafast laser hold the promise for novel tools for compact micromachining and medical usage. It is worthwhile to review some of the physical units and typical sources used for generating ultrafast laser pulses and observe where various lasers have fit in applications; see Table 20.4. The numbers are representative and can vary substantially in a range around the numbers mentioned.

Note that the table in a way charts the path to recent frontiers in R&D. Going from the R&D lab to industrial or medical applications can take much less than a decade if the technology is rugged enough and the market large enough for use outside the university laboratory. Though picosecond-class dye lasers were used in the lab ~30 years ago, the advent of more user-friendly technologies such as modelocked fiber lasers has enabled one to imagine a market in industry or medicine for these devices beyond the use in doctoral theses.

Let us now look at the key components of an ultrafast system. The complexity of the system depends on the degree of wavelength flexibility, pulse energy, peak power, pulse duration, and average power that one desires. Typical ultrafast systems have the features shown in Figures 20.6, 20.7, and 20.8. The core of the system, the basic ultrafast configuration, is a broadband laser, a modelocking mechanism or medium, and an optical pump for the broadband laser. Using Ti:sapphire as the broadband source, the pump is typically a low-power, diode-pumped, frequency-doubled-Nd:YAG laser. These lasers are quite adequate for a range of studies in molecular dynamics by researchers in the chemistry lab. For the Yb or Er fiber laser, the pump is typically a moderate-power diode array. The picosecond pulses from a compact and reliable fiber-based device are showing potential for diverse industrial and medical applications. Although a number of other broadband

Table 20.4 Comparison of various pulsed and ultrafast systems

Pulse duration	Range of energy and pulse rate	Typical peak and average power	Typical source	Comments
10×10^{-9} sec, 10 ns 10 ns	~100 mJ to 1J ~10 Hz (pps)	10–100 MW peak 1–10 W average	Q-switched Nd:YAG	Workhorse for R&D from 1975; various medical and industrial
10×10^{-12} sec, 10 ps 10 ps	0.1 μJ ~50 MHz	~ 10kW peak ~5 W	Modelocked Nd:YAG	R&D from 1975; molecular dynamics
1×10^{-12}sec, 1 ps	0.02 μJ ~50 MHz	~ 20 kW peak ~1 W	Modelocked dye lasers	R&D; dye has disadvantages
1×10^{-12} sec, 1 ps	~1 to 10 μJ ~ 1 MHz	~1 MW 10–20 W average	Modelocked, diode-pumped fiber laser	Compact and simple; entering into applications
10×10^{-15} sec, 10 fs 10 fs	0.1 micro-J ~50 MHz	~ 10 MW peak ~5 W	Modelocked Ti:sapphire	Rapidly displaced modelocked dyes. Boon to chemistry R&D
20×10^{-15} sec, 20 fs	~50 mJ ~10 Hz	2.5 TW peak ~0.5 W	Pulse-sliced, modelocked Ti:sapphire + CPA amplifier	Ultrahigh intensity possible; various physics and technology R&D
500×10^{-18} sec 500 attoseconds	~1 mJ at 800 nm at ~10 Hz and 5 fs NLO conversion to XUV with ~0.08 fs pulses	Peak powers vary by extreme UV wavelength chosen	Frequency up-converted, high-energy ultrafast	Generate X-rays for short pulse; Fundamental physics; R&D
500×10^{-15} sec 500 fs	500 J < 1Hz	1 PW (petawatt) Low pulse rate, hence low average power	Very-high-energy amplification via CPA scheme	Fundamental physics; R&D

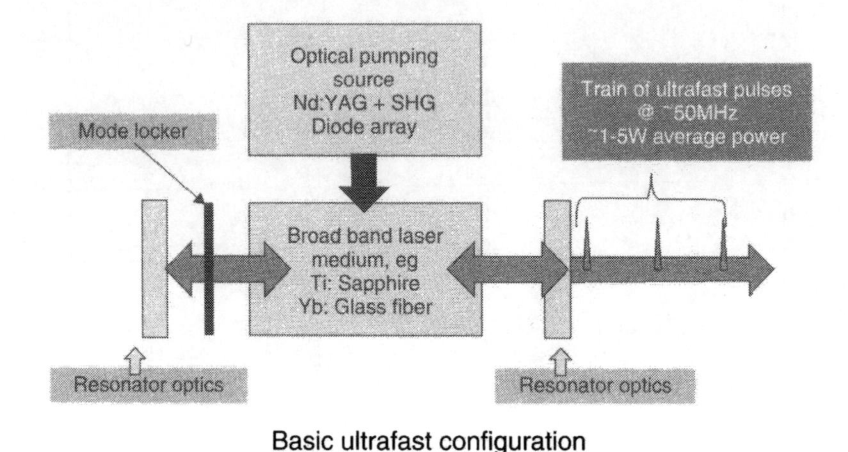

Figure 20.6 Basic layout for the core of an ultrafast laser system.

media have been used, Ti:sapphire and Yb in fiber are the most commonly encountered.

The basic ultrafast configuration is, however, just the beginning for many ultrafast instruments and applications. The accessories that any one application will require are quite diverse and application-specific. Figure 19.7 shows a block diagram for a high-performance, ultrafast system. For example, to obtain extremely high

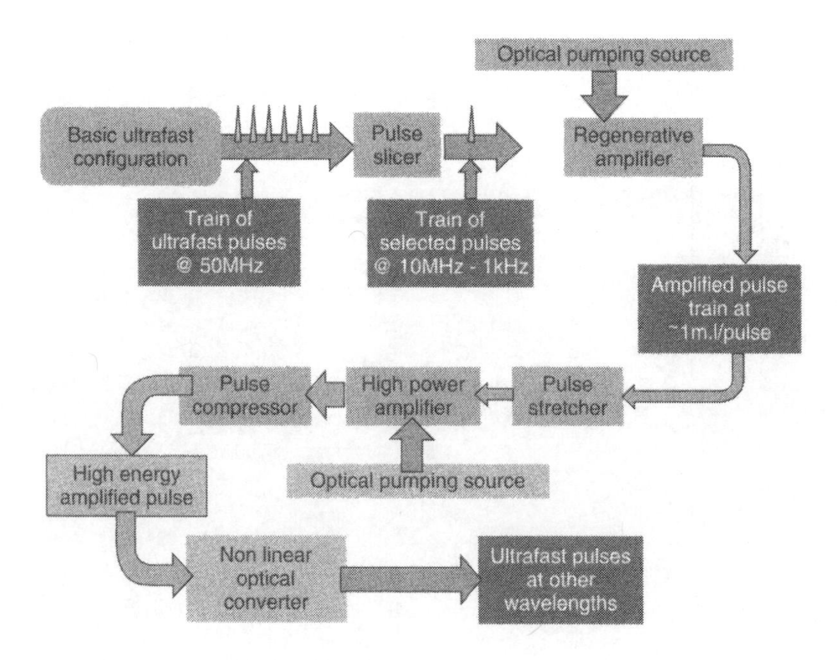

Figure 20.7 Block diagram for the components of a high-power, ultrafast laser system.

power one would slice out an individual pulse using an electrooptic switch from the train produced by the broadband source and then amplify those pulses, typically in a regenerative amplifier. Further amplification may require pulse stretching and recompression in a scheme termed chirped-pulse amplification (CPA) noted in Table 20.4. Nonlinear-frequency converters are often found in these systems to change the wavelength to much shorter [UV or extreme UV (XUV)] or longer mid-infrared wavelengths. In producing very short wavelengths such as extreme UV, by high-harmonic generation, extremely short pulses can be obtained. Pulses well under the 1 femtosecond range are now routinely generated in research labs. Such short pulses are of use to physics researchers in making freeze-frame photos of the motion of electrons in atoms and molecules. A very large and diverse set of fundamental applications derive from instrument packages such as that shown in Figure 20.7. One can see from the figure that the number of components is quite large. Obviously, the expense of such a system is not insignificant and the sale of such an instrument is a dream for any laser salesperson.

Let us now discuss some of the components of these various systems. Starting with the basic ultrafast configuration of Figure 20.6, the items of interest as shown are the medium, the pump, the resonator, and the modelocking mechanism. The approach one takes in each of these cases depends on the architecture being employed. There are significant differences for Ti:sapphire and Yb ion or Yb:Er doped glass fibers.

The prerequisite for an ultrafast system is a laser medium capable of generating and amplifying the ultrafast pulse desired. As discussed earlier, the extremely wide tuning range of Ti:sapphire, ~180 nm full width at half maximum, centered at ~790 nm could in principle sustain pulses in the sub-10 fs regime with appropriate cavity optics. Certain other solid-state lasers have large bandwidths, but the combination of low-loss materials and ease of excitation with commercial laser pumps, such as frequency-doubled Nd, make the Ti:sapphire medium the material of choice for very short pulses. Dye lasers have somewhat less of a tuning range and are not suitable for femtosecond-style applications. Parametric oscillators also can operate over a wide spectral range and, as such, one will find these devices as part of ultrafast wavelength-conversion systems. The Yb-doped glass fibers have significant broadening and, as such, the Yb:glass medium is ideal as a picosecond source for applications not requiring the ~10 fs pulses that are achieved with Ti:sapphire. A typical ultrafast cavity layout for a Ti:sapphire laser is shown in Figure 20.8. This figure serves as a reference for discussion of the resonator of an ultrafast system.

Modelocking systems found in these lasers can be active or passive. Though active modelocking was the earliest approach used in picosecond-class lasers, the passive technique now reigns supreme in most systems. The ultimate in passive modelocking can be seen in the Ti:sapphire laser, where the resonator and its beam profile in the medium and the medium itself combine to make a train of ultrafast pulses. For a TEM_{00} mode in the Ti:sapphire medium, the intensity of the light in the cavity varies across the cross section of the laser material itself. With the aid of telescopes in the cavity, the response of the medium is nonlinear in the field of the light and the medium interacting with the optical field itself induces the light to self-focus via the Kerr effect. The self-focusing effect essentially turns the medium and the surround-

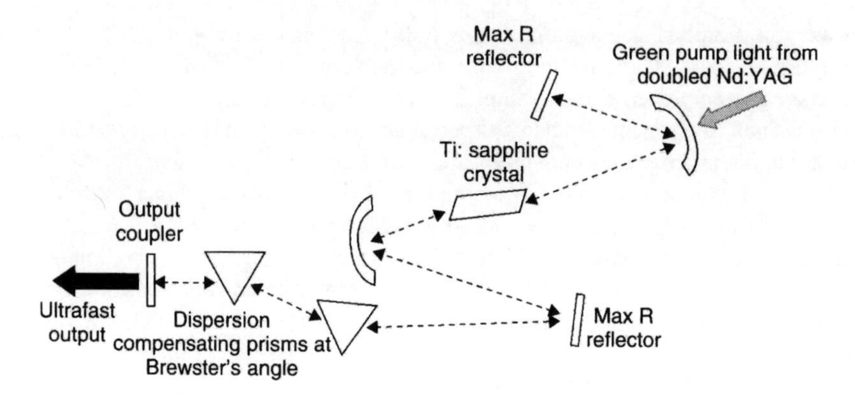

Figure 20.8 Sketch of the components in the basic building block Ti:sapphire ultrafast system. Though shown here as pumped by a frequency-doubled Nd:YAG laser, any intense source of green pump radiation can be used. Passive modelocking, using properties of the Ti:sapphire itself, is most frequently found in Ti:sapphire systems.

ing resonator optics into a saturable absorber switch. By careful design of the size of the optics, and compensation for the dispersion effects that can occur in the cavity optics, pulses as short as 5 fs can be generated.

For the fiber-laser variant of the ultrafast laser, some of these tricks do not apply. However, novel structures made with layered semiconductors can also be used as passive Q-switches for either Ti:sapphire or fiber lasers. These structures, termed semiconductor saturable absorber mirrors or SESAMs, are fabricated with alternating layers of semiconductor materials such as gallium arsenide (GsAs) and aluminum arsenide (AlAs) tailored in thickness and material for the wavelength of interest. The physical mechanism is absorption that saturates (that is, the SESAM material no longer absorbs after sufficient light exposure). The time scales for the fiber-laser-based ultrafast systems are controlled both by the passive modelocker, the SESAM, and the bandwidth of the basic medium. These can be as short as ~500 fs, or up to 10 ps for the Yb transition near one micron. The YbEr fiber laser operating near 1.56 micron is also available as a picosecond-class modelocked variant.

The optical pumps for the two different styles of ultrafast lasers are quite distinct. This leads to significant differences in both performance potential and price that a user needs to consider. The Ti:sapphire laser needs to be excited by green pump radiation, as shown in Figure 20.8. Since the storage time of Ti:sapphire is short, about 4 μs, and the excitation wavelength is in the green, excitation with diode arrays is out of the question. Excitation is typically achieved with frequency-doubled Nd lasers. Early in the development of the ultrafast Ti:sapphire laser, argon-ion lasers provided the cw green pump source. These days, one can find both lamp-pumped, frequency-doubled Nd:YAG lasers along with diode-pumped, frequency-doubled Nd systems in ultrafast lasers in the field. In contrast, the fiber lasers are directly excited by diodes or diode arrays, depending on the fiber technology used and average power desired. Yb^{3+} ions in a glass matrix have a storage time of order a millisecond, and make excellent use of diode pump radiation, typically near 970 nm. The fiber-based ultrafast laser thus has a significant advantage in size and simplicity. This will lead to the adaptation of fiber-based systems for medical and industrial applications as they develop, so long as pulse duration in the picosecond region are adequate for the application. However, the Ti:sapphire-based systems can produce much shorter pulses and, as such, are favored in the university research setting.

The resonator for the shortest ultrafast pulses is a concern in the design. The various materials in the optical train can induce a temporal spreading of the pulse due to dispersion (variation of the speed of light with wavelength) in the optics and the Ti:sapphire itself. The achievement of the shortest pulse durations requires special attention to this phenomenon, and the cavity optics are designed to compensate. One method is to use a pair of prisms inserted in the cavity with each prism at that Brewster angle, as sketched in Figure 20.8. By making the faster moving wavelengths go through a bit more quartz in the prism pair, the prisms can compensate for those wavelengths getting out of step with the slower moving wavelengths, resulting in a well-controlled shorter pulse.

To generate other wavelengths or higher pulse energy, additional components are added to the basic ultrafast system. The first step in such a process consists of boosting the energy of some of the short pulses emanating from the ultrafast laser and injecting them into an amplifier, as shown in Figure 20.7. An electrooptic switch is used to slice out a few of the ~50 nJ pulses and inject them into a second ultrafast medium. By using a regenerative amplifier scheme, the pulse can rattle around in the cavity of the regenerative amplifier and be boosted to considerably higher energy, though at a reduced pulse rate from that of the basic ultrafast laser. Typically, the pulse train coming out of the regenerative amplifier will be in the 10 Hz to 1 kHz range rather than the multi-MHZ range of the basic system. Further stages of amplification can then boost the energy to higher power.

Note that when one desires such amplified output energies from an ultrafast system, one needs to be concerned with optical damage due to the high-intensity light interacting with optical materials, coatings, and, indeed, the air inside a cavity or optical beam path. As such, the final stage or stages of amplification will utilize a scheme called chirped-pulse amplification, which uses a pair of diffraction gratings to stretch and then recompress the multiwavelength output pulse. Remember that an ultrafast laser will have a spread of wavelengths due to the uncertainty principle and the short pulse duration. A schematic diagram of a grating pair that temporarily disperses the short pulse and then recompresses it is shown in Figure 20.9. The short pulse is dispersed in wavelength by use of a diffraction grating and then sent through the amplifier with a longer pulse duration. By arranging the geometry appropriately, the wavelength dispersed pulse that comes off of the grating has a much longer pulse duration. A careful trigonometric analysis of the geometry in Figure 20.9 reveals that the path followed for the short-wavelength components of the pulse is shorter than the path for the longer-wavelength components. Thus, the short-wavelength components emerge from the grating pair slightly ahead of the longer-wavelength components, that is, the pulse has been stretched! By using separations of the grating pair of order 1 m, a 100 fs pulse can be stretched to one that is approximately 100 ps long. Since the optical damage that a pulse of light can cause

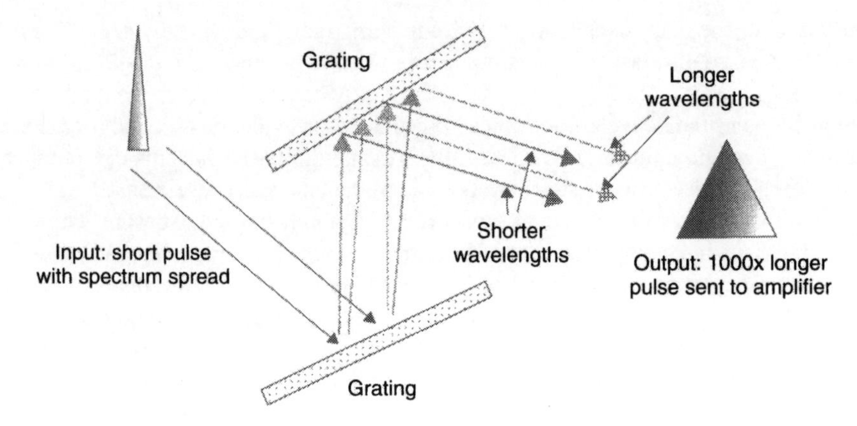

Figure 20.9 Schematic layout of the use of a grating pair to stretch the pulse of an ultrafast laser. When amplifying to high energy, a longer pulse is needed to avoid damage in some of the amplifier optics and laser materials. A second pair of gratings is then used to recompress the broadband signal to an amplified ultrafast pulse.

in a coating or an optical surface depends on the peak intensity, longer-duration pluses of a given energy have less chance of damaging an optic. After amplification the pulse is recompressed using another grating pair to effectively run the process in reverse and generate the ultrafast pulse but at the higher amplified energy. Details of the number of gratings and actual paths used vary with the particular setup, but the concept is always the same.

Such systems are used to create terawatt pulses using a fairly small table-top system. Devices producing 10 mJ in 10 fsec are such terawatt lasers. As one goes to yet higher energy, for example, amplifying to the 1 J level or beyond, the cost of the laser system and the size of the table(s) goes up. However there are efforts to build research systems with yet higher pulse energy and peak power and the ability to operate at shorter wavelengths so that the pulses shorter than 1 fsec are obtained. Longer-wavelength operation is also valuable for some work so that the infra-red fingerprint region of molecules of the spectrum can be used to diagnose molecular processes. Typically, an optical parametric amplifier is used to generate the UV emission. In certain cases, very broad-spectrum outputs are obtained to effectively provide a white light source.

There are a broad range of applications developing for such short-pulse systems. Researchers in chemical physics and materials science labs have exploited such ultrafast pulses to effectively elucidate the dynamics of reactions. Typically, two different ultrafast systems are used, tuned to different molecular resonances: one to excite or dissociate molecules, and a second to probe the species as they come apart, often termed pump-and-probe measurements. The researchers can then effectively visualize the molecular dynamics. The thrust for yet faster (attosecond) lasers in the extreme UV and X-ray regions is driven by the desire to further measure and understand the motion of electrons in atoms. As one scales the peak power and total energy, one can readily break down air. The use of ultrafast pulses to break down a column of the atmosphere has been shown with the goal of controlling lightning. The high local intensity that can be generated by focusing the output of sub-mJ class fiber laser can be used to evaporate material and effectively micromachine a material. The ultrafast laser may have significant advantage is drilling deep and narrow holes in materials like stainless steel. Micromachining of tissue may lead to a number of surgical applications based on compact, diode-driven fiber ultrafast lasers. One technique employed in development of both micromachining and surgical research is the use of the ultrafast laser to micromachine (or cut, if in surgery) beneath a surface. In surgery, one could imagine using such subsurface methods to destroy portions of a cell inside the cell wall. In laser micromachining, subsurface material ablation at the focal point of the fiber laser could allow one to change the index of a transparent material and, for example, produce a diffraction grating inside the bulk of a glass component. How commercial ultrafast applications evolve deserves further observation.

Glossary

Acousto-optic effect. The interaction of light waves and acoustic waves. A-O devices, such as Q-switches and modulators, deflect part of an optical beam passing through them.

Acronym. A word formed from the initials of its definition, for example, "laser."

Active medium. The atomic or molecular species that provides gain for laser oscillation.

Amplifier. A laser amplifier increases the power of an optical pulse generated by a laser oscillator.

Angstrom. (10^{-10} meter) Its use as a unit of optical wavelength has largely been supplanted in recent years by the nanometer (10^{-9} meter).

Autocollimator. An instrument thatuses its own collimated light to detect small angular displacements of a mirror.

Average power. The average rate at which energy flows from a laser; in a pulsed laser, the average power is equal to the product of pulse energy and pulse repetition rate.

Axial mode. A particular spatial distribution of energy stored in a laser resonator. Same as longitudinal mode.

Band gap. The energy gap between the high-energy conduction band and the lower-energy valence band in a semiconductor.

Bandwidth. The range of frequencies (or wavelengths) over which a laser operates.

Beam diameter. The diameter of a laser beam, typically measured to the points at which the power is equal to $1/e^2$ times the power at the center of the beam.

Beam divergence. The increase in beam diameter with distance from the laser's exit aperture, usually measured in milliradians.

Birefringence. The property of certain optical materials to have two refractive indices.

Birefringent filter. A spectral filter whose operation depends on its birefringent nature.

Boltzmann's law. The law of nature that describes the distribution of energy among a collection of entities.

Bragg cell. A type of acousto-optic modulator.

Note: The authors are indebted to the editors of *Laser Focus World* magazine for many definitons in this glossary.

Brewster's angle. The angle of incidence for which there is no reflectivity for one polarization of incident light.

Broadening. The increasing of a laser's bandwidth.

CARS. Acronym for "Coherent Anti-Stokes Raman Spectroscopy."

Cavity dumping. The process of storing energy in optical fields inside a resonator, then suddenly dumping it out to produce a short, powerful pulse of light.

Chemical laser. A laser in which the population inversion is produced by a chemical reaction.

Chopping. Modulating a beam by mechanically blocking and unblocking it.

Circular polarization. A polarization state in which the electric-field vector rotates as it propagates through space.

Circulating power. The power circulating between the mirrors in a laser.

Coherence. The unique property of laser light associated with a constant phase relationship among waves.

Coherence length. The distance over which waves in a laser beam retain their phase coherence. The narrower the bandwidth of a laser, the greater the coherence length (*see* bandwidth).

Collimated light. A beam of near-parallel rays of light (but the rays cannot continue to be parallel as they propagate) (*see* beam divergence).

Collisional broadening. The spectral broadening of a gas laser's bandwidth due to collisions of the atoms (or molecules). Also called pressure broadening.

Continuous wave (cw). A laser that emits a steady stream of light (as opposed to a pulsed laser).

Damage (optical). Damage to optical components caused by an intense laser beam.

Dewar. A thermos-like container used to hold cryogenic fluids such as liquid nitrogen or liquid helium.

Dielectric. A nonconducting (insulating) material such as glass.

Diffraction. The deviation of light from straight-line paths due to the wavelike nature of its propagation.

Diode. A two-terminal electronic device that lets current flow in only one direction. Most semiconductor lasers are diodes, and emit light when the current passing through them exceeds a threshold level. Diodes can also detect light, in which case they are called photodiodes.

Directionality. The property of laser light to have all its rays traveling in nearly the same direction.

Dispersion. In a dispersive medium, the refractive index depends on the wavelength of light propagating in the medium. All materials exhibit optical dispersion, but some are more dispersive than others.

Divergence (*see* beam divergence)

Doppler effect. The observed frequency shift of a wave whose source is moving relative to the observer.

Doping. The addition of trace amounts of an impurity, often resulting in great changes in the property of the substance that is doped.

Doughnut mode. The combination of transverse modes that produces a doughnut-shaped beam.

Duality. Wave–particle duality is the seeming contradiction in the nature of light; it sometimes behaves as if it were composed of particles, and it sometimes behaves as if it were composed of waves.

Efficiency. The ratio of useful output energy produced by a device to input energy consumed.

Electronic energy states. The quantum levels of atomic or molecular energy associated with the electrons' motion or position.

Electro-optic effect (linear). The dependence of a material's refractive index on an applied electric field.

Energy levels. The quantized amounts of energy that can be stored in an atom or molecule.

Etalon. An optical device composed of two mirrors precisely aligned with each other so that light can become trapped between them, reflecting back and forth indefinitely. An etalon can serve as a very narrow spectral filter.

Extraordinary polarization. In a birefringent medium, light polarized perpendicular to the ordinary polarization is defined to be in extraordinary polarization (*see* ordinary polarization).

Excimer. An excimer molecule is one that is stable in an excited state but unstable in its ground state.

Fabry–Perot interferometer. An optical device composed of two mirrors precisely aligned with each other so that light can become trapped between them, reflecting back and forth indefinitely. A Fabry–Perot can serve as a very narrow spectral filter.

Faraday effect. A rotation of a light wave's polarization as the wave passes through a magneto-optic material to which a magnetic field is applied.

Fiber optic. A thin strand of transparent material, usually glass, that guides light through an inner core that is surrounded by a lower-index cladding layer that confines the light in the core by total internal reflection.

Four-level laser. A laser in which four energy levels are involved in the lasing process.

G-parameter. An easy-to-calculate indicator of resonator stability.

Gain. A measure of the strength of an amplifier. Unsaturated gain is the gain measured with a very small input signal. As the input signal is increased, the gain becomes saturated.

Gaussian mode (*see* TEM_{00})

Grating. An interferometric device that reflects or transmits light of different wavelengths in different directions.

Ground state. The lowest energy level of an atom or molecule.

Half-wave plate. A birefringent element that (when used correctly) converts light of one polarization to the orthogonal polarization, that is, vertical to horizontal or clockwise to counterclockwise.

Harmonic generation. Production of light whose frequency is an integral multiple of the input light by means of nonlinear optics.

Hole burning (*see* spatial hole burning and/or spectral hole burning)

Hertz. A unit of frequency, one cycle per second.

Hologram. A recording of the interference of coherent light reflected from an object with light direct from the same source. The hologram can produce a three-dimensional image of the object.

Huygens principle. A simple method of visualizing wave propagation.

Index of refraction. An important optical property of a material; the refractive index is the ratio of the speed of light in vacuum to the speed of light in the material.

Infrared. Wavelengths longer than those visible to the human eye, from 700 nanometers to about 1000 micrometers.

Interference. The interaction of two or more waves to enhance or negate each other.

Interferometer. Any device that manifests the effects of wave interference.

Invar. A steel alloy with extreme thermal stability, often used in optical tables and laser resonator structures.

Junction. In a semiconductor, the layer between p- and n-doped semiconductors, which is where electrons and holes recombine to emit light in a diode laser.

Kerr cell. An electro-optic modulator in which a liquid, rather than a crystal, retards one component of polarization.

Kryton. A gas-filled, cold-cathode switching tube that conducts high peak currents for short periods.

Lamb dip. A power dip in the spectral center of a single-frequency laser tuned across its lasing bandwidth.

Laser. Acronym for "Light Amplification by Stimulated Emission of Radiation."

Lifetime. The length of time for which an atom (molecule) will remain unperturbed in a given energy level.

Lasik. Acronym for "Laser in situ keratemeulisis." Corneal surgery to correct vision using excimer or ultrafast lasers.

Linewidth (*see* bandwidth)

Longitudinal mode. A particular spatial distribution of energy along the axis of a resonator.

Loss (intracavity). The optical loss caused by elements in a laser resonator.

Maser. Acronym for "Microwave Amplification by Stimulated Emission of Radiation."

Metastable. An energy level with a long spontaneous lifetime is a metastable level. In general, "long" means tens or hundreds of microseconds, but the term is subjective.

Modelocking. Internal modulation of a laser resonator at a frequency equal to the natural spacing of the modes, causing the phases of the individual modes to become locked together. Locking the longitudinal modes of a resonator produces a train of output pulses at the modulation frequency.

Modulator. A device that imposes a cyclic variation on light passing through it. An electro-optic modulator can affect the polarization, phase, or intensity of light, and an acousto-optic modulator can affect the propagation direction or intensity of light.

Molecular laser. A laser based on transitions in a molecule rather than in an atom.

Monochromator. A device that selects light in a narrow band of wavelengths from a beam in which a range of wavelengths is present.

Multimode oscillation. Simultaneous oscillation of several resonator modes resulting in an output beam containing several distinct frequencies.

Neutral density filter. A filter that reduces the intensity of light without affecting its spectral character.

Nonlinear optics. The process of generating new frequencies of light from one (or more) original light waves.

Optic axis. The direction of symmetry in a birefringent crystal.

Optimum output coupling. The value of mirror transmission that produces maximum power from a given laser.

Ordinary polarization. In a birefringent medium, light polarized perpendicular to the optic axis is defined as ordinary polarization.

Parametric oscillator. A nonlinear device that produces tunable light at wavelengths longer than the input wavelength.

Passive modulators. Modulators that become transparent when a transition in the modulator material is saturated.

Phasematching. The process of compensating for dispersion in a nonlinear process so that different waves maintain the proper phase relationship.

Photomultiplier. A phototube in which the photoelectrons produced by light incident on the cathode are amplified by multiple stages of secondary electron emission.

Phonon. The quantized particle of a sound wave.

Photon. The quantized particle of electromagnetic radiation.

Phototube. A vacuum tube in which photons striking a light-sensitive cathode cause the emission of electrons that are collected by an anode.

Piezoelectric effect. The slight change in physical size that takes place in certain crystals when a voltage is applied to them.

Plane polarization. A light wave in which the electric field oscillates in a single plane is plane polarized.

Pockels cell. An electro-optic crystal and the electrodes necessary to modulate the phase or polarization of light passing through the device.

Polarization. The polarization describes the manner or direction of oscillation of the electric field in a light wave.

Polarizer. A device that separates one component of polarization from the other.

Population inversion. A condition in which more atoms of a species are in a given energy state than in a lower state.

Power. The rate of using energy.

Power density. Power per unit area in a laser beam.

Pressure broadening (*see.* collisional broadening)

Q-switch. An intracavity shutter that prevents laser oscillation until it is opened.

Quantum cascade laser. A nondiode semiconductor laser in which electrons passing through a series of quantum wells emit infrared light.

Quantum mechanics. The area of modern physics concerned with the behavior of nature on a very small (i.e., atomic) scale.

Quarter-wave plate. A device that can (when used correctly) convert plane polarized light to circular polarization and vice versa.

Quasi three-level laser system. A four-level laser whose lower laser level is close enough to the ground level that it is significantly populated at room temperature. Thus, at room temperature, a quasi three-level laser behaves as a three-level laser.

Refractive index. The ratio of light velocity in vacuum to light velocity in a material is the material's refractive index.

Relaxation oscillations. The oscillation of energy in a laser between the population inversion and circulating power. Relaxation oscillations appear as modulation of the output from a laser at frequencies of hundreds of kilohertz.

Resonator. The two or more mirrors of a laser that provide the feedback necessary for a laser to oscillate.

Rotational energy levels. Energy levels associated with rotational motion of a molecule.

Saturated transition. A transition that has been driven to the point where the population of the upper and lower levels are equal.

Saturated gain. The steady-state gain inside a laser medium, reduced from its initial value by stimulated emission.

Second harmonic generation. Nonlinear generation of light at exactly twice the frequency of the input light wave.

Single-mode oscillation. Oscillation of a laser in only one spatial mode. Same as single-frequency oscillation.

Small signal gain. The gain of a laser before it is reduced by stimulated emission. Also called unsaturated gain.

Spatial hole burning. The depletion of population inversion (usually due to stimulated emission) at certain places in the lasing medium.

Spectral hole burning. The depletion of population inversion (usually due to stimulated emission) at certain frequencies within the lasing bandwidth.

Spectrum analyzer. An instrument that indicates the frequencies present in an input signal.

Spontaneous emission. The natural emission of light from an atom (or molecule) as it decays from an excited energy level.

Spot size. The diameter of a laser beam at a given point.

Standing wave. The stationary wave produced by interference between two waves traveling in opposite directions.

Stimulated emission. The emission of light from an atom (or molecule) caused by an interaction between the atom (molecule) and an external light wave.

Substrate. The underlying material onto which a coating is applied. For example, many laser mirrors are fabricated by depositing a thin dielectric coating onto a glass substrate.

Synch pumping. A technique of modelocking a laser, usually a dye laser, by creating the population inversion synchronously with the passage of the intracavity pulse.

TEA laser. TEA is an acronym for "Transversely Excited Atmospheric" pressure. It is a gas laser, usually CO2, in which the exciting discharge is transverse to the optical axis of the laser. Because of the shorter discharge, these lasers operate at higher pressure (but not necessarily atmospheric) than conventional lasers. The output of a TEA laser is a fast train of high-peak-power pulses.

TEM_{00}. A designation for the fundamental, or Gaussian, transverse mode. The beam divergence of the TEM_{00} mode is lower than that of any other mode; it is diffraction limited.

Three-level laser system. A laser having three energy levels populated during the lasing cycle.

Threshold. Laser threshold occurs when the unsaturated round-trip gain in a resonator is just equal to the round-trip loss.

Transition. The changing of an atom (molecule) from one energy level to another, accompanied by the absorption or emission of energy.

Transverse mode. A particular spatial distribution of energy in a resonator.

Ultrafast. Lasers whose pulse duration is less than 1 picosecond, 10^{-12} s. Some ultrafast lasers produce light pulses that are under 1 femtosecond.

Unsaturated gain (*see.* small signal gain)

Unstable resonator. A laser resonator in which a ray is not forever trapped, as it is in a conventional resonator. Unstable resonators can be useful in obtaining the maximum possible output from a high-gain, pulsed laser.

VCSEL. Acronym for "vertical-cavity surface-emitting laser," a type of diode laser.

Waist (of a laser beam). The point along the beam where the diameter is smallest.

Wave–particle duality (*see.* duality)

Waveguide. Any device that guides electromagnetic waves along a path defined by the physical construction of the device.

Waveplate. An optical element that utilizes its birefringent properties to alter the polarization of light passing through it.

Further Reading

Ken Barat, *Laser Safety Management,* CRC Books, 2006. A thorough and readable overview of laser safety.

Michael Bass et al. (eds.), *OSA Handbook of Optics,* McGraw-Hill, 2000. Massive multivolume encyclopedia of the field.

Robert Boyd, *Nonlinear Optics,* Academic Press, 2008. A clearly written perspective on modern nonlinear optics.

Eugene Hecht, *Optics,* Addison Wesley, 2001. The standard college optics textbook.

Jeff Hecht, *The Laser Guidebook,* McGraw-Hill, 1999. Classic description of classical lasers; some parts dated.

Jeff Hecht, *Understanding Lasers: An Entry-Level Guide,* IEEE Press/Wiley, 2008. Nonmathematical introduction to lasers and their applications.

IEEE Journal of Selected Topics in Quantum Electronics, Vol. 6, No 6, Nov/Dec 2000. This issue contains many excellent review articles dealing with lasers and laser physics.

Francis Jenkins and Harvey White, *Fundamental of Optics,* McGraw-Hill, 2001. The classic reference to optics technology.

Walter Koechner, *Solid State Laser Engineering,* Springer Series in Optical Sciences, 2010. For decades, the "bible" on solid-state lasers.

Bela A. Lengyl, *Lasers (Pure and Applied Optics),* Wiley, 1971.

Anthony Siegman, *Lasers,* University Science Books, 1986. A classic reference, with lots of rigorous detail.

William Silvast, *Laser Fundamentals,* Cambridge University Press, 2004. Textbook for upper level undergraduates, explanatory.

Orazio Svelto, *Principles of Lasers, 5th ed,* Springer, 2009. A textbook that stresses intuitive explanations where math can be avoided, now in its 5th edition.

Joseph Verdeyen, *Laser Electronics,* Prentice-Hall, 1995.

Introduction to Laser Technology, Fourth edition. C. Breck Hitz, J. Ewing, and J. Hecht

Index

Introduction to Laser Technology, Fourth edition. C. Breck Hitz, J. Ewing, and J. Hecht

Made in the USA
Monee, IL
26 November 2020

aa62c9a4-1159-4463-b47f-5a70c6062f20R04